Conservation of Tropical Birds

Conservation of Tropical Birds

Navjot S. Sodhi
National University of Singapore

Çağan H. Şekercioğlu
Stanford University

Jos Barlow
Lancaster University

Scott K. Robinson
University of Florida

WILEY-BLACKWELL
A John Wiley & Sons, Ltd., Publication

This edition first published 2011

Blackwell Publishing was acquired by John Wiley & Sons in February 2007. Blackwell's publishing program has been merged with Wiley's global Scientific, Technical and Medical business to form Wiley-Blackwell.

Registered Office
John Wiley & Sons, Ltd., The Atrium, Southern Gate, Chichester, West Sussex, PO19 8SQ, UK

Editorial Offices
9600 Garsington Road, Oxford, OX4 2DQ, UK
The Atrium, Southern Gate, Chichester, West Sussex, PO19 8SQ, UK
111 River Street, Hoboken, NJ 07030-5774, USA

For details of our global editorial offices, for customer services and for information about how to apply for permission to reuse the copyright material in this book please see our website at www.wiley.com/wiley-blackwell.

Library of Congress Cataloging-in-Publication Data has been applied for

Conservation of tropical birds/Navjot S. Sodhi... [et al.]. – 1st ed.
p. cm.
Includes bibliographical references and index.
ISBN 978-1-4443-3482-1 (hardcover : alk. paper)
1. Birds–Conservation–Tropics–Case studies. 2. Biotic communities–Tropics–Case studies. 3. Climatic changes–Environmental aspects–Tropics–Case studies. 4. Global environmental change–Case studies. 5. Adaptation (Biology)–Case studies. I. Sodhi, Navjot S.
QL676.57.T76C66 2011
333.95′8160913–dc22

2010052429

A catalogue record for this book is available from the British Library.

This book is published in the following electronic formats: ePDF 9781444342581; Wiley Online Library 9781444342611; ePub 9781444342598; Mobi 9781444342604

Set in 10.5/12pt Classical Garamond BT by SPi Publisher Services, Pondicherry, India
Printed and bound in Malaysia by Vivar Printing Sdn Bhd

1 2011

Contents

Color plates appear in between pages 166 and 167

Preface

The idea for this book came about when Navjot and Cagan were having dinner in Palo Alto (California) in December 2006. With this book, we basically wanted to advance the field of conservation ornithology. Our hope was that such a book would facilitate more, and sorely needed research on tropical birds. In order to achieve comprehensive coverage in expertise and geography, we invited Jos and Scott to the project.

The resulting book is the first on the major issues in tropical bird conservation presented in a contemporary conceptual and empirical framework. We provide an up-to-date summary of the conservation problems faced by tropical birds and outline potential solutions. Because birds are the best known group and are key indicators of ecosystem health, they provide an excellent lens through which to observe and understand global conservation problems such as climate change, declines in ecosystem services, habitat loss, fires, overexploitation, and invasive species. Therefore, the book also provides an engaging synopsis of the major issues in tropical conservation and the problems also faced by other tropical wildlife. We have attempted to cover key tropical and subtropical areas around the globe, but our scope is limited mainly to terrestrial birds. The broad ecological approach should prove complementary to the detailed species' accounts provided by Birdlife International.

With millions of individuals around the world passionately engaged in observing and conserving tropical birds, this book will serve as an important resource and companion to all the people interested in birds and committed to studying and conserving them. The book is written in an accessible fashion so that it can be used by advanced undergraduate and graduate students.

Many individuals have assisted us in this project. Pei Xin Ng provided invaluable assistance in preparation of many figures and obtaining permissions. Toby Gardner, Lian Pin Koh, Alex Lees, Cristina Banks-Leite, Matthew Lim, Richard Primack, and Glendy C. Robinson commented on various chapters.

Navjot's research on tropical birds has been supported by the National University of Singapore, National Geographic, and Kim Cheng-Kim Loke and

Loke Wan Tho Memorial Foundations (through Lady Y. P. McNeice). Cagan has been supported by the Christensen Fund, National Geographic Society, and Wildlife Conservation Society. Jos was supported by Lancaster Environment Centre (UK), the Goeldi Museum, and the Federal University of Lavras (Brazil). Scott was supported by the Katherine Ordway Foundation of the Florida Museum of Natural History. We all thank folks at Wiley-Blackwell for their help, in particular, Ward Cooper, Delia Sandford, and Kelvin Matthews.

Navjot S. Sodhi, Singapore
Cagan H. Şekercioğlu, Stanford
Jos Barlow, Lancaster
Scott K. Robinson, Gainesville

1

The State of Tropical Bird Biodiversity

Overview: Tropical areas contain the largest proportion of imperiled bird species. Of 1227 bird species threatened worldwide, 79% occur in lowland and montane tropical forests. Many of the world's most threatened bird species also occur within the tropics. Tropical bird species are endangered by drivers such as habitat loss and degradation, hunting, pollution, invasive species, and disease. These drivers may act singly or in combination. Tropical bird species, especially the threatened ones, remain relatively poorly studied. Poor ornithological knowledge may hinder the design of effective conservation measures for threatened tropical birds. More ornithological research is urgently needed on tropical birds.

Lying between the Tropics of Cancer and Capricorn (i.e. 23.5° north and south of the equator), the tropics harbor two-thirds of the biodiversity in a region that is currently suffering a massive loss of native habitats (Dirzo and Raven 2003). Both mega-rich biodiversity and rapid loss of native habitats have made the tropics of high interest to conservation ornithologists. In this chapter, we highlight the state of tropical bird biodiversity by focusing on threatened bird species in the tropics. We also discuss some of the drivers of bird endangerment and highlight that tropical birds, particularly the threatened ones, need more and urgent scientific attention.

1.1 Imperiled bird biodiversity

The tropical region is at the forefront of bird extinction and endangerment. All three bird species suspected to have gone extinct since 2000 are tropical – Spix's Macaw (*Cyanopsitta spixii*), Hawaiian Crow (*Corvus hawaiiensis*), and Po'ouli

Conservation of Tropical Birds, First Edition. Navjot S. Sodhi, Çağan H. Şekercioğlu, Jos Barlow and Scott K. Robinson.

Published 2011 by Blackwell Publishing Ltd.

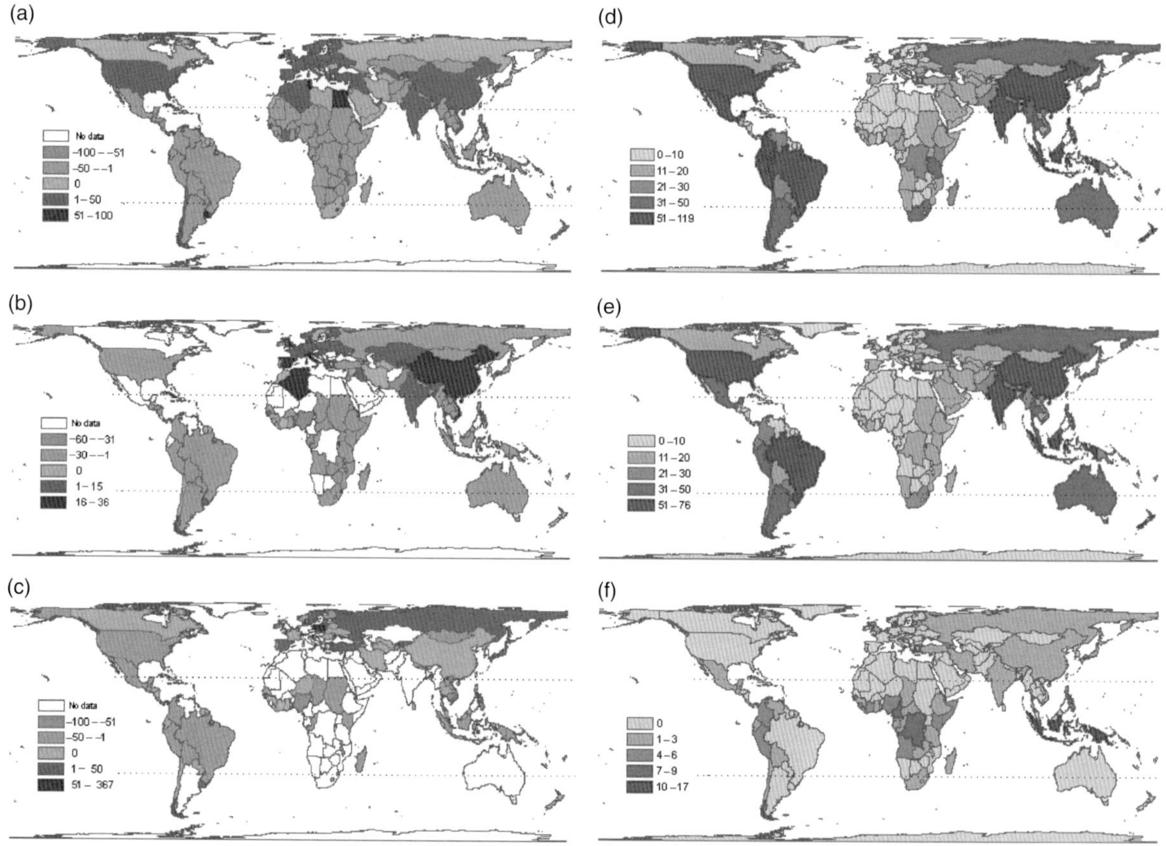

Figure 1.1 Global distribution of average annual forest cover percent change (1990–2005) and richness of threatened and data-deficient terrestrial birds among countries. Data sources for rate of forest cover change and terrestrial birds were obtained from Global Forest Resources Assessment (Food and Agriculture Organization [FAO] 2005) and IUCN Red List of Threatened Species (International Union for the Conservation of Nature and Natural Resources 2007), respectively. Definitions of (a) total, (b) natural, and (c) primary forest areas follow those of FAO (2005). (d) Globally threatened species (i.e. those classified on the IUCN Red List as "critically endangered," "endangered," or "vulnerable"). (e) Globally threatened species as defined for (d), but excluding those species listed under the Red List criteria A1c, B1b(iii), and B2b(iii) (i.e. criteria related to decline of the habitat area, in extent, quality, or both). (f) Data-deficient species (i.e. those classified on the Red List as "data-deficient"). Tropical countries are those in which all the land area is situated within ~23.5°N and ~23.5°S (indicated by the dotted lines). (After Sodhi *et al.* 2008.) (Please see Plate 1.1 for the color version of this figure.)

(*Melamposops phaeosoma*) (BirdLife International 2010a; see Chapter 3). Based on the most recent assessment in 2010, 1227 bird species (12.3%) worldwide are threatened with extinction (BirdLife International 2010a). Most of these threatened bird species occur within the tropics (Figure 1.1; Sodhi *et al.* 2008). For example, 79% of all threatened bird species are found in lowland and montane tropical forests (BirdLife International 2010a).

Overall, the endangerment of birds has been increasing steadily since 1988 (Figure 1.2; BirdLife International 2010a). In fact, birds are becoming more dramatically endangered, especially in the Indomalayan region (Figure 1.3; Butchart *et al.* 2004). The primary reason for this is massive habitat loss in this region (Sodhi *et al.* 2004a; see Section 1.2). The cause for concern is that within Indomalaya some areas such as Southeast Asia harbor not only the highest number of endemic species, but also threatened species as well (Figure 1.4; Sodhi *et al.* 2006a). Therefore, if extinctions are to happen in this region, they will represent global rather than population extinctions (see also Chapter 3). Similarly, far more threatened endemic bird species occur in tropical than in non-tropical biodiversity hotspots (Sodhi *et al.* 2007). Biodiversity hotspots are areas that contain 42% of all terrestrial vertebrates on 2% of Earth's land surface, but have lost 70% of their original vegetation due to heavy impact of human activities (Myers *et al.* 2000). Therefore, to prevent global extinctions, these areas should be immediately preserved – there are 34 biodiversity hotspots across the globe (www.biodiversityhotspots.org).

The importance of tropics for conservation is further attested by the fact that four out of five top countries housing the highest number of critically endangered bird species are tropical – Brazil, Indonesia, India, and Philippines (Figure 1.5; BirdLife International 2008a). It is not surprising that many of the world's most threatened bird species (< 60 individuals in the wild) are found in the tropics; for example, Caerulean Paradise-flycatcher (*Eutrichomyias rowleyi*; Indonesia), Bali Starling (*Leucopsar rothschildi*; Indonesia), Puerto Rican Amazon (*Amazona vittata*; Puerto Rico), Sulu Hornbill (*Anthracoceros montani*; Philippines), Tahiti Monarch (*Pomarea nigra*; French Polynesia), and Floreana Mockingbird (*Mimus trifasciatus*; Ecuador) (BirdLife International 2008a).

Birds may not be the most threatened taxonomic group – 32% and 22% of the known amphibian and mammal species are currently threatened with extinction, respectively (www.iucnredlist.org; see Chapter 3). It is unclear however if the number of threatened bird species correlates well with those of other taxonomic groups. At least one other taxonomic group shows congruence with bird endangerment: the proportion of threatened bird and mammal species correlate strongly for tropical countries (Figure 1.6; Kerr and Burkey 2002). Additionally, forests protected just for birds may also adequately protect other groups such as woody plants, moths, butterflies, and small mammals, at least in Uganda (Howard *et al.* 1998). However, there may be regional and site differences in such a complementarity (Moore *et al.* 2003).

Overall, it is clear that the tropics is the most critical region for bird biodiversity and conserving birds, and also for the conservation of other organisms.

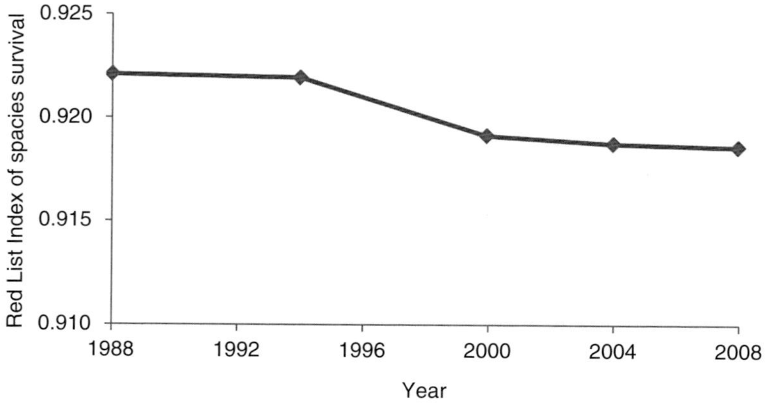

Figure 1.2 Red List Index for the world's birds. Red List Index indicates the change in birds in their threat status. (After www.birdlife.org/action/science/species/global_species_programme/red_list.html)

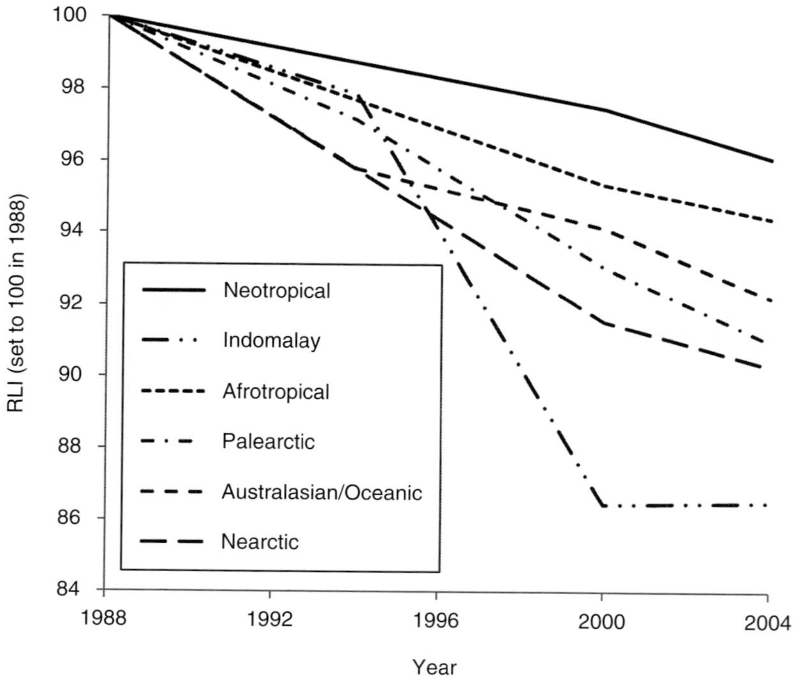

Figure 1.3 Red List Index (RLI) for birds in different biogeographic realms over time. For the definition of RLI see Figure 1.2. (After Butchart *et al.* 2004.)

1.2 Drivers of endangerment

Major causes of bird endangerment include the loss, fragmentation and degradation of native habitats (for 95% of threatened species; see Chapter 3), human use (~71%; see Chapter 7), and invasive species (~33%; see Chapter 6) (Figure 1.7;

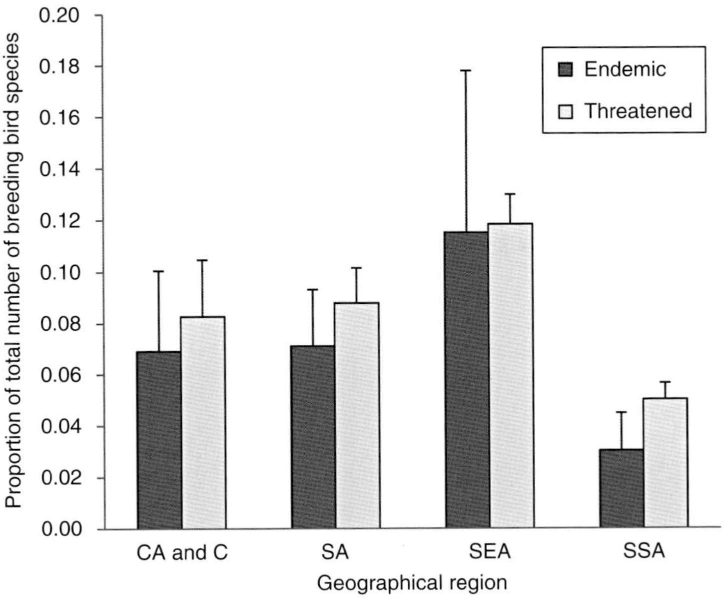

Figure 1.4 Comparison of proportion of total number of breeding bird species that are endemic and threatened among tropical regions. Data on number of breeding, endemic and threatened bird species were obtained from World Resources Institute; threatened status was obtained using IUCN Red Listing criteria. Error bars represent standard errors of mean proportion of total number of breeding bird species that are endemic or threatened. Abbreviations: SEA = Southeast Asia; SSA = Sub-Saharan Africa; CA and C = Central America and Caribbean; SA = South America. (After Sodhi *et al.* 2006a.)

BirdLife International 2008b, 2010a). Additional drivers of bird endangerment include environmental pollution, fire (see Chapter 5), disease (see Chapter 3), climate change (see Chapter 8), and indirect mortality (e.g. drowning of seabirds in longline fisheries). These noxious drivers impact species singly or in combination (Figure 1.8; Brook *et al.* 2008). Here we briefly review some of the drivers of bird decline that will not be detailed in subsequent chapters.

1.2.1 Habitat loss

Deforestation epitomizes habitat loss within the tropical region. Using the United Nations Food and Agricultural Organization's (FAO) data on forest cover change from 1980 to 1990, Laurance (1999) estimated that 15.4 million ha (hectares) of tropical forest is destroyed annually, with an additional 5.6 million ha being degraded through activities such as selective logging. According to the FAO, between 2000 and 2005, Africa and South America lost 4 million ha of tropical forests annually (FAO 2005). During the same period, Southeast Asia was deforested at one of the highest rates of 1% per year (FAO 2005). There is controversy as to whether the FAO values are accurate, as they may fail to include catastrophic events such as the vast 1997–98 forest fires in Indonesia and Amazon, and perhaps erroneously include plantations as forest cover (Achard *et al.* 2002;

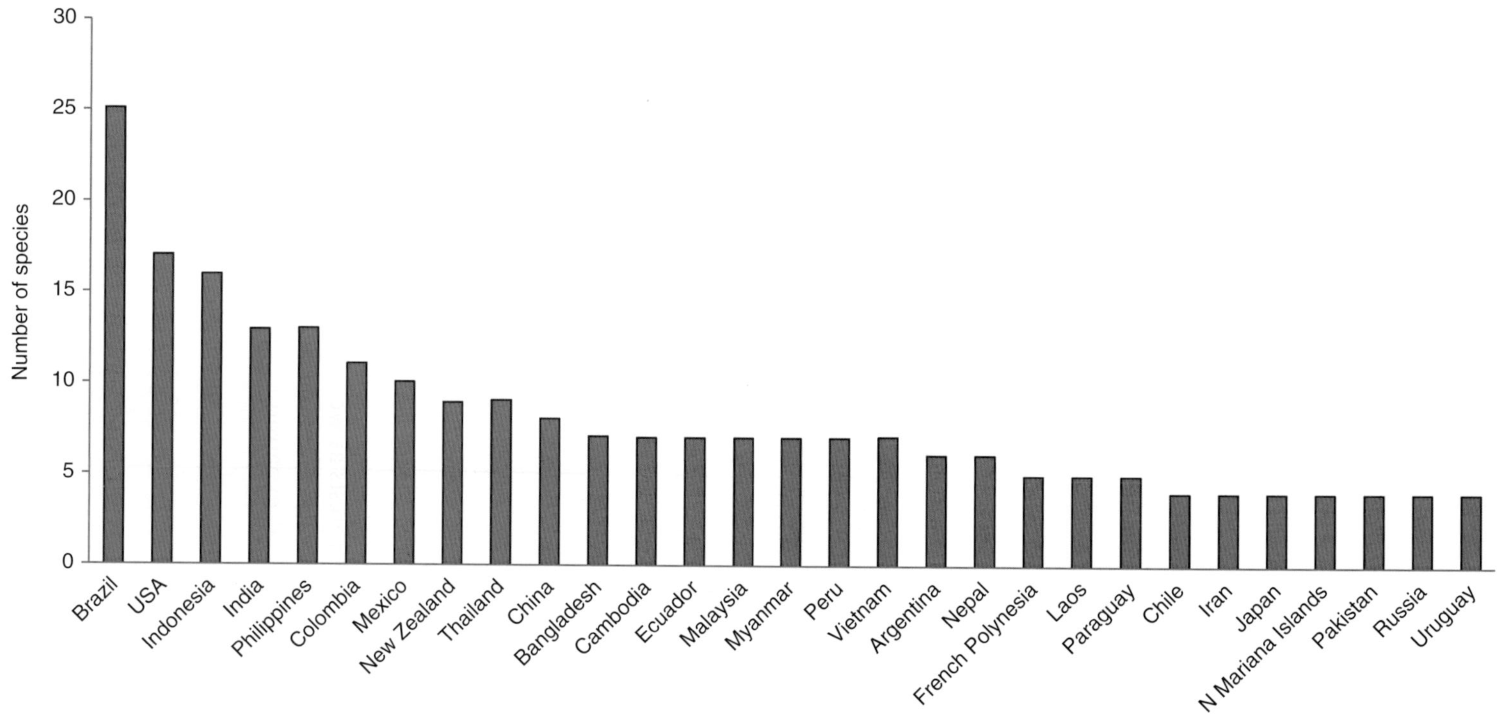

Figure 1.5 Critically endangered bird species in various countries. N = North. (After BirdLife International 2008a.)

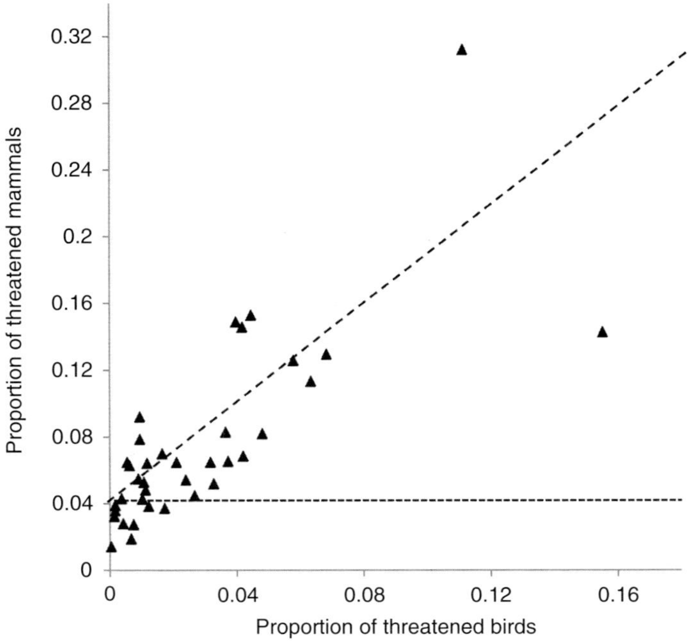

Figure 1.6 Proportions of threatened bird and mammal species per tropical country are strongly correlated. The horizontal dotted line shows the location of the *x*-intercept, located at 0.04. It indicates that in general 4% of a tropical forested country's mammal fauna is threatened with extinction before any birds. (After Kerr and Burkey 2002.)

Grainger 2008). Deploying remotely sensed satellite imagery, Hansen *et al.* (2008) reported that humid tropical forest loss was 27.2 million ha from 2000 to 2005, with the highest losses in the Americas and Asia (Figure 1.9). The two countries with the highest absolute rate of deforestation were Brazil and Indonesia (Hansen *et al.* 2008). Despite the debate about the exact levels of tropical deforestation, most would agree that the current rate is massive.

Although native forest loss seems to be decelerating over time in tropical Latin America, it continues to accelerate in tropical Asia (Matthews 2001). This trend is corroborated by a satellite imagery study conducted by Hansen and DeFries (2004). Deforestation continues at an alarming rate not only in humid but also in seasonal and montane tropical forests as well (Sodhi *et al.* 2007).

The direct causes of deforestation in rainforests (and loss of other habitats) are numerous, including agricultural expansion, logging, and urbanization. These drivers can act singly, or most likely, synergistically. The ultimate underpinnings of these causes of deforestation are complex, however, involving both socio-political and economic imperatives (Sodhi *et al.* 2004a). Devastating losses are not restricted to the tropical forests – other habitat types in the tropics such as mangroves, swamp forests, dry forests, and savannas face similar predicaments (Sodhi *et al.* 2007). For example, mangroves covered more than 200,000 km^2 of the tropical and subtropical coastlines, but are now disappearing at the rate of 1–2% annually, and may be soon wiped out from 26 of 120 countries containing

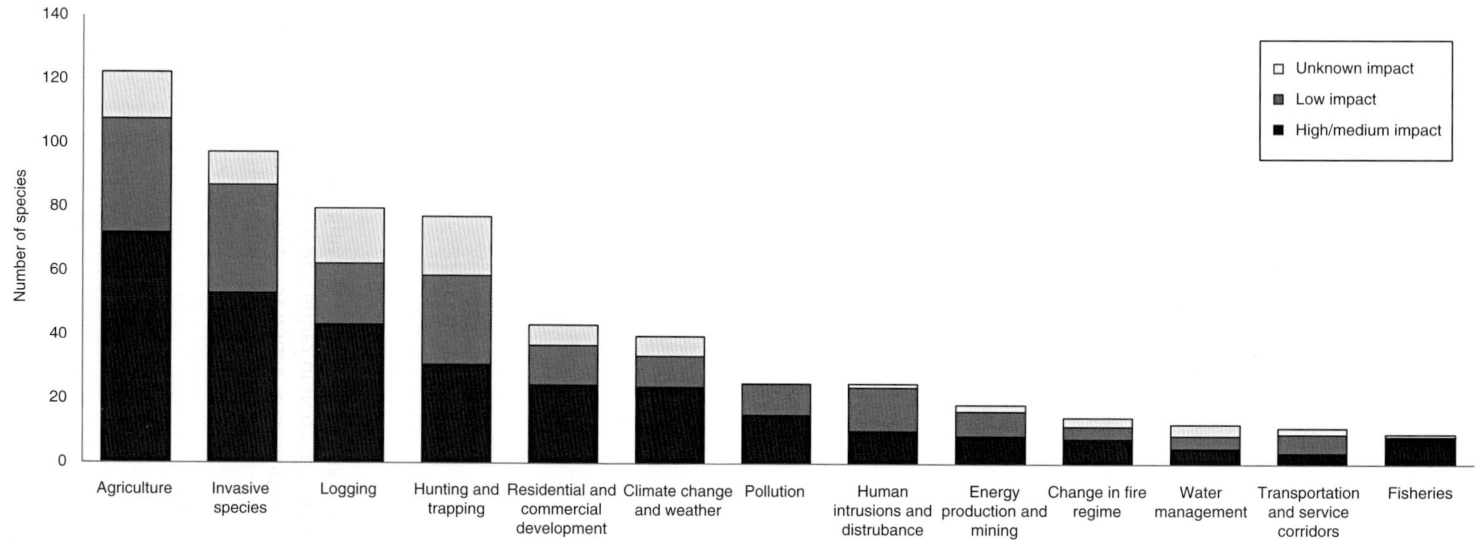

Figure 1.7 The main threats to globally threatened species. (After BirdLife International 2008b.)

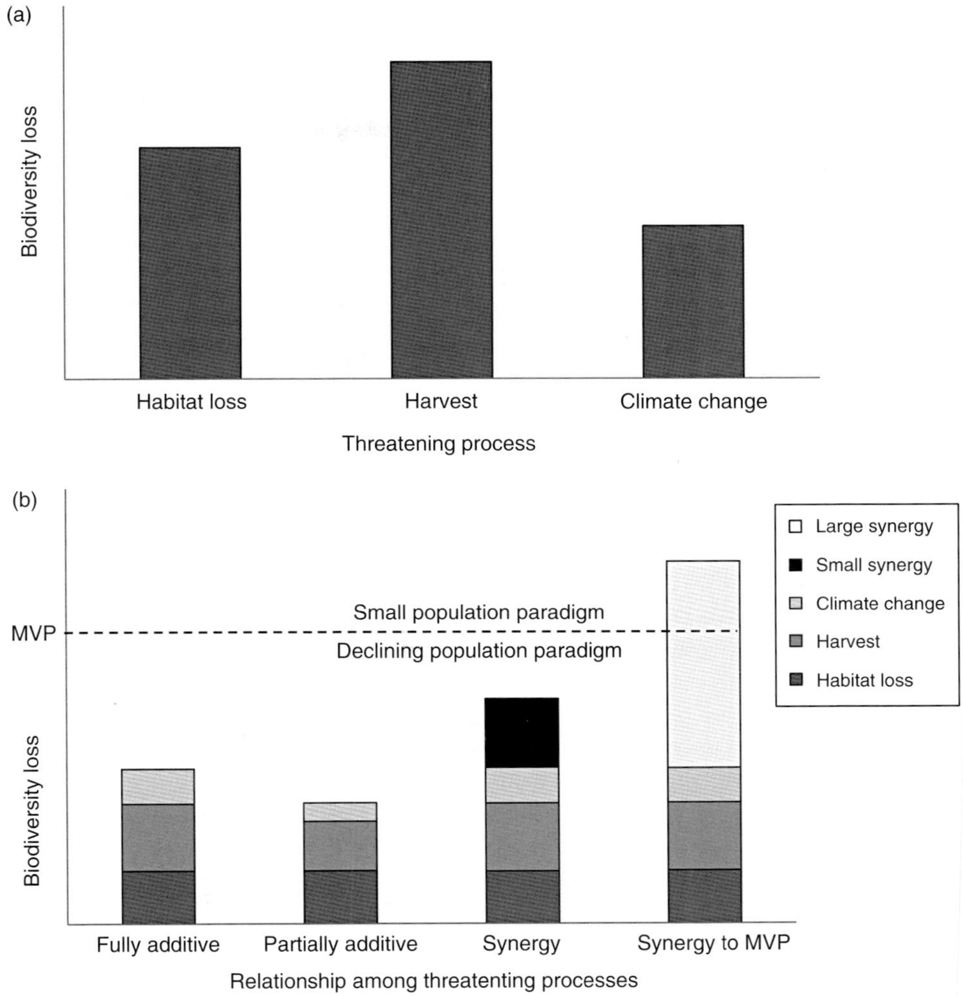

Figure 1.8 Threatening processes can cause biodiversity loss (either through declines in species' abundance or loss of species from communities), and they can act alone or in combination. The schematic shows three examples of threatening processes: habitat loss, harvest, and climate change. (a) Acting singularly, each process has varying negative effects on biodiversity such as a reduction in carrying capacity (habitat loss), direct decline in abundance (harvest), or reduction in range size (climate change). (b) Many systems show multiple threatening processes acting together. A fully additive model predicts that the combined negative effects are simply the sum of each process's individual effects, whereas combinations might result in only partial additivity (e.g. further range restrictions from habitat loss encompass those predicted from climate change). A synergy among processes implies a positive interaction (feedback) whereby the total negative effect on biodiversity is greater than the sum of each threatening process's individual contribution. Small synergistic interactions will result in population decline (i.e. the declining population paradigm); however, only when large synergies occur that push populations below their minimum viable population (MVP) size does extinction risk become non-negligible (i.e. the small population paradigm). (After Brook *et al*. 2008.)

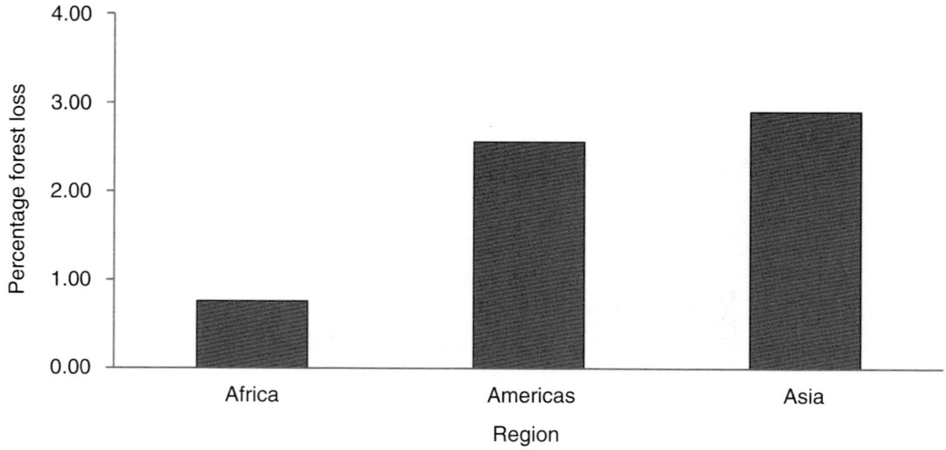

Figure 1.9 The percentage of humid tropical forests cleared in different regions. (Data derived from Hansen *et al.* 2008.)

mangroves (Duke *et al.* 2007). There has been a loss of 3.6 million ha of mangroves since 1980s, with the steepest declines in Asia (Figure 1.10; FAO 2007). Mangroves are important for bird biodiversity, with 48 bird species or subspecies restricted to mangroves (Luther and Greenberg 2009). Of these, five species are currently endangered – Madagascar Teal (*Anas bernieri*; Madagascar), Plain-flanked Rail (*Rallus wetmorei*; Venezuela), Mangrove Hummingbird (*Amazilia boucardi*; Costa Rica), Sapphire-bellied Hummingbird (*Lepidopyga lilliae*; Colombia), and Mangrove Finch (*Camarhynchus heliobates*; Galápagos Islands, Ecuador). Mangroves also provide refuge for bird species such as the Philippine Cockatoo (*Cacatua haematuropygia*) that have lost their primary habitat – inland forests (Luther and Greenberg 2009). Additionally, mangroves are enormously useful to humanity – they are important for carbon sequestration, fisheries, timber, medicines, and protection from storms, tsunamis, and sea-level rise. Therefore, numerous human communities also suffer from the loss of mangroves (Duke *et al.* 2007).

Massive habitat loss and conversion in the tropics has already heavily impacted birds and will continue to do so. Gaston *et al.* (2003) estimated that habitat conversion thus far may have reduced tropical forest bird densities at least by 15% since the pre-agricultural times, approximately 300 years ago (Ellis *et al.* 2010). Because they are apex predators, raptors are thought to be good indicators of the consequences of environmental change (see below). Habitat loss threatens 46% of diurnal tropical raptors, the highest percentage being threatened in the Austral tropics (Figure 1.11; Bildstein *et al.* 1998). The raptor decline may be caused by the high habitat loss in this region, but might also be a result of relatively better biological understanding of raptors there (Bildstein *et al.* 1998). Overall, "ecological health" (estimated as using species richness and demographic variables) of all birds declined by 18% in human-modified habitats compared with pristine forests in Southeast Asia (Sodhi *et al.* 2009b). As mentioned, habitat loss not only harms biodiversity, but also human well-being via the damaged delivery of key

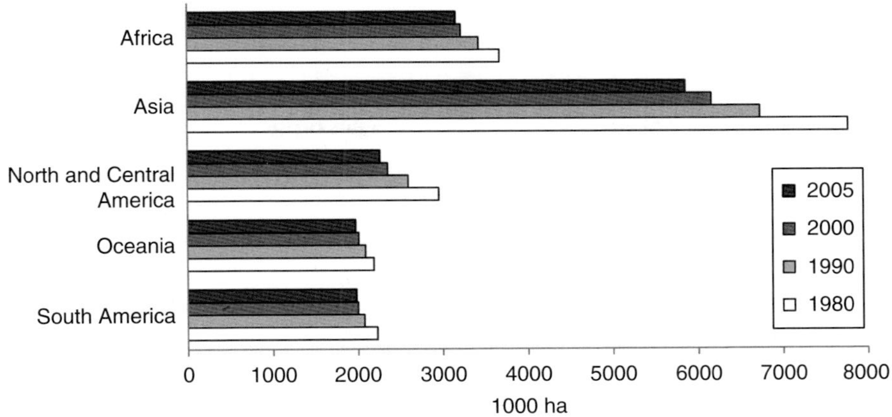

Figure 1.10 Changes in world mangrove area, 1980–2005. (After FAO 2007.)

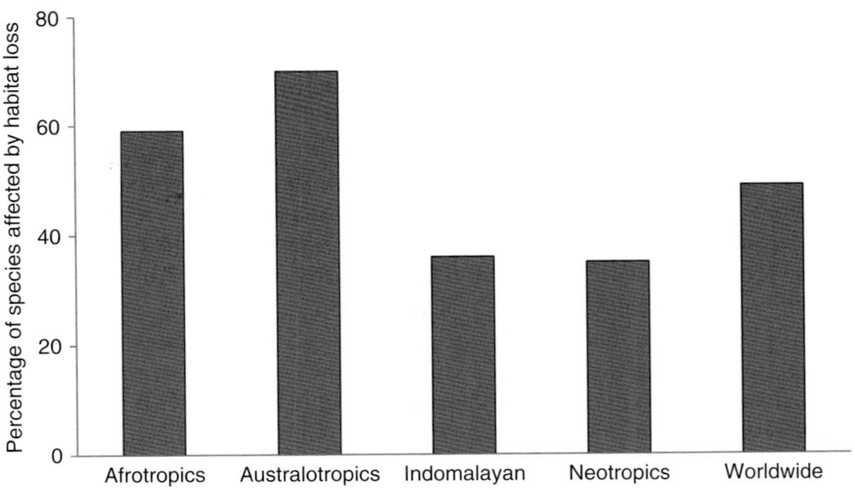

Figure 1.11 The percentage of tropical raptor species affected by habitat loss in different regions. (Data derived from Bildstein *et al.* 1998.)

ecosystem services such as pollination and flood protection (Balmford *et al.* 2002). We will further examine the consequences of habitat loss and degradation on birds in Chapters 2, 3, and 5.

1.2.2 Agricultural expansion and intensification

Agricultural areas remain the most ecologically degraded human-modified habitats (Sodhi *et al.* 2009b). However, a burgeoning human population means more mouths to feed, and agriculture remains the main factor in land conversion in the tropics. It is estimated that agricultural expansion is responsible for as much as 90% of tropical deforestation (Hardtër *et al.* 1997; Achard *et al.* 2002), 55% of which may be at the expense of intact forests (Gibbs *et al.* 2010). Globally, over the last three decades, croplands have doubled from 50 million ha to 1.4 billion ha (Niesten *et al.* 2004; Smith *et al.* 2010), and they now cover a quarter of Earth's land surface (Millennium Ecosystem Assessment 2005). By 2030, it is predicted that an additional 120 million ha of agricultural land will be needed by developing countries to support their increased populations (Jenkins 2003). Therefore, land clearing for agriculture is almost certain to continue at a rapid pace. In addition, due to factors such as low soil fertility and high levels of erosion, land conditions in many tropical areas are not particularly conducive to sustainable agriculture, thus promoting a cycle of forest destruction (Sodhi *et al.* 2007). Further, increased demands for meat production may increase the grazing by at least 10% by 2050 (Smith *et al.* 2010), further increasing competition for the land.

There are a number of quintessential examples of crop expansion in the tropics. For example, the production of soybean (*Glycine max*) has increased 100-fold since 1961 in Argentina and Brazil largely for import to China (Donald 2004). This has resulted in a severe shrinking of *Cerrado* grasslands (Donald 2004). These grasslands are home to 837 bird species, of which 3% are endemic (Klink and Machado 2005). Similar to soybean expansion, oil palm (*Elaeis guineensis*) cultivation has been expanding in the past few decades in the tropics, particularly in Southeast Asia (Koh and Wilcove 2008). In addition to cooking oil, palm oil is used in soaps, candles, cosmetics, and biodiesel. More than half of oil palm expansion in Malaysia and Indonesia has occurred at the expense of forests (Koh and Wilcove 2008). This expansion reduces the habitat for most forest birds, and studies conducted in Peninsular Malaysia and Thailand report a reduction in forest bird species richness in oil palms (Figure 1.12; Peh *et al.* 2005, 2006; Aratrakorn *et al.* 2006).

For immediate conservation, BirdLife International has identified 218 Endemic Bird Areas (EBAs) that contain the breeding ranges of at least two of 2623 restricted-range species (Stattersfield *et al.* 1998). EBAs cover about 14,500,000 km^2, mostly in the tropics and subtropics (Stattersfield *et al.* 1998). Since the 1700s, more land for crops and pastures has been used in EBAs than in the other areas (Figure 1.13; Scharlemann *et al.* 2004), suggesting that these regions remain vulnerable to future agricultural development.

Due to the expanding human footprint on the planet, conservation efforts in human-dominated landscapes or "countryside habitats" that encompass the diverse array of active agriculture, plantation or managed forests, fallow land, gardens, and small remnants of native vegetation are also needed (Daily *et al.* 2001; Gardner *et al.* 2009). About 37% of global land, some containing natural habitats, is embedded in countryside landscapes (Ellis *et al.* 2010). Surveys in these landscapes have indicated that they can harbor a substantial proportion of the regional avifauna, forest species included (Estrada *et al.* 1997; Daily *et al.* 2001; Hughes *et al.* 2002; Sodhi *et al.* 2005a; Ranganathan *et al.* 2008; see Chapter 2). For example, production forests (native forests undergoing extraction of non-timber products such as leaf litter) and areca nut plantations (*Areca catechu*; used for chewing purposes) had similar forest bird species richness, second only to intact forests in the Western Ghats (a global biodiversity hotspot) in India (Figure 1.14; Ranganathan *et al.* 2008). However, the value of the different anthropogenic land uses for maintaining bird biodiversity can vary (Peh *et al.*

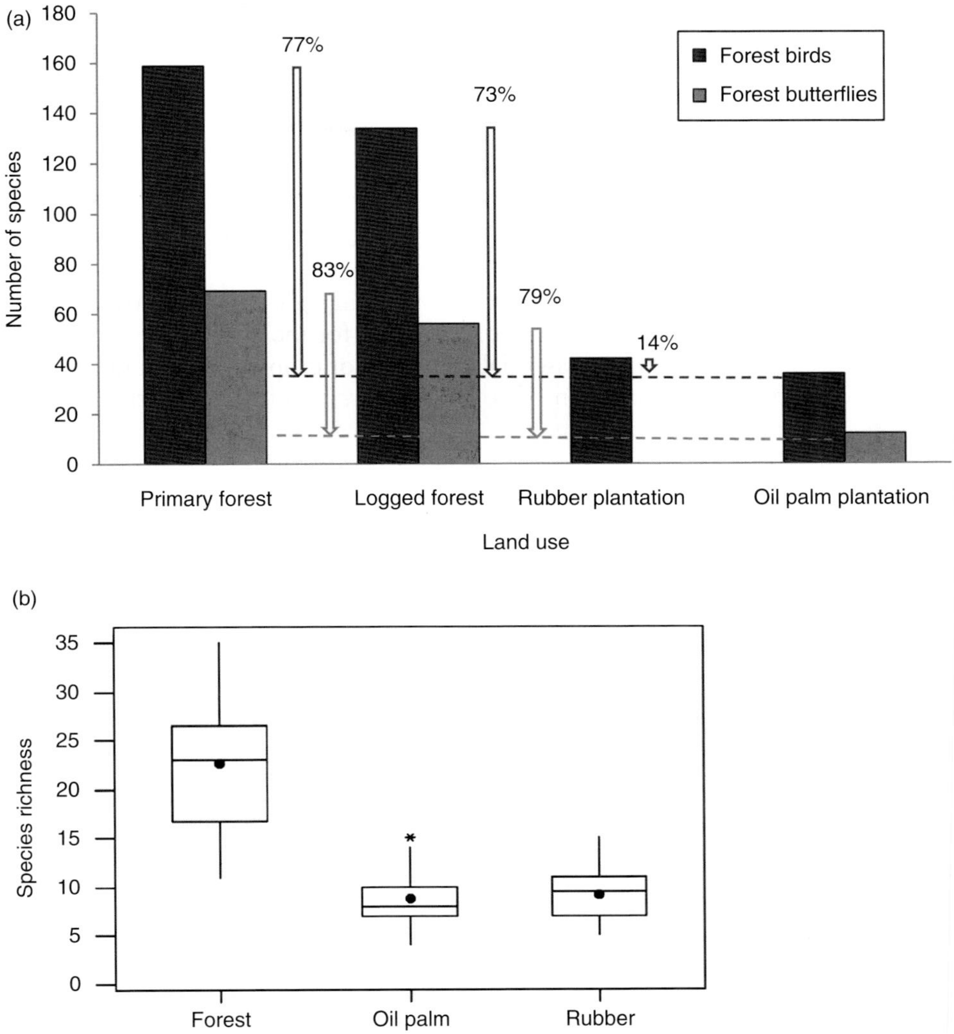

Figure 1.12 (a) Total number of species of forest birds and butterflies recorded from different land-use types in southern Malaysia and Borneo. (After Koh and Wilcove 2008.) (b) Bird species richness in forest, oil palm, and rubber in southern Thailand. The horizontal line shows the median and the box the interquartile range. Vertical lines represent the range of values between the interquartile and 1.5 times the interquartile range. Outlier is represented by asterisk. (After Aratrakorn *et al.* 2006.)

2005; Posa and Sodhi 2006; Soh *et al.* 2006). In some disturbed areas, as much as 80% of canopy cover is need to retain all forest bird species (Figure 1.15).

Low bird species richness is also observed in intensively managed plantations, especially in monocultures of non-arboreal annual crops (Estrada *et al.* 1997; Matlock *et al.* 2002; Waltert *et al.* 2004). Open country species dominate these habitats, as forest birds may be sensitive to the extreme climatic conditions present. Crops such as coffee, cacao and cardamom support a greater number of

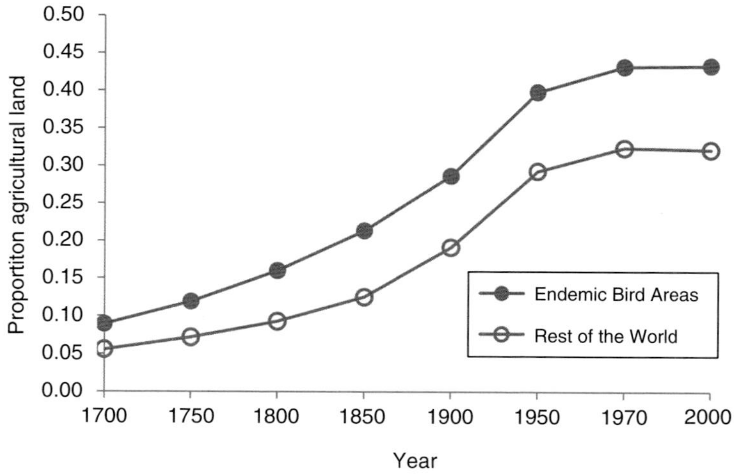

Figure 1.13 Historical changes in the proportion of land under crops and pastures in 139 Endemic Bird Areas and in the rest of the world. (After Scharlemann *et al.* 2004.)

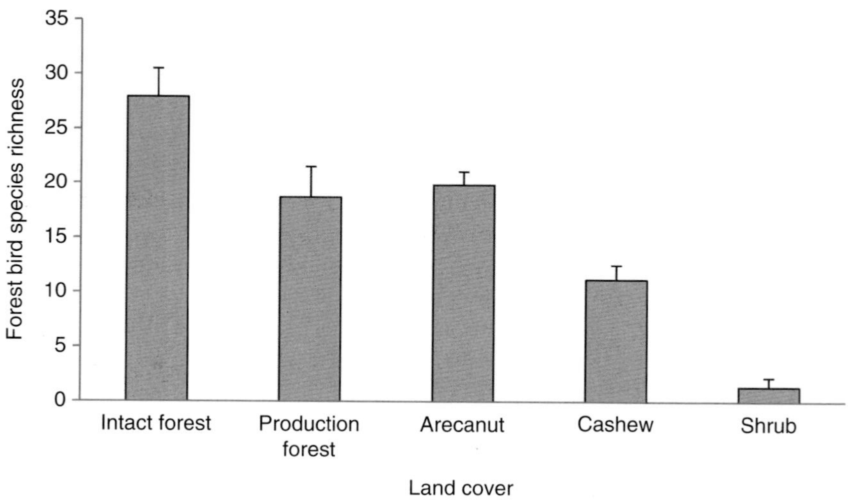

Figure 1.14 Forest bird species richness in each land cover in southwestern India. Error bars represent standard errors. (After Ranganathan *et al.* 2008, copyright National Academy of Sciences, USA.)

forest bird species, particularly if natural vegetation (i.e. "shade cultivation") is allowed to grow within cropfields (Shahabuddin 1997; Greenberg *et al.* 1997a,b; Estrada *et al.* 1997; Petit *et al.* 1999; but see Waltert *et al.* 2004). Remnant forest trees and riparian strips can be critical for the persistence of forest birds in tropical countryside (Sekercioglu *et al.* 2007; Hawes *et al.* 2008; Lees and Peres 2008a; see Chapter 2). Some primary forest birds can use older plantations of exotic trees that allow secondary growth (Mitra and Sheldon 1993) or traditional agroforests that are diverse and structurally complex (Thiollay 1995). However, species richness and diversity in these are still lower compared to primary forest

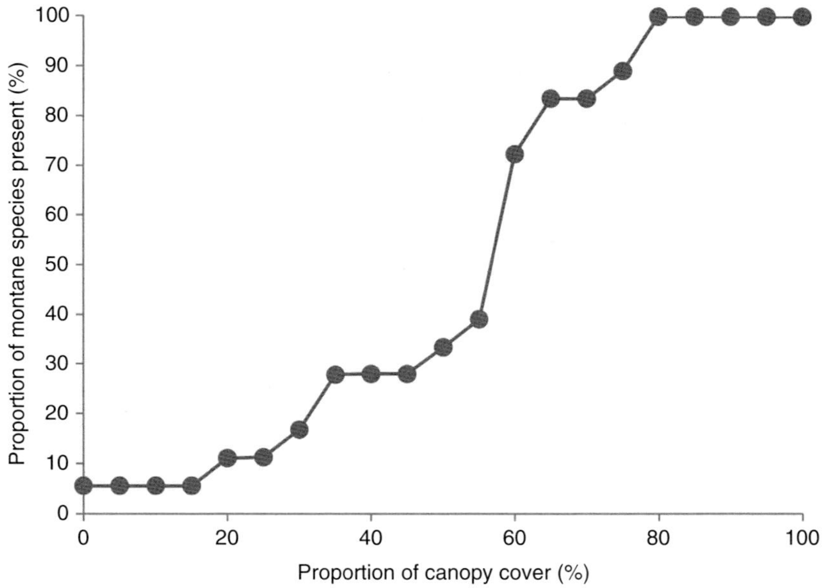

Figure 1.15 Graph showing proportion of montane species present for simulated proportions of canopy cover. (After Soh *et al.* 2006.)

(Tscharntke *et al.* 2008). Successional vegetation resulting from practices such as shifting cultivation (i.e. slash and burn) or from the abandonment of "permanent" agriculture can similarly be colonized by some forest birds. Species richness and abundance has been found to parallel vegetation recovery (Bowman *et al.* 1990, Blankespoor 1991; Andrade and Rubio-Torgler 1994; Raman *et al.* 1998). Secondary forest regrowth from agricultural fallows can contain a significant proportion of forest avifauna, as well as secondary-growth specialists (Blankespoor 1991; Raman *et al.* 1998). However these habitats are still suboptimal for forest-dependent species (Raman 2001), and over 40% of primary forest species birds are not found in biodiversity friendly matrix (areas surrounding remnant forests) habitats such as secondary forests and plantations (Barlow *et al.* 2007, 2010).

While large-scale agriculture and silviculture are generally detrimental to tropical forest birds, traditional systems of shifting agriculture practiced on small scales, with long intervals between burning and recropping, may have minimal impact on bird fauna (Zhijun and Young 2003). Nonetheless, the species richness of large forest frugivores and insectivores generally decline in agricultural areas and agroforests (Figure 1.16; Tscharntke *et al.* 2008). In contrast, nectivorous, granivorous, omnivorous, small insectivorous, and frugivorous species survive and sometimes flourish in such areas probably because of their superior abilities to track seasonal food resources. The sensitivity of insectivorous birds in agroforests may reduce crop yields as they play an important role in pest removal (Koh 2008; Van Bael *et al.* 2008). Therefore, for economic reasons alone, forest remnants should be preserved near or within agroforests (Koh 2008).

The degree of similarity between species assemblages in countryside habitats and pristine forest appear to be dependent on land-use patterns and landscape

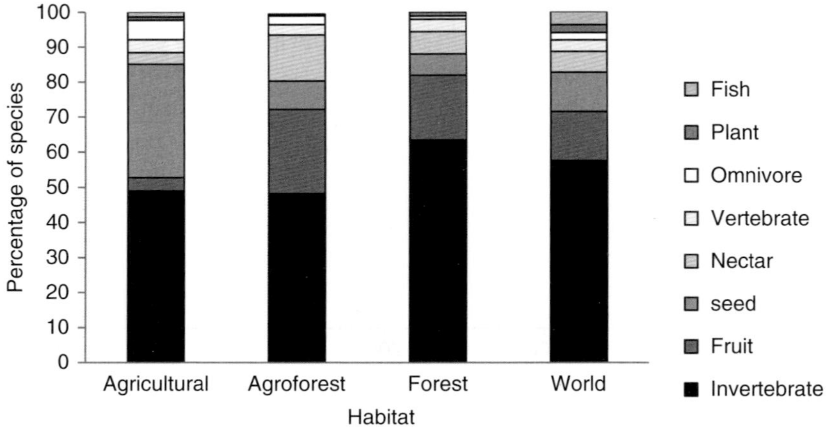

Figure 1.16 Relative bird species richness (percentage of all bird species) per habitat type based on primary diet, which is a proxy for ecological function. (After Tscharntke *et al.* 2008.)

context (Luck and Daily 2003). Pesticides adversely affect insectivores, as does the lack of leaf litter and low vegetational diversity in agricultural areas (Shahabuddin 1997), but these birds appear to benefit from insect pests in timber plantations (Mitra and Sheldon 1993). Although large frugivores generally do not benefit from the dominating crop trees of agroforests and are absent from plantations in some areas (Thiollay 1995; Shahabuddin 1997; Barlow *et al.* 2007), they have been observed in other mixed-rural habitats (Sodhi *et al.* 2005a). However, such an occurrence may be due to the close proximity of pristine forests to study areas (Figure 1.17; Barlow *et al.* 2007; Tscharntke *et al.* 2008). Overall, agricultural expansion remains a threat to tropical birds and better management of these areas is certainly needed to ameliorate this threat.

1.2.3 Selective logging

As mentioned, forests are being selective logging in many tropical countries. In selectively logged areas, forests are usually allowed to regenerate in gaps created by timber harvests, resulting in a forest structure with a mix of tree sizes and ages. Some species vacate an area when logging begins but return to it after it has been logged, but this pattern is not universal (Thiollay 1992; Dranzoa 1998). Studies indicate that many tropical forest species continue to survive in, or use, selectively logged forest. Studies from Borneo show up to 17% decline in bird species richness in logged compared to unlogged forests (Figure 1.18; Lambert 1992; Edwards *et al.* 2009; Berry *et al.* 2010). Some avian communities remain distinct from the original for periods of up to 10–15 years (Johns 1996; Thiollay 1997a, 1999). Logging can also result in significant changes in the relative abundance and composition of the avifauna, with an increase of widespread generalists or forest-edge species, compared with pristine forests (Johns 1996; Thiollay 1997a; Dranzoa 1998; Aleixo 1999). The influx of these types of species can sometimes results in high avian richness in logged areas.

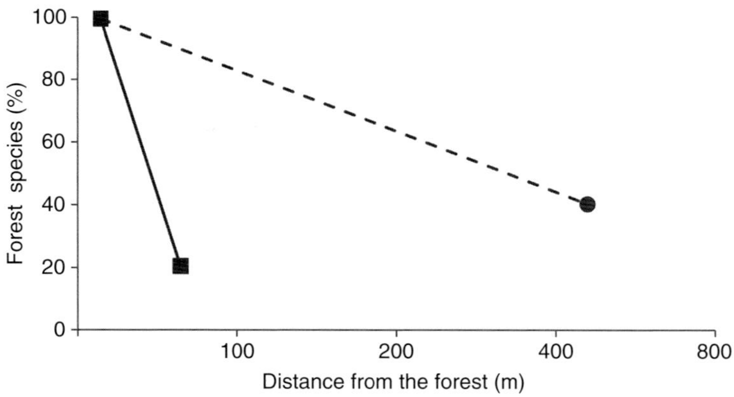

Figure 1.17 The percentage of forest species in agricultural areas decreases with increased distance from a nearby forest. (After Tscharntke *et al.* 2008.)

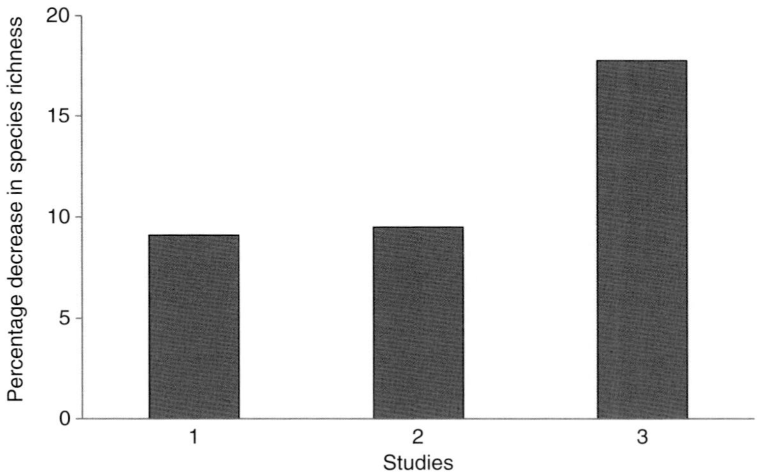

Figure 1.18 Species richness decline in selectively logged forests when compared with unlogged forests in Borneo. Figures based on three studies that are cited in the text. (Data derived from Berry *et al.* 2010.)

Selective logging affects various avian guilds differently. Some understory insectivores, as well as mixed-species flock members, are intolerant of the changes in microclimate and vegetation that occur after logging, because of their physiology and foraging-habitat specializations (Johns 1986; Mason 1996; Dranzoa 1998; Marsden 1998; Thiollay 1999). Guilds such as bark-associated insectivores and large-canopy frugivores (e.g. hornbills) also decline after large trees are lost (Johns 1989; Cleary *et al.* 2007). However, some studies report that nectarivores, generalist frugivores, omnivores, and gap or edge specialists seem to benefit from logging-related changes in vegetation (Lambert 1992; Owiunji and Plumptre 1998; Dranzoa 1998; Thiollay 1999). There is little evidence that logging disproportionately threatens rare species or those with restricted ranges (Thiollay 1997b; Marsden 1998). However, the richness and abundance of threatened bird species can be lower in logged than in unlogged forests (Edwards *et al.* 2010).

Many logging concessions are in proximity to primary forests from which birds can disperse, which may account for the minimal species loss and occurrence of some forest-dependent species in logged areas. For instance, when contiguous with unlogged forests, 75% of forest birds can persist even in forests that are repeatedly logged (Edwards *et al.* 2010). The magnitude of impact of logging on the avifauna can also depend on the management regime adopted by timber companies (Frumhoff 1995; Mason 1996; Sekercioglu 2002a). If harvest regimes do not either allow logged forests to regenerate or isolate them from unlogged areas, bird communities are unlikely to fully recover. However, one of the serious consequences of logging is increased access through roads, which can elevate hunting and further forest clearance (Laurance *et al.* 2009).

1.2.4 Infrastructure development and urbanization

Cities are expanding worldwide, with expectations that more than half of the world's total human population will be living in them by 2030 (Palmer *et al.* 2004). It is postulated that 70% of Earth's land surface will be impacted by infrastructure development in the next 30 years (UNEP 2002). Some regions are likely to be more affected than others, with more than 75% of all land area affected by such developments in the Caribbean and Asia-Pacific (UNEP 2002).

Many threatened bird species are projected to be negatively impacted by infrastructure development in the Asia-Pacific region (Figure 1.19). Unlike other land-uses, it is difficult, if not impossible, for vegetation to recover via natural succession in urban areas. Thus, impacts on biodiversity are likely to be long term (Pautasso *et al.* 2010). Urbanization increases biological homogenization, causing the extirpation of native species and promoting the establishment of non-native, urban-adaptable species that are becoming increasingly widespread and locally abundant across the planet (McKinney 2006). There is a near-complete absence of forest species in many tropical urban areas with human commensals such as Rock Doves (*Columba livia*) and House Crows (*Corvus splendens*) flourishing (Sodhi *et al.* 1999; Lim and Sodhi 2004; Pauchard *et al.* 2006; Posa and Sodhi 2006). However, some frugivorous species that can feed on fruit-bearing ornamental plants are able to persist in city parks and low-density housing areas (Petit *et al.* 1999; Lim and Sodhi 2004; Posa and Sodhi 2006). The presence of remnant forests may be the most important determinant of forest bird diversity in tropical cities (Sodhi *et al.* 1999; Lim and Sodhi 2004). Therefore, it is unfortunate that cities in developing tropical countries typically do not maintain natural forests in the urban environment (Pauchard *et al.* 2006; Posa and Sodhi 2006). Our understanding of the effects of urbanization in regions of high avian diversity, such as the tropics, is still rudimentary (Chace and Walsh 2006), and measures for urban conservation will be crucial in the coming decades as urban sprawl is set to replace native and rural habitats.

Infrastructure does not need be massive to affect birds. Linear infrastructure such as roads, power lines, gas and oil pipelines are common features of human

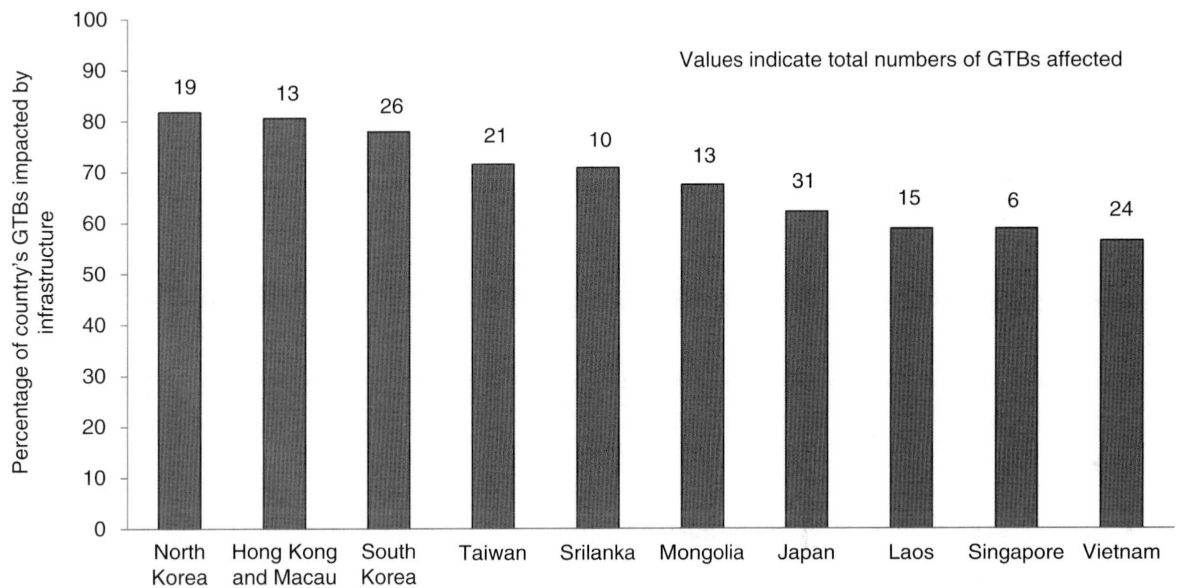

Figure 1.19 The ten countries in Asia with the highest percentage of their GTBs (Globally Threatened Birds) impacted by infrastructure development. (After http://www.birdlife.org/action/science/sowb/case_studies/p38-39.pdf)

Figure 1.20 Road-killed vulture in Panama. (After Laurance *et al.* 2009.)

activity. These are becoming ubiquitous in many tropical areas due to logging, mining, agriculture, and oil and gas development activities (Laurance *et al.* 2009). Tropical bird species become susceptible to hunting, road kill, elevated predation and invasions near these linear disturbances. Slower-flying bird species, and those whose activities coincide with heavy traffic (e.g. crepuscular species), remain vulnerable while crossing roads (Figure 1.20; Laurance *et al.* 2009). Movements of forest-dependent insectivorous bird species (e.g. those following army-ants or formed mixed-species flocks) were interrupted by road clearing in Amazon (Laurance *et al.* 2004). Generally, species avoiding the forest edges also avoided the road clearings.

In addition to linear infrastructure, hydroelectric projects also threaten terrestrial and aquatic tropical birds. Increasing energy prices and shrinking stocks of oil will facilitate more dams, and it is likely that large dams will eventually affect almost all major tropical rivers (Junk 2002). Dams lead to: (i) deforestation, especially of highly productive floodplain habitats and the loss of disturbance-dependent species of river edges; (ii) reduction in aquatic biodiversity; (iii) degradation of downstream habitats such as forests, estuaries, and deltas; and (iv) reduction in water quantity and quality (Junk 2002; www.birdlife.org/action/science/sowb/case_studies/p38–39.pdf). In Asia, existing or planned dam projects will endanger threatened bird species such as Masked Finfoot (*Heliopais personata*) and Indian Skimmer (*Rynchops albicollis*) (www.birdlife.org/action/science/sowb/case_studies/p38–39.pdf).

1.2.5 Environmental pollution

Pollution remains a threat to about 9% of globally threatened bird species, mainly by causing either elevated mortality or reduced reproductive success (BirdLife International 2008b). However, effluents released through agriculture, forestry, and industry further degrades habitats of 11% of globally threatened species (BirdLife International 2008b). Tropical ecotoxicology is still in its infancy and thus impacts of pesticides on tropical biotas are poorly understood (Lacher and Goldstein 1997). Nonetheless, fertilizers and pesticides (i.e. herbicides, insecticides, and fungicides) used in agricultural areas adjacent to tropical forests can take a toll on tropical forest biotas. Because chemicals tend to accumulate up the food chain, top predators are likely to be the most affected by toxic substances used in pesticides and fertilizers. For instance, severe reduction in the range of Mauritius Kestrels (*Falco punctatus*) in the 1970s may have been brought about due to heavy DDT (dichlorodiphenyltrichloroethane; an organochlorine) usage on the island to control malaria-spreading mosquitoes (Safford and Jones 1997). Raptors have been among the most prominent victims of organochlorine insecticide use because they are at the top of their food web and accumulate organochlorines through the food chain.

Environmental pollutants, for example, have been implicated in the dramatic declines of the breeding populations of various raptors in the 1970s and 1980s in North America and Europe. Impacted species included the Peregrine Falcon (*F. peregrinus*), Bald Eagle (*Haliaeetus leucocephalus*), Golden Eagle (*Aquila chrysaetos*), and Osprey (*Pandion haliaetus*) (Wiemeyer *et al.* 1984, 1988; Ratcliffe 1993; Nygåard and Gjershaug 2004). These species were exposed to a "cocktail" of organochlorines that included DDT, aldrin, dieldrin, and heptachlor, as well as PCBs (polychlorinated biphenyls; non-insecticides with many uses, such as coolant) (Fry 1995). Negative effects included severe declines in reproductive success (Newton 1979). Reproductive success was compromised because of the production of thin egg-shells, which could break under the weight of an incubating parent (Newton 1979). These effects were exacerbated by high embryo and nesting mortality. Declining reproductive success eventually resulted in epic population declines in some raptors in Europe and North America (Newton 1979). DDT was banned in various countries in the 1970s and 1980s but it is still being used for malaria control in tropical countries such as India (www.who.int/malaria/en/). Since the ban of DDT in North America, populations of many raptors have recovered, aided by aggressive captive breeding and release programs, such as the one for Peregrine Falcons (Ratcliffe 1993).

In most countries, DDT and related organochlorine insecticides have been usually replaced by organophosphate and carbamate chemicals. However, sadly these chemicals also remain a threat to birds. In 1995, 700 Swainson's Hawks (*Buteo swainsoni*) perished in Argentina, presumably because of an insecticide

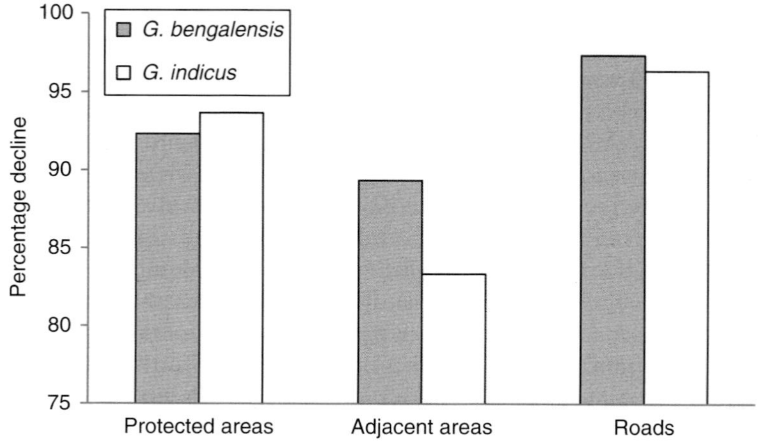

Figure 1.21 The percentage of decline in the number of two vulture species in India. Declines are based on surveys conducted in 1991–1993 and 2000. (Data derived from Prakash *et al.* 2003.)

(organophosphate monocrotphos) extensively used to eliminate grasshoppers (Woodbridge *et al.* 1995). As a result of pesticide-mediated mortality in wintering grounds, Swainson's Hawks have been suffering population declines in parts of its breeding range in North America (Sarasola *et al.* 2008). As in northern areas, raptors such as Grey Falcon (*F. hypoleucos*) and Eastern Grass-owl (*Tyto longimembris*) in tropical Australia (Mooney 1998) remain vulnerable to pesticides. Environmental contaminants threaten 11% of diurnal tropical raptors, with the highest risk in the Austral tropics (Figure 1.11; Bildstein *et al.* 1998). Raptors in other tropical regions also remain susceptible. For example, fenthion used to eradicate Red-billed Quelea (*Quelea quelea*) causes the mortality of raptors when they feed on treated individuals (Keith and Bruggers 1998).

The quintessential example of the effects of environmental contaminants on birds comes from tropical vultures. Since 1990s, there has been over 92% decline in vulture populations in India (Figure 1.21; Prakash *et al.* 2003). Veterinary use of the anti-inflammatory drug diclofenac appears to be behind these declines (Oaks *et al.* 2004; Green *et al.* 2006). Vultures are likely exposed to diclofenac when they feed on the carcasses of livestock that were treated with this drug. Diclofenac residues in vultures result in renal failure and visceral gout (build up of uric acid in internal organs) (Oaks *et al.* 2004). Possibly because of diclofenac-related effects, breeding populations of the White-backed Vulture (*Gyps bengalensis*) have been obliterated from the Keoladeo National Park (India) (Figure 1.22; Prakash *et al.* 2003). To reverse the decline of vultures, the Indian government banned diclofenac in 2006 (www.birdlife.org/news/news/2006/05/vulture_update.html). Pharmaceutical companies in the Indian subcontinent are now promoting meloxicam as an alternative to diclofenac, which is considered to be safe for vultures (Swarup *et al.* 2007). However, it is unclear if measures taken have reversed the vulture declines.

Figure 1.22 The decline in the number of active nests of the White-backed Vulture at Keoladeo National Park (India). (Data derived from Prakash *et al.* 2003.)

Noxious chemicals used by humanity also wash into local waterways, affecting aquatic birds (Junk 2002). Despite the negative effects of pesticides, their exports to tropical countries have been steadily increasing. For example, in 2006, US$73–100 million worth of these chemicals were exported to these countries (www.fao-statistics.org). International pesticide companies facilitate pesticide use in tropical developing countries by selling cheap pesticides with expired patents (Wanger *et al.* 2010).

1.2.6 Other drivers that threaten tropical birds

In addition to the drivers described above, and those that will be discussed in the following chapters, tropical birds remain vulnerable to factors such as drought, hurricanes, and longline fisheries. Fisheries bycatch remains a formidable threat to tropical birds: an estimated 200,000 birds perish annually in fishing nets (Zydelis *et al.* 2009). Long-lived seabirds such as albatrosses remain particularly vulnerable to mortality through fishing nets because the chances of population recovery are low due to delayed and slow reproduction (Veran *et al.* 2007). For example, annual adult survival of the Galapagos Waved Albastross (*Phoebastria irrorata*) declined between 1999 and 2005, likely due to high mortality caused by fisheries (Awkerman *et al.* 2006). It is now endangered with extinction (BirdLife International 2010a). Significant longline fishing operations are currently being conducted in tropical waters (see Huang 2009), so the fishing industry should be required to devise and employ seabird avoidance methods. Tori lines (streamer lines) and establishing protected areas near seabird colonies may be effective measures to reduce seabird mortality through fisheries (Gilman *et al.* 2005; Mancini *et al.* 2009). In addition to fisheries, tropical seabirds remain susceptible to oil-spills (Gonzalez-Solis *et al.* 2007).

Other disasters such as hurricanes, cyclones, and typhoons can also be catastrophic for tropical birds. Soon after a hurricane, there can be heavy mortality of forest birds (Will 1991). Nectivorous and frugivorous bird species (e.g. toucans, tinamous, parrots, and hummingbirds) seem to be severely impacted by hurricanes (Lynch 1991; Tanner *et al.* 1991; Wunderle 1991), probably because of factors such as lack of fruits and flowers in hurricane-damaged forests (Askins and Ewert 1991). However, generalist foragers usually become more numerous in hurricane-damaged forests (Tejeda-Cruz and Sutherland 2005). Migratory birds also remain vulnerable to hurricanes. Two neotropical migrant species, the Black-and-white Warbler (*Mniotilta varia*) and Northern Parula (*Parula americana*), had lower numbers on St John (US Virgin Islands) four months after than in the two years before Hurricane Hugo (Askins and Ewert 1991). Even two years after the hurricane, hurricane-damaged forests lack typical lowland rainforest birds such as woodcreepers, forest antbirds, and furnaiids (ovenbirds) (Will 1991). Therefore, hurricane-damaged forests may require active management (e.g. reforestation) to attract and retain forest bird species.

1.3 The enigmatic

Ornithological knowledge in the tropics remains limited. In every tropical region, birds remain poorly studied compared to non-tropical birds (Figure 1.23; Sodhi *et al.* 2007). This is attested by the fact that there are high numbers of

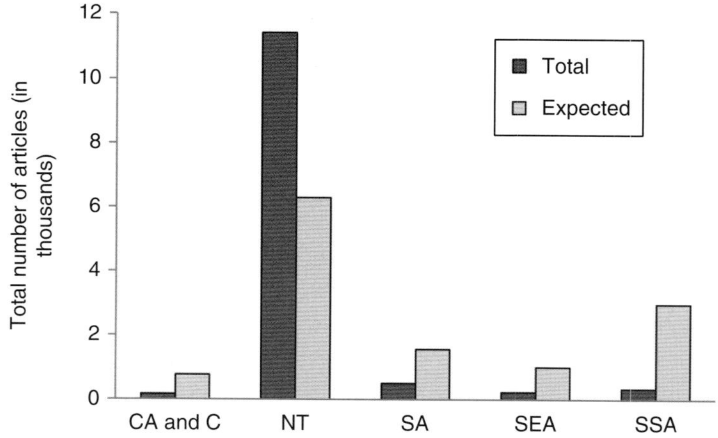

Figure 1.23 Total number of published (dark bars) and expected (light bars) articles on biodiversity-related issues of birds among tropical regions and non-tropical regions. All comparisons were based on the number of internationally peer reviewed research articles (excluding marine studies) published between 1986 and 2005 extracted from the database BIOSIS Previews. For each taxonomic group, the expected number of publications for each geographical region was calculated by dividing the number of publications evenly among geographical regions, weighted by the total number of recorded species per region for that particular taxonomic group. Abbreviations: SEA = Southeast Asia; SSA = Sub-Saharan Africa; SA = South America; CA and C = Central America and Caribbean; NT = non-tropical regions.

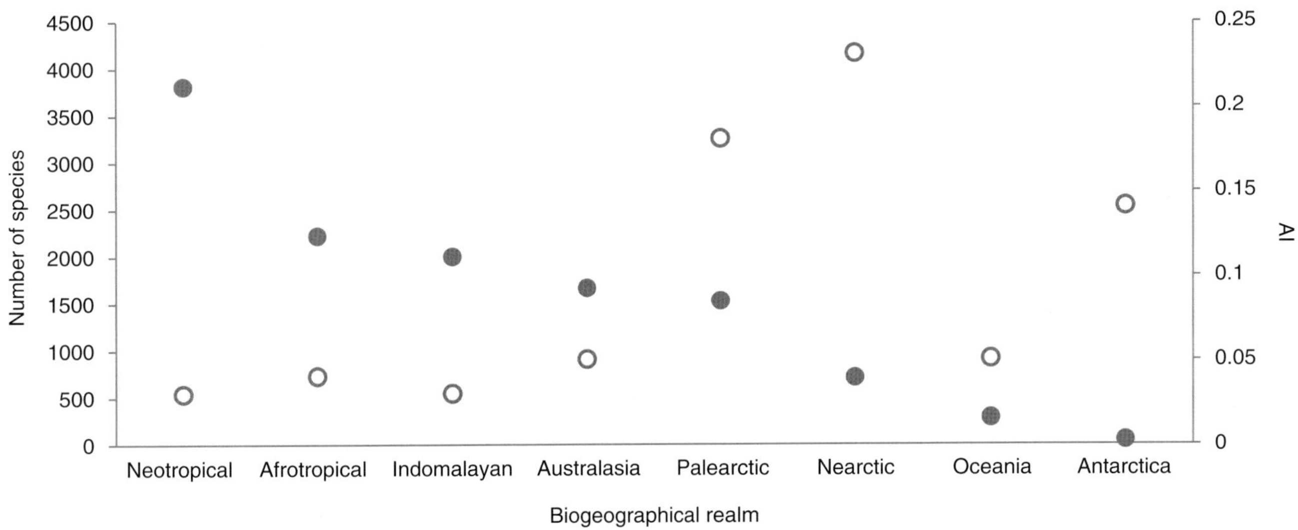

Figure 1.24 Biogeographical attention index (open dots) and total number of bird species (filled dots) for each biogeographical realm. AI = attention index was obtained by dividing the number of papers by the number of species. (After Brito and Oprea 2009.)

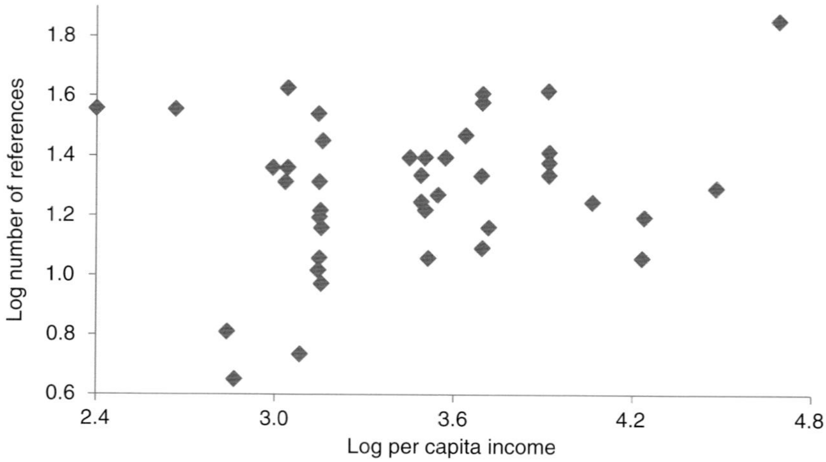

Figure 1.25 The relationship between numbers of references for a random sample of 50 threatened bird species and per capita income in their range countries. (After Brooks *et al.* 2008.)

data-deficient birds in the tropics than other regions (Figure 1.1; Sodhi *et al.* 2008). Despite higher species richness and endangerment in tropical areas, the highest number of conservation-related scientific articles on birds comes from the Nearctic and Palearctic regions (Figure 1.24; Brito and Oprea 2009). Lack of adequate research funds and personnel may hinder research in the tropics. In fact, more studies on threatened bird species are conducted in richer than in poorer countries (Figure 1.25; Brooks *et al.* 2008), suggesting that there is a greater need to develop ornithological research in poor developing tropical countries. It is disconcerting that 600 threatened bird species do not have a single scientific article published on their biology (Brooks *et al.* 2008). Another problem is that the academic establishment tends to value and fund global meta-analyses and modeling papers based on existing data more than field-based papers on the biology of individual species, which results in a disincentive for young scientists to do long-term field research under difficult conditions. Patently, more research on tropical birds, especially the threatened species, is urgently needed.

2

Effects of Habitat Fragmentation on Tropical Birds

Overview: Habitat fragmentation is a characteristic feature of most terrestrial ecosystems, and the long-term prospects of many tropical bird species ultimately depends on their ability to persist in fragmented landscapes. We discuss the theoretical premises that underpin much of the work on habitat fragmentation, and then examine the empirical evidence regarding the conservation importance of fragment area, edge effects, and the role of the matrix (the land surrounding the patches of native habitat). Finally, we examine temporal changes in forest fragments, which suggest the full consequences of present-day fragmentation may take decades or centuries to become apparent.

2.1 Introduction

Habitat fragmentation occurs when a large expanse of a particular, broadly defined habitat type is reduced to smaller fragments (also called "remnants," "patches," and "habitat isolates") that are isolated by surrounding, but different habitats (generally termed "matrix" habitats – Wilcove *et al.* 1986). Although some ecosystems are naturally fragmented, the vast majority of observed habitat fragmentation is anthropogenic in origin (Fahrig 2003).

Habitat fragmentation invariably occurs when areas of forest, or other native vegetation types, are subdivided and patches become isolated by expanding urban areas, agriculture, and other types of land uses. Millions of hectares of tropical forests currently exist as fragments (Wright 2005), and they continue to be created at a rapid rate. For example, Broadbent *et al.* (2008) estimate that deforestation created over 5500 new fragments in the

Conservation of Tropical Birds, First Edition. Navjot S. Sodhi, Çağan H. Şekercioğlu, Jos Barlow and Scott K. Robinson.

Published 2011 by Blackwell Publishing Ltd.

Brazilian Amazon between 1999 and 2002 alone, while around 80% of all patches of the Atlantic Forest are smaller than 50 ha (Ribeiro *et al.* 2009). In heavily modified areas these remnant fragments play a critically important role in preventing a catastrophic loss of species, and habitat fragmentation has become one of the most studied issues in conservation biology (Fisher and Lindenmayer 2007). It is generally assumed that fragmentation effects are currently stronger in tropical than in temperate systems (Baldi 1996; Fahrig 2003; Lindell *et al.* 2007), possibly because temperate regions were fragmented so long ago that they have already lost the vulnerable area-sensitive species (Henle *et al.* 2004).

Because habitat fragmentation has become such a dominant area of ecological research, the term has come to be used in a broad way that encompasses many different areas (Fisher and Lindenmayer 2007). For example, habitat fragmentation is often cited as being one of the greatest threats to terrestrial biodiversity, but this is conceptually misleading as it does not distinguish from the direct effects of habitat loss *per se* (Tilman *et al.* 1994), which can often exert a much greater effect on species presence in a landscape (Fahrig 2003; Zurita and Bellocq 2010). To clarify the terms we employ in this chapter, we follow the review of the "State of the World's Birds" (Birdlife International 2008b) where fragmentation is recognized as the splitting apart of habitat as a consequence of expanding and intensifying agriculture (and other land-use change), rather than a threat to biodiversity *per se*.

We first discuss the theoretical premises that underpin much of the work on habitat fragmentation. Next, we assess evidence relating birds with the major consequences of fragmentation, examining in turn the importance of area-effects, edge effects and degradation, and issues relating to connectivity (isolation, the composition of the matrix, and corridors). We then evaluate the time-lags associated with many fragmentation effects which are fundamental for assessing the longer-term consequences for biodiversity. Finally, we examine the synergistic interactions between different fragmentation-related processes, and discuss the general conservation implications that can be drawn from over four decades of fragmentation research.

2.2 Theoretical premises of fragmentation

The theoretical basis for research on habitat fragmentation can be traced back to the equilibrium theory of island biogeography (MacArthur and Wilson 1967). The principle premise of this theory states that the number of species on an island would reflect a balance between the rate at which populations of established species are lost and the rate at which new species and populations colonize. The conservation implications of this theory are twofold, and relate to area and isolation. First, island biogeography theory (IBT) predicts that all else being equal larger fragments will always hold more species than smaller fragments due to the lower rate of species loss. Second, less isolated (better connected) fragments will hold more species than more isolated fragments, due to the increased chance of colonization.

Despite being reassuringly intuitive, the relevance of theory for conservation and reserve design was soon questioned by Simberloff and Abele (1976), who argued that proof that large reserves were always better for biodiversity was inconclusive. This developed into the "SLOSS" debate (Single Large or Several Small), which centred on whether a single large or several small reserves would be more effective for preserving species subject to fragmentation. Although the SLOSS debate spawned many interesting studies, it has been suggested that the scientific minutiae may have hampered adequate reserve/fragment planning and management, for which size is only one of many factors that need to be considered to guarantee effective conservation (Saunders *et al.* 1991).

More recently, the limitations of island biogeography theory for conservation have been addressed in a number of excellent reviews (e.g. Haila 2002; Fahrig 2003; Ewers and Didham 2006; Laurance 2008a,b). Overall, it seems that a theory based on islands surrounded by a very hostile matrix (IBT was based on patterns of species evolution and dispersal across oceanic island archipelagos) is difficult to apply to human-modified terrestrial landscapes that are comprised of a complex patchwork of vegetation types (Gilpin and Diamond 1980; Haila 2002; Prugh *et al.* 2008). The influence of area and isolation are also confounded by the role of edge effects and habitat degradation (Ewers and Didham 2006; Laurance 2008a) (see Section 2.4).

Other important limitations of island biogeography theory relate to the fact that habitat conversion (to agriculture or urban uses) is almost always non-random, and habitat characteristics such as rainfall, soil quality, and elevation all exert a strong influence on the likelihood that certain areas will be converted for agricultural, urban or industrial use (Seabloom *et al.* 2002). As a result, protected areas and other remaining patches of native habitat are almost always situated on marginal land, characterized by qualities such as steep terrain, a propensity to flood, poor soils, or their remoteness (Margules and Pressey 2000; Laurance 2008a). At the same time, species distributions are also non-random, and are often influenced by similar underlying habitat characteristics. The combination of these non-random patterns has resulted in accelerated predictions of species-loss in plant communities in California (Seabloom *et al.* 2002). The fragmented tropics are likely to have been similarly affected, but our very limited biological knowledge of the areas that have already been converted mean it is almost impossible to test for these effects.

2.3 Area effects in tropical birds

Area effects are probably the most studied aspect of habitat fragmentation. Reduction in forest area can result in losses in habitat diversity and resources, with ensuing declines in bird populations that make them more vulnerable to local extinction (Brook *et al.* 2003a). Small populations are more vulnerable to extinction due to their heightened susceptibility to stochastic events (Gilpin and Soulé 1986) and the loss of genetic variability (Spielman *et al.* 2004). We review some of the empirical evidence for area effects in tropical birds, and examine the issue of thresholds in habitat loss at the landscape level.

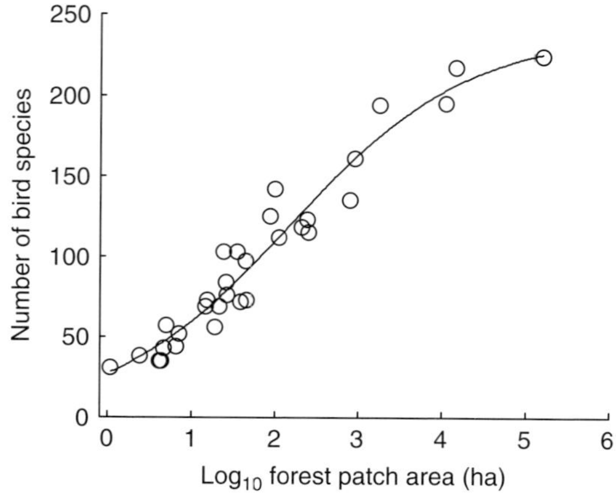

Figure 2.1 Log-linear species–area relationships for all bird species occurring in the 31 forest patch sites in the Brazilian Amazon, ranging from 1.2 ha to 14,476 ha. (Adapted from Lees and Peres 2008a.)

2.3.1 Empirical evidence for area effects, and some caveats

Many studies show that area is a good predictor of the species richness of forest birds in fragmented tropical landscapes. For example, Lees and Peres (2008b) found an almost perfect fit between area and the species richness of the entire bird community in 31 forest patches in the Brazilian Amazon along the "Arc of Deforestation" (Figure 2.1). Similar area-dependent results have been observed from a variety of tropical environments. The species richness of forest-dependent avifauna increased with fragment size in the Cerrado region of central Brazil (Marini 2001). Watson *et al.* (2004) found that remnant area was the main predictor of forest bird species richness in Madagascar. Newmark (1991) found that smaller fragments held fewer species of understory birds than larger control fragments in the Usambara Mountains in Tanzania. Over one-third of West Africa forests birds were shown to be area sensitive in 35 fragments in Ghana (Beier *et al.* 2002). Similarly, Van Balen (1999) sampled birds of 19 lowland forests (6 to 50,000 ha) on the island of Java and found that 30 forest species (i.e. those found in forest edges and interior habitats) occurred only in forests > 1000 ha in area, and that 13 of these were restricted to forests > 10,000 ha.

These studies illustrate the general rule of thumb in conservation ecology, that protection of larger patches is better for maintaining bird species richness and species of high conservation value. They also seem to provide strong empirical support for the species–area relationship, and the argument that many species of forest bird require large patches of relatively undisturbed habitat to persist. However, in some cases species–area relationships can be partially explained by sampling effort, where more species are observed in the larger patches simply because the sample effort is often higher in these patches (Cam *et al.* 2002). Furthermore, strong species–area relationships may also be explained by factors that are collinear with changes in patch area, including edge effects, isolation, connectivity, and degradation (Ewers and Didham 2006). For example,

Ferraz *et al.* (2007) found that both area and isolation affected forest birds in an experimental fragmentation landscape near Manaus, Brazil. Furthermore, also in the Amazon, Lees and Peres (2006) showed that forest quality (canopy disturbance from logging or burning) was also an important predictor of the most sensitive forest-dependent species. Studies in the Brazilian Atlantic forest have found that area effects were much weaker when edge effects were taken into account (Banks-Leite *et al.* 2010), and that connectivity can be more important than fragment area for most community metrics (Martensen *et al.* 2008). We will assess the relative importance of these effects in detail in the rest of the chapter, as they are critical to our understanding of the importance of fragmentation for tropical bird conservation.

2.3.2 Thresholds at the landscape level

One particular issue regarding area relates to the possibility of threshold fragmentation effects, which were defined by Andrén (1994) as "rapid changes in the size and isolation of patches at critical proportions of habitat in the landscape" that have "important consequences for species persistence in fragmented landscapes." The search for general thresholds in landscape change (e.g. percent cover of native vegetation) is appealing, and would provide clear guidelines for the management of native habitats. However, most research on this subject has been theoretical and there is only limited empirical support for thresholds (Lindenmayer *et al.* 2005), although a recent study in the Brazilian Atlantic forest suggests that measures of bird community similarity may show a sudden decline when the forest cover falls below 20–40% of the landscape (Zurita and Bellocq 2010; also see Figure 1.15).

The overall utility of thresholds for developing general management rules is likely to be limited, as different species in an assemblage are unlikely to share common traits such as home range, territory size and reproductive capacity, and many species could be lost above established thresholds (Mönkkönen and Ruenanen 1999). The importance of species-specific traits were demonstrated by comparing two theoretical species of songbird (With and King 2001), where the level of habitat required for species persistence varied from as little as 5% to as much as 90%, depending on the sensitivity of the species to edges and the pattern and distribution of the remaining habitat in the landscape.

2.4 Edge effects

The loss and fragmentation of native habitat generates many new edges. For example, the combined effect of deforestation and logging created as much as 38,000 km of new edge each year in the Brazilian Amazon between 1999 and 2002 (Broadbent *et al.* 2008). The amount of edge created during habitat loss is dependent on the size and shape of the resulting fragments. Larger fragments tend to have a greater immunity to the negative influences of fragmentation due to their larger core area (i.e. the region unaffected by edge effects) and relatively smaller proportion of edge.

2.4.1 Physical, biological and human driven edge effects

The consequences of edge creation are known to include a mixture of physical, biological and social phenomena that occur at the interface between contrasting habitats. Physical edge effects have been well studied where tropical forests adjoin agricultural lands, where forests closer to the edge are exposed to increased solar radiation (Saunders *et al.* 1991), higher air temperatures and increased soil heating (Camargo and Kapos 1995; Murcia 1995). In addition, edges are more exposed to wind (Laurance and Curran 2008), resulting in elevated vegetation mortality, either through direct physical damage or by increased drought stress (Saunders *et al.* 1991).

These physical effects are accompanied by a suite of biological phenomena. One of the most noticeable in tropical forests is the change in vegetation structure and composition, as edges become dominated by ruderal and light-demanding pioneer species (Gascon *et al.* 2000; Tabarelli *et al.* 2008). The tropical forest bird community is strongly linked to variables such as canopy openness and the density of vegetation in the understory (e.g. Barlow and Peres 2004a), and any changes in forest structure are likely to result in the loss of primary forest specialists and an increase in species associated with secondary forests and non-forest habitats.

Edge effects are also thought to affect bird communities by increasing the number of "forest-avoiding" generalist predators. For example high nest predation is thought to have been one the principal factors explaining the loss of forest birds from Singapore during the 20th century (Wong *et al.* 1998). Because predation events are difficult to observe, artificial nest experiments have been used to compare predation pressure among sites. Burkey (1993) reported higher predation on chicken eggs in rain forest edges than in forest interiors in Belize and Mexico. Artificial nest predation has also been reported to be higher in fragments than in continuous forests (Loiselle and Hoppes 1983; Sieving 1992). In Singapore, 80% (n = 328 nests) of artificial ground nests were depredated, with primary forest experiencing at least 12% less predation than other fragments (Wong *et al.* 1998). Sixty-two percent of arboreal nests (n = 110) were also depredated (Sodhi *et al.* 2003). Despite the evidence, doubts still remain about the importance of these nest predation events, as artificial nest experiments may mimic natural predation events poorly because of factors such as lack of parental defence and differences in the characteristics of native and "exotic" eggs that are used in most experiments. A review of 120 studies examining nest predation in relation to fragmentation concluded that the responses of nest predators to fragmentation is complex, taxon-specific, and context-dependent (Chalfoun *et al.* 2002).

There are also a suite of edge effects that can be attributed to increased human activity around forest fragments. Many of these degrade forest structure close to the forest edge. Tropical forest edges are also more vulnerable to fire, especially when surrounding matrix such as pastures are frequently burned (Alencar *et al.* 2004; see Chapter 5), with devastating impacts on the specialist tropical forest bird community (Barlow and Peres 2004a). Forest edges are often associated with invasion by livestock, which can impact directly on the composition of bird species found at the edges of tropical forests (Kruger and Lawes 1997; Lees and

Peres 2008a). Selective logging often occurs at such broad scales that it is not considered as an edge effect (see Chapter 1), although it normally follows that areas closer to the forest edge are more likely to be logged that those much further away, where physical access is difficult. Selective logging impacts the forest bird community in a similar way to fire (see Chapter 5), reducing the number of forest specialists adapted to the dark understory of primary forest (Mason 1996; Barlow *et al.* 2006; Felton *et al.* 2008).

Some human-induced edge effects are more cryptic, as they do not have a direct influence on forest structure. For example, fragments may also provide easy access to hunters and poachers. Peres (2001) argued that in heavily fragmented areas exposed to hunting, most mid-sized to large vertebrates may be driven to local extinction in the Brazilian Amazon, including larger game birds such as curassows (*Crax* and *Mitu* spp.), guans (*Penelope* and *Pipile* spp.), and large tinamous (*Tinamus* spp.) (see also Chapters 3 and 7). Although very poorly studied in the tropics, chemical compounds such as fertilizers and pesticides could also have negative impacts on forest birds as they drift from surrounding agricultural areas into forest edges (Murcia 1995; Weathers *et al.* 2001; see Chapter 1).

For physical and biological edge effects, the magnitude and distance of any impact is likely to depend upon the contrast in structure and composition between the adjacent systems, extreme temperatures and high winds, as well as the presence of invasive species (Harper *et al.* 2005). Within a tropical context, edge effects are likely to be most pronounced where dense tropical forests meet open agricultural land or water, and are likely to be less apparent in more open tropical dry forests or Savanna systems (Harper *et al.* 2005). They can also be buffered by the surrounding matrix (see Section 2.5.2). However, human-mediated edge effects may be just as important in these more open tropical systems, which can be strongly modified by altered fire regimes in the surrounding landscape (see Chapter 5) and hunting (see Chapter 7).

2.4.2 Can edge effects explain area effects?

Edge effects may be so important that they can even swamp the area-related effects of fragmentation (Ewers and Didham 2006; Laurance 2008a). A study by Banks-Leite *et al.* (2010) demonstrates the potential importance of these edge-related effects for tropical forest birds. The researchers collected data on edge-related differences in vegetation structure and bird community composition at 31 sites in fragmented and continuous forest in the Atlantic forest of Brazil. Fragments ranged from 3 to 145 ha in size, and were all located in the secondary forest patches that typify this critically threatened biome. The response of birds to edge effects and area effects were positively correlated (Figure 2.2). Furthermore, the magnitude of edge to interior differences on bird community composition declined markedly with patch size, and community composition was no longer affected by patch area after controlling for edge effects. These results indicate that many of the area-related effects on birds in fragmented landscapes may in fact be explained by edge effects, and demonstrate the importance of maintaining larger and more regularly shaped fragments wherever possible.

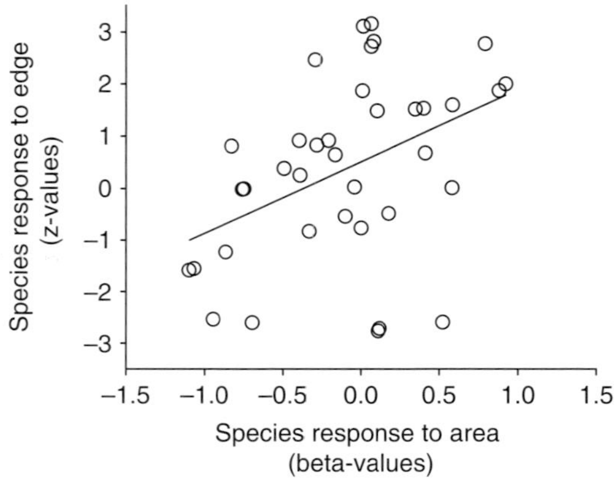

Figure 2.2 Area effects can be mediated by edge effects, and species-level responses to area and edge are often correlated. (Adapted from Banks-Leite *et al.* 2010.)

Figure 2.3 A narrow and degraded riparian forest strip in the São Felix region of the Brazilian Amazon. Evidence suggests these strips need to be at least 400 m wide to have a chance of conserving most tropical forest species (Lees and Peres 2008b; figure 2.4). (Photograph by Toby Gardner.) (Please see Plate 2.3 for the color version of this figure.)

2.4.3 Fragment shape and the conservation value of linear forest remnants

Fragment shape is an important determinant of the importance of edge effects, especially within smaller fragments where relatively small increases in fragment irregularity greatly reduce the remaining interior habitat (Saunders *et al.* 1991). Linear remnants of habitat are one of the most extreme shapes, but are a common feature of many managed tropical landscapes (Figure 2.3).

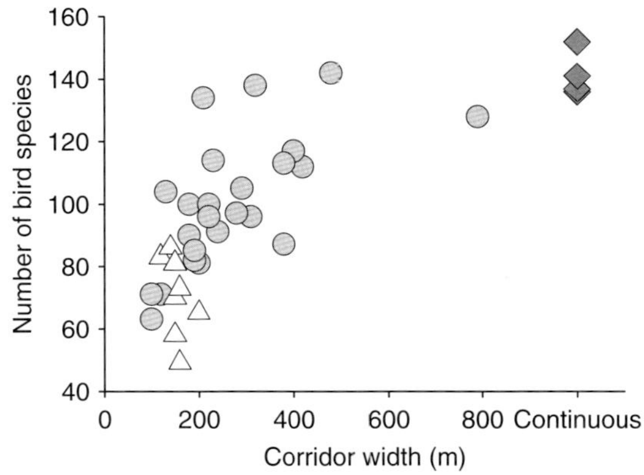

Figure 2.4 The relationship between the mean width of riparian forest corridors and bird species richness in the Brazilian Amazon. Forest corridors were either connected (grey circles) or unconnected (open triangles) to large forest patches. Five control sites were sampled within continuous forest (dark gray diamonds). (Adapted from Lees and Peres 2008b.)

They are often found alongside watercourses (riparian habitat is often afforded a high level of legal protection) or as a way of delineating land-use boundaries. Given their prevalence, these linear remnants may provide a particularly cost-effective and politically acceptable biodiversity conservation strategy (compared to changes in production system or set-asides of large blocks of native habitat).

These forest strips could have a valuable conservation role by aiding dispersal and increasing connectivity (see Section 2.5.4). They can also provide a useful habitat for forest birds. Four recent studies examine how their conservation value be maximized, and provide insights about the importance of strip width, the role of degradation, and isolation. Lees and Peres (2008a) sampled birds in 37 riparian forest sites in southern Brazilian Amazon. The width of the forest strip was a key factor in determining the presence of forest species, and the data suggests that riparian forests need to be at least 400 m wide to ensure the persistence of the majority of the regional avifauna (Figure 2.4). The importance of width was supported by a similar study in Honduras where more forest bird species and individuals were recorded in belts 50 m or wider (Arcos *et al.* 2008).

Degradation of forest structure within forest strips was also highlighted as an important factor in the study by Lees and Peres (2008a), although it was also strongly affected by their width. Cattle intrusion had a major influence on the forest structure, and occurred in 70% and 89% of all connected and unconnected corridor plots. Substantial conservation benefits could be accrued by fencing these riparian strips, to prevent cattle from reaching the stream, and providing alternative sources of water for cattle (Lees and Peres 2008a).

Isolation distance (from larger "source" forest patches) is also an important factor influencing the conservation value of forest strips for tropical forest birds. Unconnected riparian strips held fewer forest specialists than connected strips in a study in Costa Rica (Seaman and Schulze 2010), and isolation distance was one of most important variables affecting bird community composition and the abundance of disturbance-sensitive species in a study examining the role of *terra firme* (dry land) and riparian forest strips in the Brazilian Amazon (Hawes *et al.* 2008). In this latter study, the strips embedded more than 2 km into the *Eucalyptus* matrix contained species increasingly similar to secondary forests in the same region, even though the structure of the forest remained similar to control sites in continuous areas of forest (Hawes *et al.* 2008). Lees and Peres (2008a) also find weak support for the importance of isolation, but these effects were generally masked by the collinear influence of width and degradation.

Despite their extreme edge to interior ratio, forest strips can play an important role for the conservation of tropical forest birds, and these studies demonstrate that their value as a habitat *per se* can be increased by maximizing their width, maintaining connectivity, and preventing degradation. Of course, riparian and dry land forest strips can also deliver additional benefits, such as increasing connectivity (see Section 2.5.4) and helping maintain water quality and the integrity of the hydrological system. As such, they should be a fundamental part of any landscape-wide management plan in tropical landscapes.

2.5 Fragment isolation and the importance of connectivity

2.5.1 Theoretical and conceptual issues regarding connectivity

There are many different ways of conceptualizing connectivity. It is important to distinguish between connectivity from a species-specific perspective, and human perceptions of connectivity based on our own biased visualizations and measurements. To help clarify the difference, Fischer and Lindenmayer (2007) use "habitat connectivity" to define the connectedness of different patches of habitat from the perspective of a particular species (which is similar to the concept of "functional connectivity" used by other authors). Similarly Fischer and Lindenmayer (2007) use "landscape connectivity" to represent the way humans visualize and measure connectivity in a landscape, which would include the use of metrics such as the percentage cover of native vegetation. Landscape connectivity is often used as a proxy of functional connectivity as it is easy to measure and requires no a priori ecological knowledge. Studies on broad species assemblages tend to focus on landscape connectivity, while those looking at individual species movements are measuring habitat connectivity.

In classical island biogeography theory, the connectivity between habitat fragments is a function of their degree of isolation, which is often measured as either (i) the distance to the nearest habitat patch, (ii) the distance to the nearest large habitat patch (such as continuous forest), or (iii) the distance to the nearest patch occupied by the same species (Prugh *et al.* 2008). Of these, the distance to the nearest patch occupied by the same species has proved to be the most effective at predicting whether or not a species will be present (Prugh *et al.* 2008). This is likely to be because populations in patchy habitats can function as a metapopulation, where patch extinctions are offset by dispersal and subsequent colonization from so-called "source" populations in larger patches. At the theoretical point of

Figure 2.5 Understory insectivores such as the White-plumed Antbird (*Pithys albifrons*) are among the most vulnerable species to forest fragmentation. (Photograph by Jos Barlow.) (Please see Plate 2.5 for the color version of this figure.)

population equilibrium, the rate of extinction is equal to the rate of colonization (Levins 1970; Hanski 1989).

Although the metapopulation concept is theoretically appealing, most of the support for this in fragmented landscapes comes from studies on arthropods (Ewers and Didham 2006). Furthermore, the metapopulation concept may not be applicable to many tropical forest birds, which despite their obvious mobility are surprisingly poor at dispersal. Many understory specialists such as antbirds (Figure 2.5) are adapted to the dark interior of tropical forests and are reluctant to cross gaps (Stouffer and Bierregaard 1995; Sodhi *et al.* 2004b; Laurance and Gomez 2005; Lees and Peres 2009) and may be physically incapable of flying over distances greater than a few hundred meters (Moore *et al.* 2008).

As a subset of metapopulation theory, source–sink theory can also be applied in the fragmentation context. Source–sink theory assumes a consistent net flow of migrants from populations in superior habitats to those in inferior habitats (Pulliam 1988). Fragmented landscapes may be partitioned into population sources and sinks. "Sources," due to their superior habitat quality, may have reproductive surpluses that engender consistent net exportation of organisms.

Large patches of continuous forests can serve this purpose in fragmented landscapes. In "sinks," local reproductive success cannot offset local mortality, so net importation of individuals is required to avoid extinction. This immigration process is also known as the "rescue effect" (immigration by unrelated individuals into isolated populations; Brown and Kodric-Brown 1977). Small patches in fragmented landscapes may often act as sinks.

2.5.2 The importance of the matrix

Simple measures of isolation based on distance are misleading in most landscapes as they ignore the important role of the matrix habitat, the land that lies between the fragments (Ewers and Didham 2006; Laurance 2008a; Prugh *et al.* 2008). In reality many fragments are not true isolates, and the agricultural matrix often contains structural features that resemble the remaining habitat patches, including hedgerows, riparian vegetation, corridors set-aside for conservation purposes, low contrast land-uses such as agroforestry, and even "stepping stones" such as isolated standard trees.

The benefits of a structurally complex matrix are twofold, as it may provide a resource and living space for a nested subset of the forest birds, and could facilitate the dispersal of forest birds between nominally isolated fragments (Lindenmayer and Franklin 2002). We first assess evidence regarding the importance of the matrix in determining dispersal limitation in tropical forest birds, before examining the importance of the matrix and corridors for bird conservation.

2.5.3 Can dispersal ability predict species persistence in fragments?

Different species will be affected differently by fragmentation, with traits such as large body size, low mobility, high trophic level, and matrix tolerance frequently hypothesized to affect a species' vulnerability (Ewers and Didham 2006; see Chapter 3). For birds, it seems that an ability to use or disperse through the matrix appears to be one of the most important factors in determining a species' ability to persist in fragmented landscapes (Antongiovanni and Metzger 2005). Experimental evidence from Amazonia suggests that many understory tropical forests birds avoid even small gaps, such as the narrow gaps created when roads are constructed (Develey and Stouffer 2001; Laurance *et al.* 2004), but similar species fair much better when fragments are surrounded by a more permeable matrix (secondary forests or exotic tree plantations) rather than pastures (Stouffer and Bierregaard 1995; Gascon *et al.* 1999; Antongiovanni and Metzger 2005). In contrast, highly mobile taxa such as parrots and other families are relatively insensitive to large gaps (Lees and Peres 2009; see Chapter 3).

The importance of dispersal has been observed in many different tropical forest regions. High dispersal ability is one of the best determinants of the persistence of birds in fragmented landscapes in Southeast Asia (Sodhi *et al.* 2004b), Costa Rica (Sekercioglu *et al.* 2002) and the Taita Hills in Kenya (Lens *et al.* 2002a). Recent evidence from a dispersal study in Panama strongly suggests that the ability to fly even short distances (less than 100 m) across water varies dramatically among species of understory forest birds, and is strongly correlated with the history of species extinction and their current distribution across the archipelago (Moore *et al.* 2008).

Overall, the message regarding dispersal is remarkably consistent, and species with a limited ability to disperse across the matrix tend to decline or disappear from fragments following isolation. However, we still lack a complete picture of the dispersal limitation of birds in fragmented tropical landscapes. Most observations of gap-crossing movements involve experimental manipulations (Develey and Stouffer 2001; Laurance *et al.* 2004; Moore *et al.* 2008) which are unlikely to reflect normal dispersal conditions. For example, Van Houtan *et al.* (2007) examined bird movements in the experimental fragmentation project in Manaus. Contrary to expectations based on experimental studies, they show that many forest birds dispersed widely before landscape fragmentation occurred, and that these same species often emigrated from smaller to larger fragments following isolation.

Another problem with dispersal studies is that most observational studies are likely to miss once-in-a-lifetime gap-crossing events (such as natal dispersal events) that could occur when birds disperse (e.g. Lees and Peres 2009). While these rare one-off dispersal events are very hard to study, they could be one of the most important determinants of colonization and species persistence in fragmented landscapes (Lees and Peres 2009). Genetic approaches may shed further light on dispersal in tropical bird populations in the future.

2.5.4 Increasing connectivity for tropical birds

Given the importance of dispersal for maintaining bird populations in fragments, it is important that we understand which landscape features can facilitate dispersal, and at what spatial and temporal scales. One of the most important and studied ways of increasing the connectivity of tropical landscapes is through forest corridors. As discussed in Section 2.4.3, narrow, degraded and isolated forest strips hold an impoverished bird fauna with few forest specialists (Hawes *et al.* 2008; Lees and Peres 2008a; Seaman and Schulze 2010). However, these same strips may play a very important role in the landscape by facilitating movement between fragments and increasing habitat connectivity (Beier and Noss 1998; Haddad 2003).

Most of our knowledge on the use of forest strips by birds stems from translocation experiments that involve capturing birds, moving them away from their territories, and radio-tracking their efforts to return through the landscape (Sekercioglu 2009). A study from highly fragmented tropical dry forest of Costa Rica moved two species of bird 0.7–1.9 km from their territories, and mapped their return route across three alternative landscape elements: 150 m wide riparian corridors, 15–30 m wide fencerows, and open pasture (Gillies and Clair 2008). The study confirms the importance of maintaining wide forest corridors to link remaining patches of forest, as the Barred Antshrike (*Thamnophilus doliatus*), which was considered a forest specialist in these dry tropical forests, restricted its return path to the wide riparian corridors, ignoring the fencerows even when these provided the most direct route back to their territory. These results are likely to be conservative, as many species are likely to require wider and less disturbed forest strips than the Barred Antshrike (Sekercioglu 2009).

Overall, dispersal is clearly favored by a matrix composed of vegetation types and landscape features that minimize the contrast with native habitat. However, from a management perspective it is also important to understand at what scale will these features actually provide functional connectivity for the species of conservation concern in that landscape. While this will obviously be species specific, some basic rules of thumb may exist. For example, in their study of the bird fauna of Atlantic forest fragments in Brazil, Martensen *et al.* (2008) conclude that many species either require direct linkages (corridors) or between-fragment distances of less than 30 m (which is particularly alarming, as the average distance between fragments is over 1400 m in the Atlantic Forest; Ribeiro *et al.* 2009).

2.5.5 The matrix and edge effects

The matrix can also play an important role in buffering edge effects, and protecting fragments from the most extreme conditions of the matrix. For example, regenerating native forest can help avoid the edge-related tree mortality in fragments that occurs when the matrix is dominated by pasture (Mesquita *et al.* 1999). In turn, birds typical of the forest interior (including species using mixed-species flocks, terrestrial species, and army-ant followers) have been found to show less edge-avoidance when fragments are surrounded by secondary forest than when they are surrounded by pasture (Stouffer and Bierregaard 1995; Laurance *et al.* 2004). A more favorable matrix can also buffer against species loss. In a study carried out in Andean subtropical wet forest, 20% of the forest species that declined in fragments surrounded by cattle pasture persisted in the fragments surrounded by tree plantations (Renjifo 2001).

2.5.6 Countryside conservation in tropical forest landscapes

As well as aiding dispersal and buffering forest remnants, the matrix of agricultural land uses and associated landscape elements such as hedgerows and standard trees can help support a significant portion of the regional diversity in fragmented tropical forest landscapes (Gardner *et al.* 2009; see Chapters 1 and 10). Gascon (1999) found that between 40% and 80% of primary forest frogs, birds and ants were detected in a matrix composed of either pasture or regenerating forest during the experimental fragmentation project in Manaus. Hughes *et al.* (2002) detected 144 bird species from 29 families in agricultural areas in southern Costa Rica, many of which were observed in standard trees and hedgerows far from remaining forest patches. These 144 species accounted for 46% of the regional pool of species, although it is important to note that most of these species are likely to be of least conservation concern, and shifting baselines means that many forest specialists may have already been lost from these heavily deforested landscapes (Gardner *et al.* 2009).

While some species typical of forest patches are able to breed and feed within the matrix (Sekercioglu *et al.* 2007), it is not clear if others require the remaining fragments for at least some part of their diet or life cycle (such as roosting or nesting). Ruiz-Gutiérrez *et al.* (2010) assessed occupancy dynamics in a fragmented Costa Rican landscape, and concluded that the protection of remnant forest habitats is the key for the long-term persistence of all members of the bird community. Furthermore, a study in the Brazilian Amazon compared the bird fauna in extensive primary forests with two favorable matrix habitats (secondary forests and *Eucalyptus* plantations) and provides a more pessimistic evaluation of these matrix habitats for primary forest specialists. Obligate ant-following and dead-leaf gleaning insectivores were never recorded outside of primary forest (Barlow *et al.* 2007), and the study suggests that between 40% and 68% of the species observed in the primary forest would be lost if primary forests were converted to landscapes dominated exclusively by secondary forests and tree plantations (Barlow *et al.* 2010). While the matrix should not be ignored in conservation strategies, there is no substitute for maintaining large patches of intact habitat wherever this is possible (Peres 2005).

2.6 Temporal change in forest fragments

Most ecological studies have assessed fragments at a single point in time. However, a fundamental question for biodiversity conservation is whether these fragments will conserve tropical species in the long term? This is very difficult to answer accurately, as many of the patterns can take years to manifest themselves. For example, many tropical trees are long-lived and may take decades or even centuries to perish following fragmentation (e.g. Taberelli *et al.* 2008) and landscape history can strongly affect the present distribution pattern of species in fragmented landscapes (Metzger et al. 2009). Here we examine the empirical and theoretical evidence regarding the likely long-term consequence of forest fragmentation for bird communities.

Temporal dynamics of fragmentation have been observed immediately after isolation, when fragments go through a short period of crowding due to individuals displaced by the removal of surrounding vegetation moving into the remaining habitat. This "crowding effect" has been frequently observed in fragmentation studies (Debinski and Holt 2000), and can alter intra- and inter-specific interactions through increased competition and other density-dependent effects (Saunders *et al.* 1991). As the forest was felled around the experimental fragments in the Brazilian Amazon, many bird species found sanctuary in the remnant forests (Bierregaard *et al.* 1992). After isolation, bird capture rates at least doubled in newly isolated fragments. These elevated bird numbers persisted for about 200 days until population sizes fell below pre-isolation levels (Bierregaard *et al.* 1992), with the extent of avifaunal collapse being greater in smaller fragments.

In the years, decades or even centuries after fragmentation, small patches of remnant habitat are likely to harbor more species than they are capable of maintaining. These species are eventually lost over time in a process known as "species relaxation." The term "extinction debt" is used to describe the time lag between habitat loss and the local extinction of populations and species (Tilman *et al.* 1994).

As examined in Chapter 3, Brooks *et al.* (1999b) found that the half-life (the time taken to lose half of the species within a defined area) for avifaunal extinctions in fragmented forests in Kenya was approximately 50 years on average, but also that this half-life decreased with decreasing fragment size (see Figure 3.2).

This study is supported by evidence from the Brazilian Amazon, where 100 ha fragments lost half of its bird species in less than 15 years (Ferraz *et al.* 2003). Leck (1979) also reported a loss of 25 bird species from 87 ha forest fragment in Ecuador after just 5 years of isolation, and a review of 20 fragmentation studies across the world concluded that the three longer-term studies (14 years) revealed strong patterns that would have been missed in short-term investigations (Debinski and Holt 2000).

Few fragments have been monitored for more than 50 years. The avifaunal changes over 100 years (1898–1998) in a 4 ha patch of rain forest in Singapore (Singapore Botanic Gardens) revealed that 49% of forest species were lost during this time. By 1998, 20% of the birds observed were introduced species including the House Crow (*Corvus splendens*) (Sodhi *et al.* 2005b). Sodhi *et al.* (2006c) reported avifaunal turnover in a 86 ha tropical woodlot (Bogor Botanical Gardens, Indonesia) containing 54% native and 46% introduced plant species and receiving a mean of over 80,000 human visitors per month. Since its isolation in 1936, subsequent surveys have shown a gradual reduction in avifaunal species richness. By 2004, the original richness of this woodlot declined by 59% (from 97 to 40 species) and its forest-dependent avifauna declined by 60% (from 30 to 12 species). All seven forest-dependent bird species that attempted to colonize this woodlot by 1987 perished thereafter (see Chapter 3). Both of these studies show that the conservation value of small fragments for forest birds declines over time, and that they are vulnerable to competition from invasive species (see Chapter 6).

Ferraz *et al.* (2003) developed a scaling rule from a 13-year-long dataset from following the experimental creation of 13 fragments in the Brazilian Amazon. They estimate that 100 ha forest fragments lose one-half of their species in <15 years, and that a 10-fold decrease in the rate of species loss would require a 1000-fold increase in area. These scaling rules provide a bleak assessment of the long-term conservation potential of fragmented habitats, as predictions suggest that even fragments as large as 10,000 ha would lose many species (likely the ones of most conservation concern) once they have been isolated for 100 years. Time lags can also help explain why there have been no proven extinctions from the heavily deforested and highly fragmented Brazilian Atlantic forest, with an analysis based on species–area relationships suggesting that between 30% and 50% of the endemic forest species will go extinct as a result of the deforestation that has already been carried out (Brooks *et al.* 1999a).

2.6.1 Fluctuating asymmetry may provide evidence of longer-term consequences

Given the eventual effects of fragmentation may take decades or even centuries to become apparent, some authors have attempted to look for ways of assessing stress in bird populations as a kind of early warning system that could help predict extinctions. One ingenious way to achieve this is to look for fluctuating asymmetry (FA), which is a measure of the deviation from perfect symmetry that can be related to the developmental stability of an individual (Lens *et al.* 1999). By measuring the tarsus length of seven forest-restricted bird species, Lens *et al.*

(1999) showed that individuals from the most degraded fragment showed four- to sevenfold higher asymmetry levels compared to those from the least degraded one. Fluctuating asymmetry was also able to predict reduced survival probability of the critically endangered Taita Thrush (*Turdus helleri*) in the same region (Lens *et al.* 2002b), demonstrating its potential use for identifying species that would be unlikely to persist in fragments in the longer-term. These results were also found in fragmented Brazilian Atlantic forest, where fluctuating asymmetry was greater in birds from forest fragments than in those found from continuous forest, and was negatively correlated with fragment size. Interestingly, fluctuating asymmetry also differed across foraging guilds, and was most evident among the understory insectivores, which are frequently cited as being among the most vulnerable to fragmentation (Anciães and Marini 2000).

2.7 Conclusion

Our knowledge regarding the consequences of habitat fragmentation for tropical birds has been informed by many different studies, most of which have been stimulated by the theoretical premises of the theory of island biogeography (MacArthur and Wilson 1967). The conclusions of these studies regarding the conservation of fragmented landscapes are mostly intuitive, and suggest that the conservation prospects of birds in tropical forests landscapes can be improved by: (i) increasing or maintaining the area of remaining patches of native habitat; (ii) minimizing human-induced edge effects by managing the use of fire, livestock, agrochemicals, and other activities that can degrade the quality of the remaining patches; (iii) minimizing physical and biological edge effects and maximizing dispersal between patches by reducing the contrast with the matrix; and (iv) increasing landscape connectivity through corridors and other schemes.

While scientists may argue about the relative importance of each of these factors, differences between the conclusions of different studies can be easily explained by the configuration of landscape where the experiments took place. For example, experiments are unlikely to find strong isolation effects if fragments are very close together, adjacent to large tracts of native forest, or nested within a hospitable matrix (Ferraz *et al.* 2007), while edge effects are likely to be less prevalent in older fragments, or in fragments that are heavily degraded, as many of the edge-sensitive species are likely to have already been extirpated (see Banks-Leite *et al.* 2010). Moreover, regional difference in the historical frequency of disturbance may mean that some tropical systems are more likely than others to suffer the negative consequences of habitat fragmentation (Gardner *et al.* 2009).

Although most tropical fragments are biologically impoverished, in many cases they are the only remaining patches of native habitat, and they therefore play a crucial role in protecting the last populations of many threatened species (Barnett *et al.* 2005). They also act as reservoirs for some globally endangered bird species and can be used to restore extirpated or perishing avifauna (Renjifo 1999; Turner and Corlett 1996). Evidence suggesting there is a strong temporal basis to the ecological effects of fragmentation paints a bleak picture for prospects for biodiversity conservation in these human-dominated tropical habitats (Gardner *et al.*

2009). While small fragments may lose species gradually over time, they are also much more vulnerable to stochastic environmental disasters, such as severe droughts and wildfires (see Chapter 5). These impacts are likely to worsen in a world where increasing human dominance further isolates the remaining patches, and where climate change threatens the stability of many tropical regions (see Chapter 8).

3

Tropical Bird Extinctions

Overview: The current bird extinction rate may be at least 30 times more than expected. However, bird conservation initiatives may have resulted in some alleviation of the bird extinction rate in the last century. It may take decades for bird species to go extinct following the loss, degradation, and fragmentation of native habitats. Nonetheless, a number of studies report bird losses following habitat alteration in the tropics. Specialized, large-sized, and rare species, and those with restricted ranges, are especially vulnerable to extinction. Loss of seed-dispersing birds may hamper forest regeneration in human-dominated landscapes. Some forest bird species may be able to use disturbed landscapes but it is not clear if they are able to sustain their populations there.

In this chapter we first discuss current and future extinction rates. We also report on the concept of extinction debt and whether birds are the most endangered taxa. We briefly highlight some of the case studies of bird extinction in the tropics. The bird species that are vulnerable to extinction and ecosystem effects of bird losses are also discussed. We end by reporting on possible persistence of tropical birds in degraded landscapes.

3.1 Extinctions over time

Since 1500, 154 bird species are believed to have gone extinct (Pimm *et al.* 2006). Indicating no halting of avian extinctions, 18 species have been recorded to be extinguished in the last quarter of the twentieth century, and at least three more

Conservation of Tropical Birds, First Edition. Navjot S. Sodhi, Çağan H. Şekercioğlu, Jos Barlow and Scott K. Robinson.

Published 2011 by Blackwell Publishing Ltd.

species have been likely lost since 2000. As mentioned in Chapter 1, the last known Spix's Macaw (*Cyanopsitta spixii*) in the wild disappeared in Brazil in 2000, and the last wild individuals of Hawaiian Crow (*Corvus hawaiiensis*) disappeared in 2002. Likewise, the last known individual of Po'ouli (*Melamprosops phaeosoma*) on the Hawaiian Islands died in captivity in 2004 (Figure 3.1; BirdLife International 2010a). The Po'ouli may have survived only 31 years after its scientific description in 1974 (Pimm *et al.* 2006), and it is possible that some bird species have been lost before scientific discovery.

(a)

(b)

Figure 3.1 (a) Spix's Macaw. (b) Hawaiian Crow. (c) Po'ouli. (Photographs by: (a) Mickey Muck; (b) Mark Sweep; (c) courtesy of San Diego Zoo.) (Please see Plate 3.1 for the color version of this figure.)

(c)

Figure 3.1 (Continued)

To estimate the rate of extinctions, the number of extinctions per species per year is calculated (Pimm *et al.* 2006). The possible background extinction rate is one extinction per million species-years (MSY) (Pimm *et al.* 1995b). Background extinctions occur as a consequence of environmental change, newly established competitive interactions through evolution or invasion, and occasional catastrophes such as storms, fire, or disease. Therefore, based on approximately 10,000 reported bird species, the background extinction rate would be just one bird extinction per 100 years (Pimm and Jenkins 2010). However, reported avian extinctions since the 1500s thus far indicate extinctions are occurring at least 30 times this rate (Pimm *et al.* 2006).

The bird extinction rates in the past two centuries have been high (19th century: 115 extinctions [E]/MSY, and 20th century: 65 E/MSY) but seem to dropping possibly due to conservation interventions (Pimm *et al.* 2006; see Chapter 10). However, if 182 critically endangered species expired over the next 30 years, the extinction rate could jump to 500 E/MSY (Pimm *et al.* 2006). This future extinction rate may be an underestimation because about half of 300 newly discovered bird species since 1950 are either already extinct or threatened (Pimm and Jenkins 2010). Recent projections for the 21st century estimate 400–1000 bird species committed to extinction on average, although this can be much higher depending on the amount of climate change and habitat loss (Sekercioglu *et al.* 2004, 2008).

Stork (2010) argues that species extinctions may be less frequent than expected due to factors such as increased availability of regenerating forests. He suggested that population or local extinctions (also called extirpations) may be more prevalent. Hughes *et al.* (1997) estimated that millions of populations may be going extinct largely due to habitat loss. Sodhi *et al.* (2006a) projected that due to heavy deforestation, Southeast Asia could lose up to 2761 of its national bird populations by 2100. Indonesia stands to lose the maximum bird populations due to deforestation – a country with the highest number of resident and endemic bird species in Southeast Asia (929 and 408 species, respectively; Sodhi *et al.* 2006a). Unfortunately, the predictions by Sodhi *et al.* may be optimistic as they do not consider the likely cumulative effects of other drivers of biodiversity change such as fire, over-harvesting, invasive species, and climate change (Brook *et al.* 2008; see below and other chapters). Similar, and perhaps better, projections need to be developed for other tropical areas. Furthermore, we have undergone a massive loss of bird individuals: Gaston *et al.* (2003) estimated that currently there are about 100 billion breeding bird individuals on Earth, which may at least represent one-fifth or one-fourth loss of those existing during pre-agricultural times, i.e. approximately 300 years ago.

3.2 Extinction debt

Possibly because of comparatively low reproductive effort and high environmental stability, tropical bird species are considered to have low adult mortality. Adult mortality of only 10–30% per year has been estimated for birds from various tropical locations (Trinidad, Sarawak (Malaysia), and Barro Colorado Island; see Willis 1974). This fact suggests that it may take some time for a tropical bird species to disappear following habitat loss unless such habitat loss decreases adult longevity. The Puerto Rican Parrot (*Amazona vittata*, longevity in captivity is 23 years), for example, had about 2000 individuals around 1900 in the Caribbean National Forest. Their numbers declined to 20 over the next 60 years (Brash 1987), indicating that it may take decades for the demise of some species. Supporting this postulation, Brooks *et al.* (1999b) estimated that a half-life (time taken to lose half of the species) for bird extinctions in fragmented forests in Kenya was 50 years (Figure 3.2; see Chapter 2). However, half-life decreased to 23 years for a 100 ha fragment. Smaller fragments may contain small populations, which may have a higher probabability of extinction (see below). Further, many tropical species are reluctant to cross non-forested habitats (Stratford and Robinson 2005), thus the possibility of the "rescue effect" (immigration by unrelated individuals into isolated populations; Brown and Kodric-Brown 1977) is minimal. Half-lives may be further shortened in fragments by indirect factors such as high predation, parasitism, and damage by catastrophic events such as fires and storms (Brooks *et al.* 1999b). In fact, 100 ha fragments in Amazon can lose half of their species rapidly, i.e. in less than 15 years (Figure 3.2; Ferraz *et al.* 2003). Overall, fragments < 100 ha may lose their sensitive species within one or two decades, while those > 100 ha may take decades or perhaps a century to lose such species (Ferraz *et al.* 2003). Nonetheless, the species may differ in their ability to tolerate habitat deterioration depending upon traits such as habitat flexibility and dispersal capabilities (Lens *et al.* 2002a; see Section 2.6.1).

The time taken to lose all the sensitive species following habitat perturbation is also called "relaxation time" (Diamond 1972). Therefore, long-term data are needed to determine the exact extent of avian extinctions following initial habitat loss. The extinction lag times can result in "extinction debt" (Tilman *et al.*

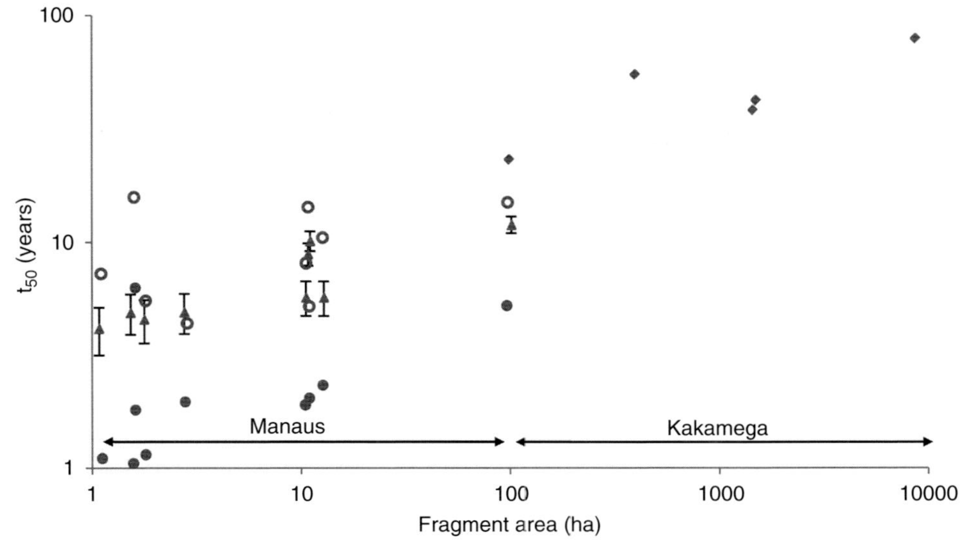

Figure 3.2 Time to lose 50% of the bird species from forest fragments in Manaus (Brazil; circles and triangles) and Kakamega (Kenya; diamonds) (After Ferraz *et al.* 2003, copyright National Academy of Sciences, U.S.A.)

1994) because some species may have unviable populations (these species are also called the "living dead") and take decades to be wiped out. Time to extinction should be longer for species that have high productivity and survival (Saether *et al.* 2005). Fragmentation can reduce survival and thus persistence, as Ruiz-Gutierrez *et al.* (2008) found in apparent annual adult survival of White-ruffed Manakin (*Corapipo altera*) which was reduced in small-sized fragments (0.9–25 ha) in Costa Rica.

3.3 Are birds the most endangered taxa?

The widespread interest in birds and conservation inititaives have likely prevented two-thirds of bird extinctions in the past 30 years (Pimm *et al.* 2006; see Chapter 10). However, birds may not be the most threatened taxa (Figure 3.3; Stork *et al.* 2009). Within the tropics, 1512 (32%) of 4728, 907 (13%) of 7128, and 714 (22%) of 3296 species of amphibians, birds, and mammals are currently threatened (Laurance and Useche 2009). It is plausible that many other tropical taxa could be more vulnerable than birds (see Chapter 1). Birds' high mobility give them a big advantage, and the more mobile long-distance migrants are 2.6 times less threatened than sedentary species (Sekercioglu 2007). As few other groups are capable of flight and most organisms are less mobile than birds, only protecting birds may not be sufficient to prevent widespread extinctions in the tropics (Pimm *et al.* 2006). Nevertheless, it should be noted that most bird species found in the understory of tropical forests are actually very reluctant to cross gaps (Lees and Peres 2009; see Section 2.5.3).

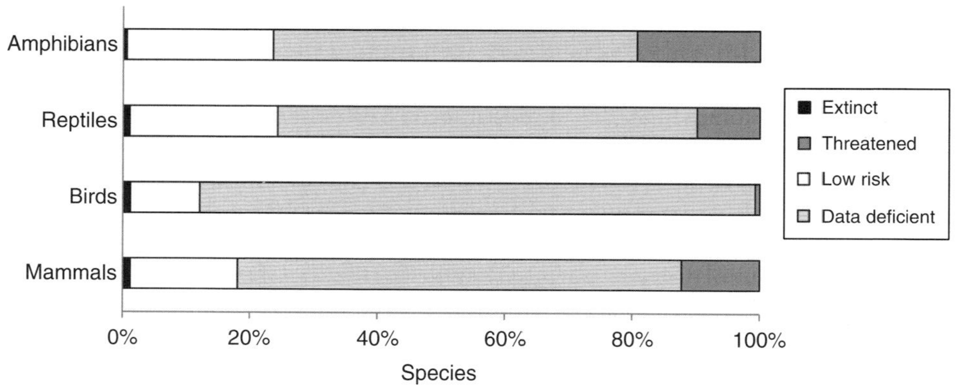

Figure 3.3 Proportions of species by threat category for four taxonomic groups. (After Stork *et al*. 2009.)

3.4 Case studies of recent bird extinctions

Various studies show that bird species are lost following tropical deforestation and degradation. We present some of the key studies below.

Since the British first established a presence in Singapore in 1819, more than 95% of the estimated 540 km^2 of original vegetation cover has been entirely cleared, initially from the cultivation of short-term cash crops (e.g. gambier *Uncaria gambir* and rubber *Hevea brasiliensis*) and subsequently through urbanization and industrialization (Corlett 1992). As a consequence, Singapore has lost 65 of 203 bird species, with 61 (67%) of the 91 original forest species going locally extinct (Castelletta *et al.* 2000; Brook *et al.* 2003a). Some areas in Southeast Asia with less deforestation than Singapore have also suffered bird extinctions. Lying within Southeast Asia is the small (198 km^2) oceanic island of Damar (Indonesia). Over the past century, 75% of the island's primary forest has remained intact. Therefore, at first glance, Damar Island might appear to represent a conservationist's dream. However, despite retaining most of its original habitat, a study by Trainor (2007) shows that by 2001, this island had lost six resident forest-dependent bird species (16%) out of 38 recorded in 1890s. While the exact cause(s) of these birds' demise is unknown, forest loss and disturbance seem to be the main culprits. Trainor's study is striking as it shows that the loss of as little as a quarter of forest may spell doom for some of the forest avifauna. The link between deforestation and extinction has also been demonstrated by a study on six Southeast Asian landbridge islands, ranging from 2.6 to 1179 km^2. Depending upon the amount of forest loss, zero (Pulau Malawali and Pulau Mantanani) to 15 (Pulau Bintan) diurnal resident landbird species were apparently extirpated since the early 1900s (Sodhi *et al.* 2010; see below).

Bird extirpations have also been reported from mountainous tropical areas. Prior to European colonization in 1885, the Eastern Usambara Mountains (Tanzania) were almost completely forested, but by the 1990s, over 40% of the forest had been cleared (Newmark 1991). A large portion of remaining forest is

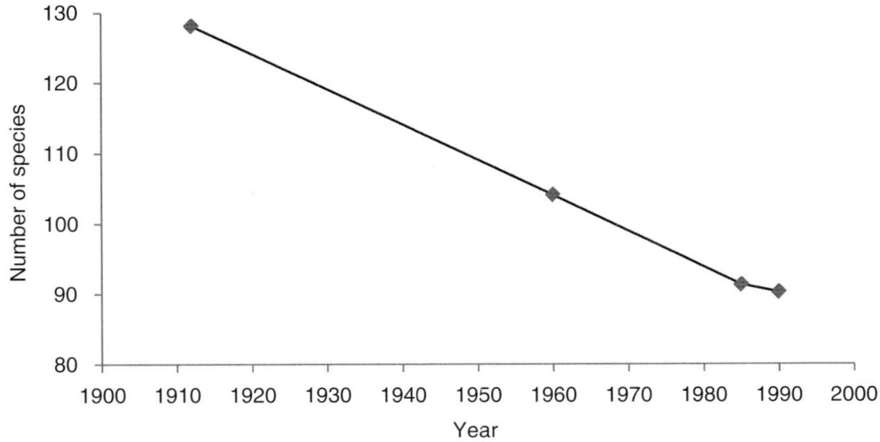

Figure 3.4 Loss of bird species at San Antonio (Colombia) through years 1911 to 1990. (After Kattan *et al.* 1994.)

fragmented. Newmark (1991) surveyed 10 fragments (0.1 to 571 ha) for understory forest birds. Eight species (e.g. Yellow-throated Wood Warbler, *Phylloscopus ruficapilla*, and Tiny Greenbul, *Phyllastrephus debilis*) were only present in the largest fragment, suggesting that they may have gone extinct from the other fragments. Similarly, in a fragmented landscape containing cloud forests in the western Andes of Colombia, 40 (31%) of the original 128 forest bird species disappeared between 1911 and 1990 (Figure 3.4; Kattan *et al.* 1994).

Additionally, there are studies showing that small remnant forests in deforested landscapes have poor conservation value for forest birds. Barro Colorado Island (BCI) in Panama is a 1562 ha lowland forest that was isolated between 1911 and 1914 to create a Gatun Lake as a part of the Panama Canal (Sigel *et al.* 2010). Birds on BCI have been inventoried since 1929; 70 species (28%) have been extirpated, possibly because of habitat deterioration and lack of recolonization of forest species (Robinson 1999; Sigel *et al.* 2010). In other words, about nine forest bird species were lost each decade. A number of extant species such as Slate-colored Grosbeaks (*Pitylus grossus*) have suffered more than 95% population decline (Robinson 1999), and may be extirpated in the future. In contrast, the 1611 ha Le Selva Biological Station (Costa Rica) suffered a comparatively lower loss of eight bird species (3%) probably because of its more recent isolation (15 years) and corridor connection with a larger forest (44,000 ha) (Sigel *et al.* 2010). In the Atlantic forest of Brazil, forest fragmentation and loss may have resulted in the local extinction of 28 bird species in the past 70 years (Ribon *et al.* 2003). Further supporting the vulnerability of forest avifauna in fragments, three forest fragments near Lagoa Santa (Brazil) lost 13 forest species between 1870 and 1987 (Christiansen and Pitter 1997).

A number of other studies also highlight the vulnerability of birds in isolated fragments. Between 1973 and 2007, a tropical lowland rainforest fragment (2 km^2) in Ecuador possibly lost 101 bird species, 74 (73%) of which were forest species (Figure 3.5; Pearson *et al.* 2010). A tropical woodlot, the Bogor Botanic

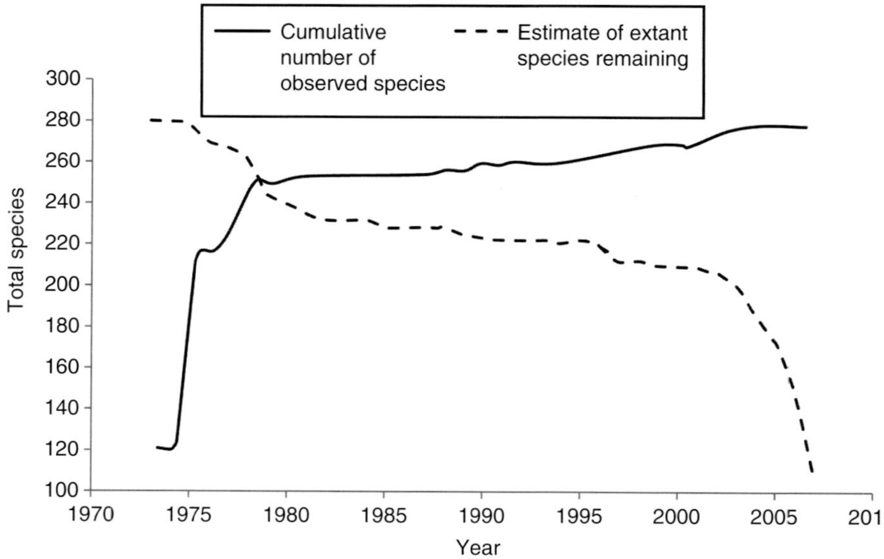

Figure 3.5 The cumulative number of non-migratory, forest-associated bird species added to the Rio Palenque Science Center (Ecuador) species list annually compared with the cumulative annual number of species lost under the extreme assumption that the last year a species was observed represents the year it became extinct. (After Pearson *et al.* 2010.)

Gardens (86 ha; 54% native and 46% introduced plants; mean 83,649 visitors/month), has been surveyed several times before and after its isolation in 1936 (Sodhi *et al.* 2006b). By 2004, the original avifaunal richness of this woodlot declined by 59% (from 97 to 40 species) and its forest-dependent avifauna declined by 60% (30 to 12 species). All seven forest-dependent bird species, probably originating from nearby shops, that attempted to colonize this woodlot by 1987 perished thereafter (Figure 3.6; Sodhi *et al.* 2006b). This study shows that area reduction, isolation, intense human use, and perverse management (e.g. understory removal) of this patch have probably negatively impacted the long-term sustainability of its forest avifauna.

Forest fragments may gain disturbance-tolerance or invasive species (see also Chapter 2). In a semideciduous forest fragment (1451 ha) in southeast Brazil, 25% of endemic bird species were lost over 25 years (Antunes 2007). However, 28% of gap- and edge-using bird species increased in abundance during the same time. Similar results are reported from tropical Australia. Because of grazing by livestock, changes in the fire regime, and vegetation clearing (both in and around), 9% of bird species were extirpated (e.g. Paradise Parrot, *Psephotus pulcherrimus*) from a subtropical woodlot in central Queensland (Australia) between 1873 and 1999 (Figure 3.7; Woinarski and Catterall 2004). The greatest loss was of species preferring dry rainforest, grasslands, and swamps. The woodlot also gained 6% bird species, mostly widespread species preferring open habitats (e.g. Australian Brush-turkey, *Alectura lathami*) (Woinarski and Catterall 2004). Similarly, Sodhi *et al.* (2005b) studied the faunal change over 100 years (1898–1998) in a 4 ha

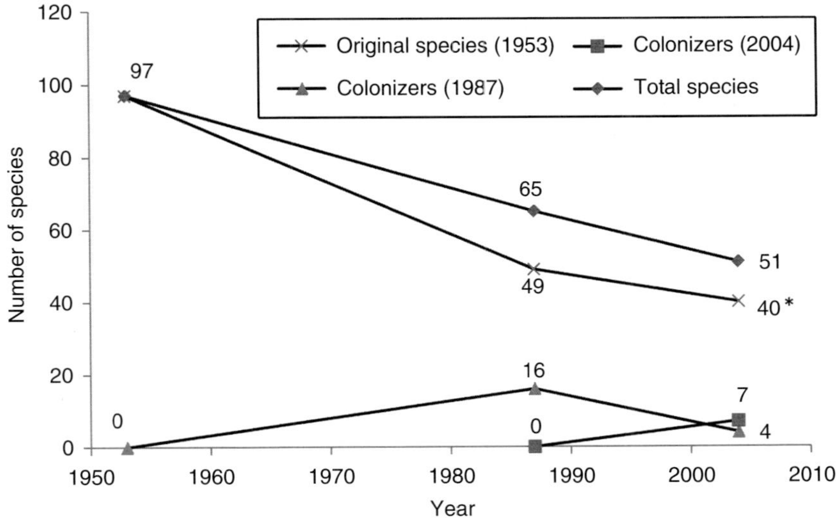

Figure 3.6 Bogor Botanical Gardens (BBG; Indonesia) avifauna extinctions and colonizations from 1953 to 2004. Numbers along the lines are number of bird species. An asterisk indicates the number and includes a species previously recorded as extinct in 1987 but that recolonized BBG in 2004. (After Sodhi *et al.* 2006b.)

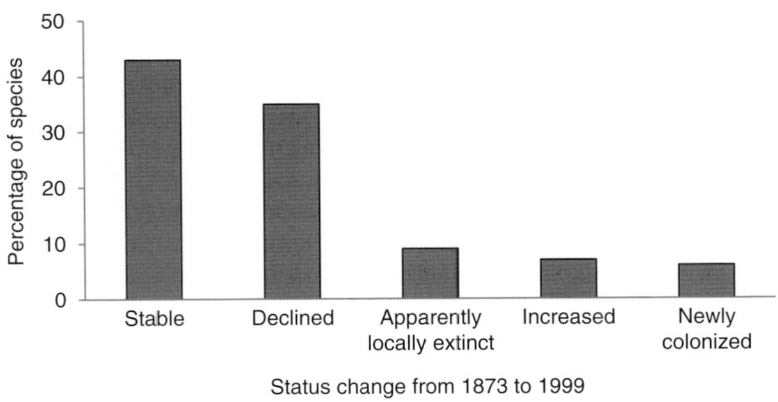

Figure 3.7 Percentage of bird species by status change in Coomoobolaroo (Queensland, Australia) from 1873 to 1999. (Data derived from Woinarski and Catterall 2004.)

patch of rainforest in Singapore (Singapore Botanic Gardens). Over this period, many forest species (e.g. Green Broadbill, *Calyptomena viridis*) were lost, and replaced with introduced species such as the House Crow (*Corvus splendens*). By 1998, 20% of individuals observed were from introduced species, with more native species expected to be extirpated from the site in the future through competition and predation. This study showed that small fragments decline significantly in their value for forest birds over time, and are vulnerable to competition from invasive species.

Most of these studies on species loss suggest that larger fragments are better for forest birds than smaller ones (Ferraz *et al.* 2007, and see Section 2.3). This is supported by strong species–area relationships (see Figure 2.1) and the results from an experimentally fragmented forest near Manaus (Brazil), where 37 bird species went extinct at least from one fragment between 1992 and 2001 (Stouffer *et al.* 2009). About 30% of species were extirpated from 1 ha fragments, but only 5% did so in 100 ha fragments. Fragments of 100 ha usually loose half of their species within 15 years following isolation – a 10-fold decrease in bird losses requires a 1000-fold increase in fragment area (Ferraz *et al.* 2003). These results suggest that larger fragments may be able to retain more species. However, some fragments gained species, possibly due to bird movements through regenerating matrix (areas surrounding fragments) (Stouffer *et al.* 2009).

3.5 Drivers of extinctions

Ultimately, decreased survival, fecundity, or both may result in the extinction of a species. Mechanisms causing extirpations may act independently or synergistically (see Figures 1.8 and 3.8; Brook *et al.* 2008). Therefore, pinpointing a single cause of an extinction may be difficult (Pimm 1996, Reed 1999). However, habitat destruction and degradation is likely to be the major cause of avian

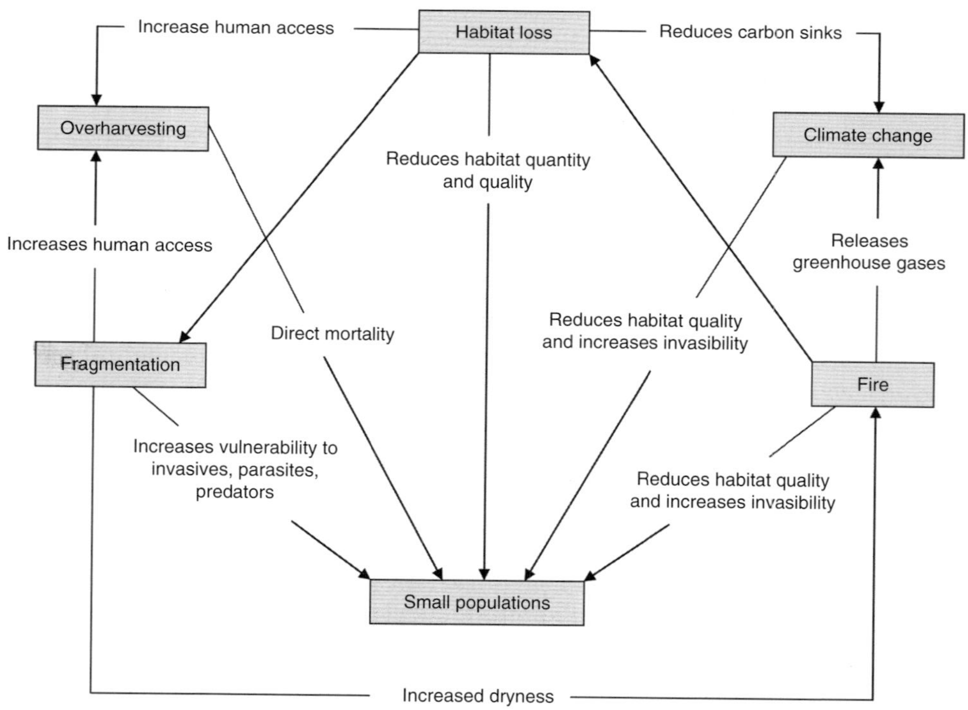

Figure 3.8 An example of the synergistic feedbacks that threaten species in disturbed tropical rainforests. (After Brook *et al.* 2008.)

extinctions (BirdLife International 2008a; Pimm and Jenkins 2010; see Chapter 1). For example, habitat conversion and destruction has been and will be the principal cause of tropical bird endangerment in Australia (Figure 3.9; Garnett and Brook 2007). Along similar lines, the bird extinctions on Southeast Asian islands correlated with the level of forest loss (Figure 3.10; Sodhi *et al.* 2010).

Deforestation reduces both habitat quality and quantity. For instance, habitat conversion can result in the loss of preferred habitats of some forest bird species. Only one or two woodpecker species occurred in fragments smaller than 2500 ha in Java, possibly because of the lack of suitable nesting trees (Van Balen 1999). Similarly, the loss of forest streams in the western Andes might have caused the

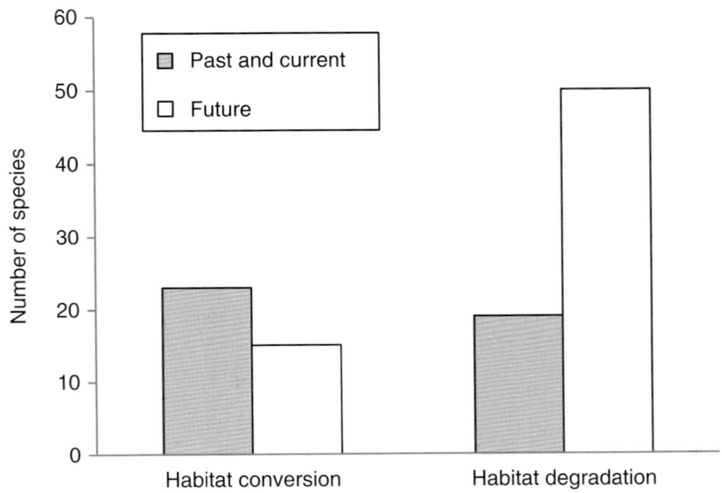

Figure 3.9 Past and anticipated stresses on tropical Australian birds. (Data derived from Garnett and Brook 2007.)

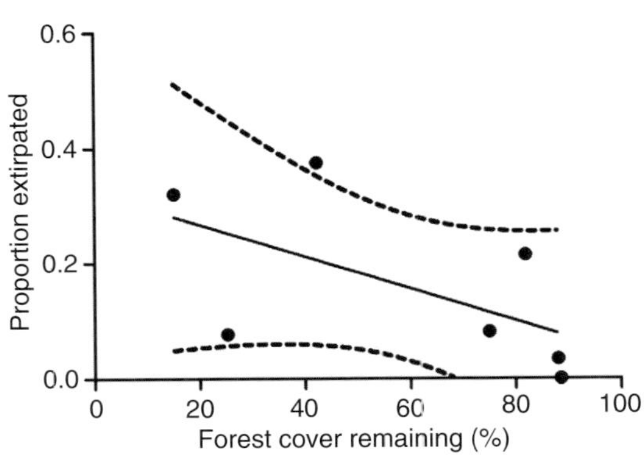

Figure 3.10 Relationship between the proportion of birds extinct (extirpated) that are forest species on seven Southeast Asian islands and the percentage of forest cover remaining. The linear regression (solid line; dashed lines, 95% confidence intervals of the regression fit) explains 34.8% of the variation in the data. (After Sodhi *et al.* 2010.)

local extinction, for instance, of the Andean Cock-of-the-rock (*Rupicola peruviana*) and the Crested Ant-tanager (*Habia cristata*) (Kattan *et al.* 1994). Mesic forests are also preferred by forest-dependent endemic and quasi-endemic bird species in the cerrado region of Brazil (Marini 2001), suggesting that loss of such forests may be detrimental to them.

As mentioned, various drivers causing extinctions may act synergistically. For example, as well as being a direct cause of some extinctions, habitat loss can also alter fire regimes (see Chapter 5), facilitate colonization by invasive species (see Chapter 6) and infiltration by disease vectors and parasites, and open up habitats for hunters (see Chapter 7) and other human-induced edge effects (see Chapter 2). Climate change is also emerging as an important driver of bird extinctions (see Chapter 8). Although intermediate climate change is projected to result in the loss of 400 to 550 bird species by 2100 (Sekercioglu *et al.* 2008), land-use changes may lead to greater bird species losses in the tropics (Figure 3.11; Jetz *et al.* 2007). Specifically, habitat loss resulting from agricultural expansion and hunting remain the main threats to tropical forest birds (Figure 3.12; Laurance and Useche 2009). Both of these threats may affect tropical forest birds synergistically (Figure 3.13; Laurance and Useche 2009). The reasons for this may be because agricultural expansion either increases access to poachers, or forest birds may enter agricultural areas where they may be especially susceptible to hunting (see Chapter 7). Habitat loss also interacts with climate change and makes it difficult or impossible for birds to shift their distributions in response to changing climates.

3.6 Extinction vulnerability

Why are some bird species more extinction prone than the others? We list the main hypotheses that can be used to explain extinction vulnerability of birds in Table 3.1. Specialization is one of the traits that may make a species sensitive to disturbance (Norris and Harper 2004; Colles *et al.* 2009) and increases bird extinction likelihood (Sekercioglu *et al.* 2004). Specialization precludes species to cope with habitat alteration and novel conditions (e.g. turnover in prey species). Bird species adapted to dark microhabitats (e.g. closed canopy) may be susceptible to increased light levels in disturbed forests and may disappear (Pearson *et al.* 2010). Indicating the vulnerability of forest bird species, in Singapore, substantially more extinctions occurred among forest birds (61 of 91 species) than non-forest-dependent birds (13 of 127 species) (Castelletta *et al.* 2000). Similarly, 73% species extirpated from a forest fragment in Ecuador were forest species (Pearson *et al.* 2010).

Dietary specialization also makes a species vulnerable to habitat disturbance. In Singapore, 76% of insectivorous birds disappeared likely because of deforestation (Castelletta *et al.* 2000). Similarly, 50% of bird species that declined in abundance at La Selva Biological Station (Costa Rica) between 1960 and 1999 were insectivores (Sigel *et al.* 2006). Insectivores may be more extinction prone because of impoverishment of insect fauna coupled with their inherently poor dispersal abilities (Sekercioglu *et al.* 2002). Studies suggest that dispersal abilities can affect

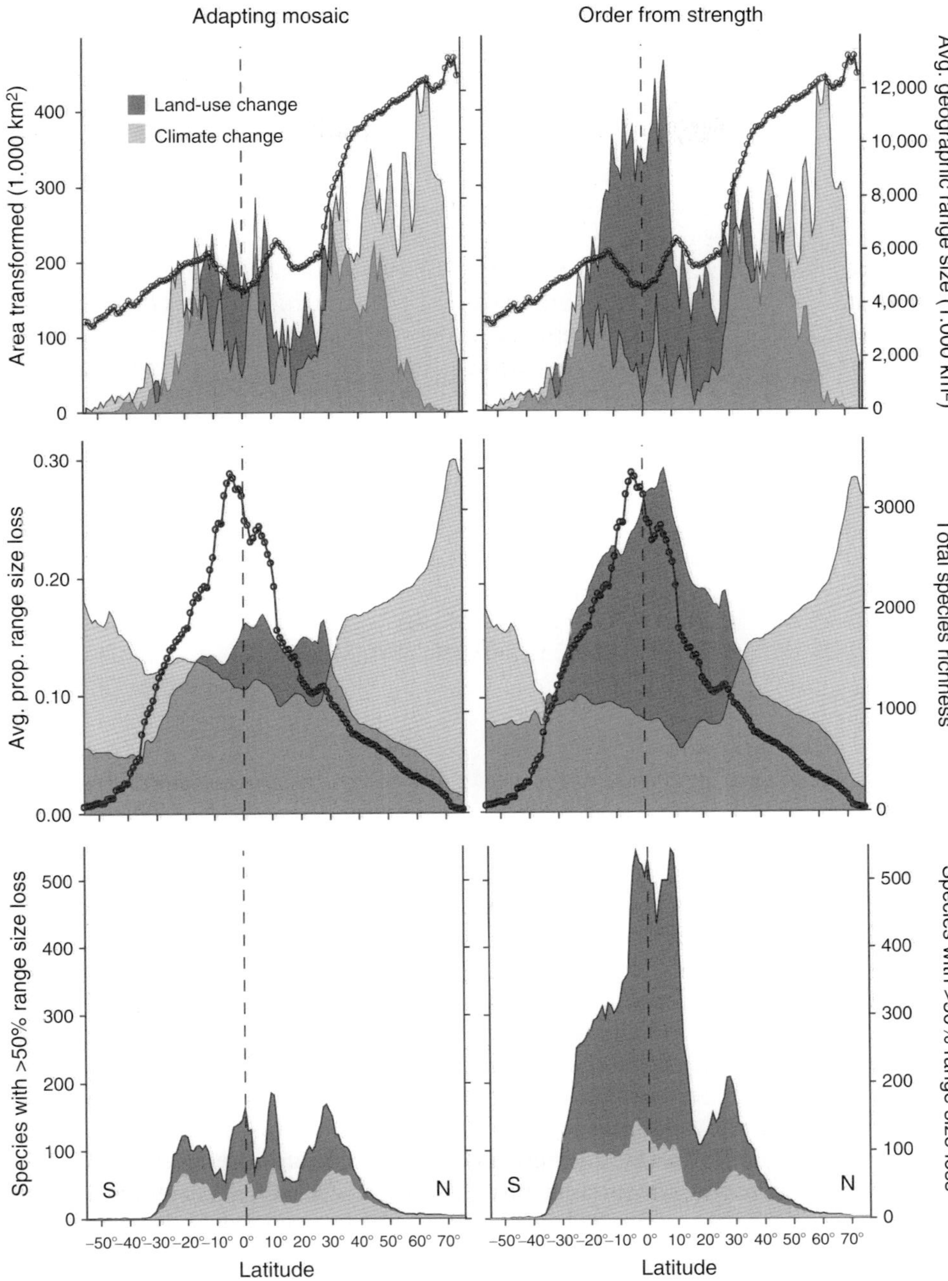

Figure 3.11 Climate and land-use changes and loss in avian species richness and range sizes. Changes are evaluated for two different scenarios – 'Adaptive Mosaic' and 'Order from Strength'. (After Jetz *et al.* 2007.) (Please see Plate 3.11 for the color version of this figure.)

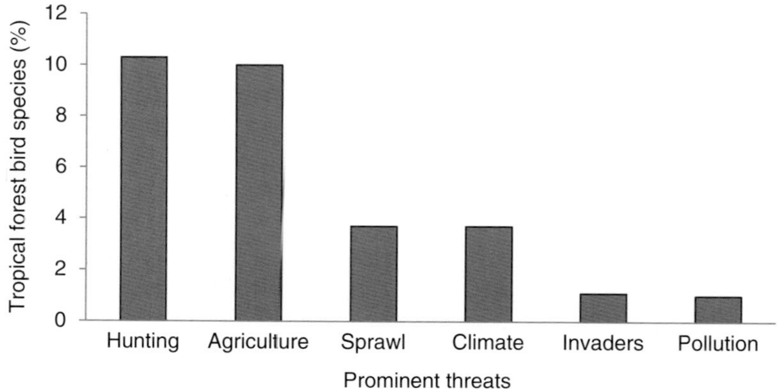

Figure 3.12 Prominent threats to tropical forest birds. (Data derived from Laurance and Useche 2009.)

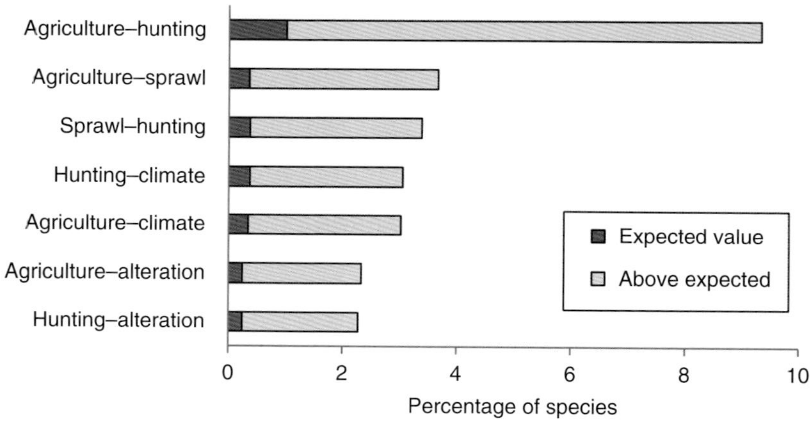

Figure 3.13 Expected and observed (expected + above expected) frequencies of combinations of environmental threats to imperiled bird species. (After Laurance and Useche 2009.)

persistence in fragments (e.g. Van Houtan *et al.* 2007; Lees and Peres 2008b; Moore *et al.* 2008). In general tropical forest birds are thought to have poor dispersal abilities and may be psychologically inhibited to cross forest gaps (Newmark 1991). For example, some forest birds are unlikely to cross non-forested habitat even 30 m wide (Stratford and Robinson 2005) or disperse between fragments separated by > 30 m (Martensen *et al.* 2008). Similar to insectivorous birds, frugivores may also be vulnerable in degraded or reduced forests because such areas may no longer provide a sufficient volume of fruits year-round (Leck 1979). Frugivores also appear to be more affected by hunting than other avian groups (Peres and Palacois 2007).

Habitat loss also impacts some behaviorally specialized birds (Van Houtan *et al.* 2007; Lees and Peres 2008b). For instance, species forming mixed-species

Table 3.1 Main hypotheses that explain bird losses following habitat perturbations. These hypotheses are non-mutually exclusive

Hypothesis	*Prediction*	*Possible mechanisms(s)*
1. Habitat specialization influences extinction	Forest species disappear disproportionately	Loss in habitat quality and quantity
2. Foraging specialization influences extinction	Guilds such as insectivores and frugivores will suffer losses	Reduction in food availability, poor dispersal
3. Body size affects extinction	More large-sized species will be lost	Low reproductive output, large area requirement
4. Range size influences extinction	Small-ranged species are extirpated	Low abundance, high specialization
5. Abundance affects extinction	Rare species disappear	Low population replacement, high specialization
6. Home range sizes influences extinction	Species with large home ranges disappear	Lack of adequate habitat
7. Bird dispersal abilities affect extinction	Species with poor dispersal abilities are lost	Poor chances of recolonization
8. Bird physiology affects extinction	Light-sensitive birds disappear	Alteration of microhabitat
9. Complex social behavior influences extinction	Species such as mixed-flock members and those following army ants are lost	Loss of affiliate, loss of suitable habitat

flocks disappeared frequently from small fragments (1 and 10 ha) in Manaus (Brazil; Stouffer and Bierregaard 1995). Some of the mixed-species flock members (e.g. Cinereous Antshrike, *Thamnomanes caesius*) may have high foraging success in mixed flocks or have large territories and high territory fidelity (Munn and Terborgh 1979; Powell 1985). There may not be enough space for large territories in small fragments. Mixed-species flocks had half the species richness in urban areas than in submontane forests of Peninsular Malaysia (Lee *et al.* 2005b). Flocking species sensitive to habitat disturbance were members of the Families Corvidae, Nectariniidae, and Sylviidae, had narrow altitudinal ranges, and were exclusively dependent on primary forest and understory (Lee *et al.* 2005b).

Similarly, obligate army-ant followers disappeared from small fragments, probably because of these birds' large area requirements (Bierregaard and Lovejoy 1989; Bierregaard *et al.* 1992). These species (e.g. Figure 2.5; White-plumed Antbird, *Pithys albifrons*) may have home ranges between 1 and 5 km in diameter (Willis and Oniki 1978; Bierregaard and Lovejoy 1989; Harper 1989), an area not available in small fragments (≤10 ha). Along the same lines, the abundance of five ant-following species declined by 52–100% in fragments < 115 ha in western Kenya (Peters and Okalo 2009).

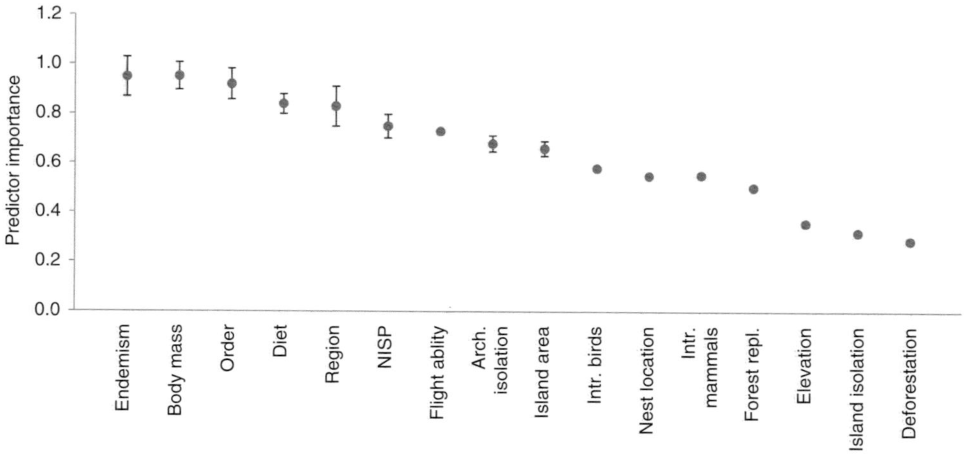

Figure 3.14 Relative importance (with standard deviation) of predictor variables over a random forest for cumulative risk. Importance score reflects the impact of predictor removal on overall classification accuracy. arch. = archipelago; intr. = introduced; repl. = replacement; NISP = measure of fossil sampling effort. (After Boyer 2010.)

Body size also affects extinction vulnerability – larger birds are usually more extinction-prone than smaller birds (Castelletta *et al.* 2000; Brook *et al.* 2003a; Gage *et al.* 2005; Keane *et al.* 2005). Perhaps due to allometric scaling laws (Peters 1983), large-sized animals have naturally lower population sizes, lower reproductive rates, larger area requirements, and/or higher food intake than small animals (Sodhi *et al.* 2004b). Larger animals also usually suffer high hunting pressure (Peres and Palacois 2007). Endemic and large-sized species have been disproportionately lost from the Pacific Islands since 3500 years (Figure 3.14; Boyer 2010). However, flightless bird species were vulnerable to extinctions that happened between 3500 and 200 years ago (Boyer 2010). Selective extinctions of large-sized birds have altered the distribution of existing birds on Hawaiian Islands (Figure 3.15; Boyer 2008). Nonetheless, body mass may not always predict bird extinctions, and in fact small-sized species may be more vulnerable in some sites (Woinarski and Catterall 2004; Sigel *et al.* 2010). This result could be due to past extinctions. Extant heavier tropical birds in Australia are not threatened by extinction, possibly because of the "extinction filter," i.e. extinctions of vulnerable heavier species during prehistoric times (> 200 years ago) (Garnett and Brook 2007).

In addition to body size, other morphological variables also affect extinction proneness. Populations under sexual selection pressure (e.g. sexual dichromatism) may suffer costs such as increased parasitism and predation (Doherty *et al.* 2003). Further, high investment in sexual traits may render some species less adaptable to changing environments. However, Jones *et al.* (2001) found that sexual dimorphism was not one of the predicators of forest bird abundance on the islands of Sumba and Buru (Indonesia). The effect of sexual dimorphism on the extirpation of forest bird species remains poorly studied.

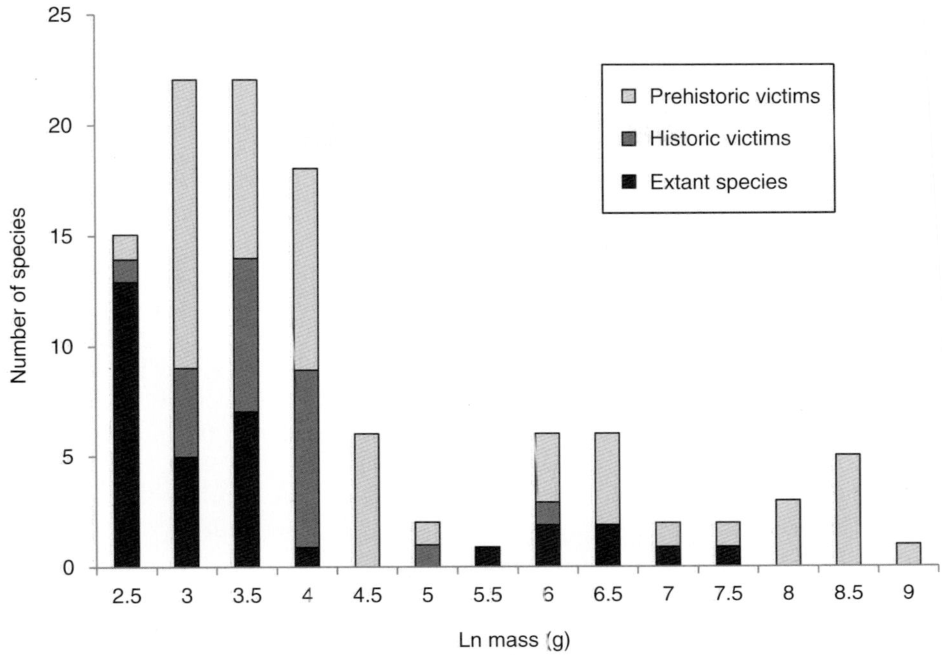

Figure 3.15 Body size frequency distributions of Hawaiian bird species. Ln = log. (After Boyer 2008.)

Biogeographical variables can be important in species' vulnerability to extinction. Populations of forest birds at the edges of their ranges are more vulnerable to forest disturbance (Christiansen and Pitter 1987). Populations may be at their physiological and ecological limits at their distributional edge. In addition, because of low densities, populations at the edges of their ranges could be particularly vulnerable to genetic (e.g. inbreeding depression) and/or environmental stresses (e.g. pesticides; see Møller 1995). Nineteen of 29 species at the upper or lower limit of their elevational distributions were extirpated from the western Andes (Kattan *et al.* 1994). Similar instances have been reported elsewhere (Christiansen and Pitter 1997; Gillespie 2000). However, in the Colombian Andes, Renjifo (1999) found that species beyond their usual elevational limits were not disproportionately likely to become extinct, suggesting, in some cases, resilience even at distributional limits or that these species may be adapting to climate warming (see Chapter 8).

Extinction proneness can also be affected by the size of a species' geographical range (Feeley *et al.* 2007a; Harris and Pimm 2008; Pearson *et al.* 2010), and is one of the most important criteria for determining the threat status on the Red List (BirdLife International 2008b). Tropical forest species generally have smaller geographical ranges than compared to species residing elsewhere (Stork *et al.* 2009). Species with small geographical ranges are usually specialized and have low local abundance (Williams *et al.* 2009). Therefore, small ranges may make species more vulnerable to stochastic perturbations. Proportionally more passerines with small geographical ranges in the Americas are threatened than those with large geographical ranges (Manne *et al.* 1999). Harris and Pimm (2008) estimated that tropical forest species with ranges < 11,000 km^2 should be considered threatened with extinction. In the Australian tropics, Williams *et al.* (2009) found that specialized and restricted range veterbrate species had high and uniform local abundances, which may bet hedge them against extinction.

Deforestation may reduce habitat sizes for species requiring large home ranges. Four canopy frugivores likely requiring large home ranges (e.g. the Scaly-naped Parrot, *Amazona mercenaria*) disappeared from San Antonio, Ecuador (Kattan *et al.* 1994). Assuming that home range scales positively with body mass, Beier *et al.* (2002) determined that threshold patch size in Ghana for a 24 g and 920 g species was 10 ha and 8000 ha, respectively. However, Renjifo (1999) argued that for some species large home ranges may not be a constraint as such species could be adapted to track variable food resources and thus be able to travel even through the matrix (areas surrounding the forest).

Rare or uncommon bird species may be more extinction prone than common species (Feeley *et al.* 2007a), but rarity is not an independent factor and correlates with variables such as body size, habitat specificity, and geographical distribution (Kattan 1992; Goerck 1997). In Ecuador, some rare species such as the Tiny Hawk (*Accipiter superciliosus*) and Crane Hawk (*Geranospiza caerulescens*) were possibly extirpated (Leck 1979). Small initial population size, requirement for large territories, and diet specialization may be critical factors in their elimination. Similarly, rare and uncommon species were more likely to be extirpated than common species in the Colombian Andes and Usambara Mountains of Tanzania (Newmark 1991; Renjifo 1999). In tropical moist rainforests of eastern Queensland (Australia), rare species were negatively affected by fragmentation (Warburton 1997). However, rare species are not always more extinction prone than common species. Karr (1982) sampled the understory fauna of Limbo Hunt Club (Panama) 9 km east of Barro Colorado Island (BCI), presumed to be the source fauna for BCI. The species that were extirpated from BCI were not less abundant at Limbo Hunt Club (e.g. Song wren, *Cyphorhinus phaeocephalus*). Species with higher annual population variability in the Limbo Hunt Club were the ones that disappeared from BCI. Karr hypothesized that such species track variable food resources and may require larger areas for their local movements. Similarly, apparently widespread and common species are sometimes extirpated, for example, the Brown-hooded Parrot (*Pionopsitta haematotis*) and Red-capped Manakin (*Pipra mentalis*) in Ecuador (Leck 1979).

Extinction of a species entails a permanent loss of evolutionary history. Bird extinctions can be phylogenetically non-random (Van Euler 2001). Few specific tropical case studies deal with phylogenetic effects. Thiollay (1997b) suggested that the presence of congeners may lead to competitive pressure, narrower niches, and lower abundance of at least some potential competitors. However, Terborgh and Winter (1980), with data from Trinidad and Venezula, and Jones *et al.* (2001), with data from Sumba and Buru, suggest that congeners did not affect survivorship of their relatives. Indicating the vulnerability of some families to extinction, Terborgh and Winter (1980) found that families containing disproportionately more susceptible species in Fernanado Po, Trinidad, Hainan, Sri Lanka, and

Tasmania (temperate island) were Bucerotidae, Cracidae, Falconidae, Phasianidae, Picidae, Timaliidae, Tinamidae, and Ramphastidae, although patterns for other families were mixed (containing similar numbers of both susceptible and resilient species). In the Colombian Andes, only Icteridae, as a family, was more prone to extinction (Renjifo 1999). In Las Cruces (Costa Rica), Sekercioglu *et al.* (2002) found a positive correlation between the number of species of a bird family present in non-forested habitats and their presence in small fragments. They seemed to suggest that family-level dispersal characteristics and abilities to exploit deforested habitats may assist in species persistence in small fragments.

As mentioned, it is highly likely that different variables act synergistically to extirpate a bird species. Supporting this thesis, factors such as range sizes, specialization, body size, and foraging specialization acted together to extirpate forest bird species in Ecuador (Pearson *et al.* 2010). Additionally, many of the variables driving extinction are correlated. For instance, body mass correlated with rarity for tropical dry forest birds in Central America (Gillespie 2000). Therefore, while studying bird extinctions, the affect of all possible variables should be considered.

3.7 Ecosystem resonance of bird extinctions

Some bird losses may have repercussion for ecosystem functioning in the tropics (see Chapter 4). Declining bird populations have great implications for ecosystem processes, especially given that extinction threat is not uniformly distributed among avian functional groups, and some key groups such as scavengers, frugivores, and tropical forest insectivores are more threatened than the global average (Sekercioglu *et al.* 2004). Disruptions of ecological processes through species loss in degraded forests may also lead to cascading and catastrophic co-extinctions (Koh *et al.* 2004; Dunn *et al.* 2009). Seed dispersal is one of the key ecological processes in tropical landscapes. Because many tropical trees produce large, lipid-rich fruits adapted for animal dispersal (Howe 1984), demise of bird frugivores may have consequences for forest regeneration (Wright *et al.* 2007). Indicating that avian frugivores may be important seed-dispersing agents, seed removal by them was found to be higher than those by primates in Afrotropical forests (Kirika *et al.* 2008). Similarly, one of the dominant fruiting trees (*Dacryodes excelsa*) in Puerto Rico failed to re-establish in areas it previously occupied, possibly because of the extinction and decline of avian frugivores (Brash 1987). Some late successional montane tree species (e.g. *Canarium asperum*) rely heavily on frugivorous bird species such as hornbills and fruit pigeons for seed dispersal (Hamann and Curio 1999). These bird species are under heavy hunting pressure and their likely extinction may disrupt the recruitment of the trees.

A similar predicament has been found for an endemic tree (*Leptonychia usambarensis*) in Tanzania that relies on birds to disperse its seeds (Cordeiro and Howe 2003). Bird species that dispersed the seeds of this tree in a continuous forest were absent or rare in fragments. This resulted in fewer seeds removed from trees in fragments and lower juvenile recruitment there (Cordeiro and Howe 2003).

Similarly, seed removal of a keystone palm (*Euterpe edulis*) was 24% lower on a frugivore-impoverished landbridge island than mainland Atlantic forest in Brazil (Fadini *et al.* 2009). It is disconcerting that reduction in fruit removal by forest specialists due to forest disturbance may not be compensated by forest generalists and visitors (Kirika *et al.* 2008). Further, frugivorous bird needs to be sufficiently abundant to adequately disperse seeds in human-dominated tropical landscapes (Pejchar *et al.* 2008).

As mentioned in Chapter 6, the introduced brown tree snake has resulted in the loss of numerous forest bird species on Guam. Among these demises are important pollinators and seed dispersers. As a consequence, no bird species visited two species of plants (*Bruguiera gymnorrhiza* and *Erythrina variegata*) that are bird pollinated on Guam (Figure 3.16; Mortensen *et al.* 2008). Due to lack of avian seed dispersers, seed dispersal was also compromised with significantly low saplings of *B. gymnorrhiza* on Guam than in Saipan where frugivores still persist in the absence of the tree snake (Mortensen *et al.* 2008).

The biological control of herbivorous insects by birds may be of value in both anthropogenic and natural forest settings (Tscharntke *et al.* 2008). Although it is clear that insectivorous birds play an important role in controlling outbreaks of herbivorous insects in agroforests (Perfecto *et al.* 2004), there are fewer studies with comparable results in natural forest stands (Van Bael *et al.* 2003; Sekercioglu 2006a,b). Generally, insect herbivores inflict high damage in both the canopy and understory of forests that lack avian insectivores (Van Bael *et al.* 2003; Van Bael and Brawn 2005; Dunham 2008); this ability of birds to control herbivorous insects complements that of other insectivorous predators (Greenberg *et al.* 2000). As discussed above, understory insectivores and frugivores decline with increased disturbance and fragmentation of tropical forests (Sekercioglu *et al.* 2002; see Chapter 2). Thus, the potential implications for tropical forest productivity of the loss of birds in these two key trophic guilds need to be carefully examined. Although frugivory and insectivory are becoming better understood, more data are needed on how forest disturbance affects other avian functions such as pollination, vertebrate predation, and scavenging.

3.8 Extinction resistence

Some tropical bird species are able to cope with habitat disturbance. For instance, despite the loss of almost of its forest, forest species such as the Cebu Flowerpecker (*Dicaeum quadricolor*) still persist on the Cebu Island (Philippines) (Posa *et al.* 2008). The species–area relationship leads to a rule of thumb that a 90% loss in habitat leads approximately to a 25–50% loss of species (Simberloff 1992a). The predictive power of this relationship may be weak because it does not account for either habitat diversity, fragmentation, or surrounding matrix (Simberloff 1992b; Koh and Ghazoul 2010; see Chapter 2), but it is the only such existing model and is still used to make crude predictions of the extent of biotic extinctions. Thus far, tropical bird extinctions in some areas seem to be fewer than predicted by the species–area equation (Brooks *et al.* 2002). This fact indicates

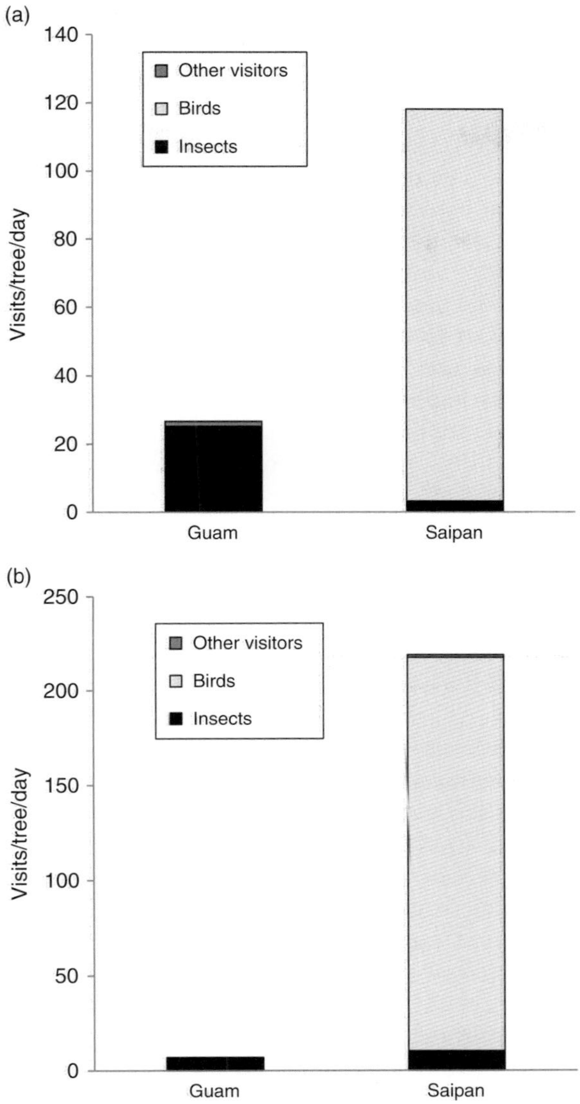

Figure 3.16 Animal visits to the flowers of (a) *Bruguiera gymnorrhiza* and (b) *Erythrina variegate* var. *orientalis*. (After Mortensen *et al*. 2008.)

temporary or permanent resilience in some of the species and/or a time lag in the extinction process.

Fragmented landscape containing mature forest can retain as much as 96% of the original avifauna even several decades after isolation (Renjifo 1999). Similarly, arecanut palm (*Areca catechu*) plantations retained 90% of extant forest species in Western Ghats (India) despite > 2000 years of agricultural activities (Ranganathan *et al*. 2008). However, because of lack of historical studies, it is unclear if all extinctions were adequately documented in this area. Nonetheless, countryside landscapes (landscapes devoted to human uses such as plantations or managed forests) can retain as much as 50% of native bird species (Ranganathan and Daily 2010; see also Figure 1.14). Nevertheless, bird diversity of countryside landscapes can be augmented by practicing less intensive agriculture and preserving native forests within and near the landscapes (Ranganathan and Daily 2010).

Many other studies show that a proportion of forest avifauna can be found in disturbed habitats such as regenerating forests (e.g. Thiollay 1992; Warkentin *et al.* 1995; Peh *et al.* 2005; Sheldon *et al.* 2010). However, the amount of native forest cover in a disturbed landscape can determine the presence of forest bird species there (Sodhi *et al.* 2005a). The minimum forest cover required to retain all forest bird species can be site-specific but does vary between 20% and 80% (Sodhi *et al.* 2005a; Posa and Sodhi 2006; Soh *et al.* 2006; Mordecai *et al.* 2009; Rompré *et al.* 2009; see Figure 1.15).

However, most studies of countryside landscapes are of short duration and we have limited understanding of the long-term population dynamics and persistence of forest birds in countryside landscapes (see Section 2.6). Few data are available, on whether persisting species in degraded areas are reproducing and surviving as well as their counterparts in pristine areas. However, Yap *et al.* (2007) found no differences in proportion of juveniles and breeding and molting adults between the understory of selectively logged (30 years ago) and unlogged forests in Peninsular Malaysia. Therefore regenerating forests may have conservation potential, especially when they are adjacent to unlogged forests, as was the case in this study.

Long-term studies to understand bird persistence in degraded landscapes are rare. To determine the biotic sustainability of plantations (Acacia, *Acacia mangium*, and Albizia, *Paraserianthes falcataria*), Sheldon *et al.* (2010) surveyed the same ones in 1982 and 2005. Despite complete isolation from the primary forests and managing activities such as tree cropping, the number of primary forests species only declined from 90 to 81 in the plantations over the years. However, the plantations were depauperate in forest trogons, barbets, hornbills, and pittas (Sheldon *et al.* 2010). This study shows that plantations can sustain at least some primary forest species over decades. Similarly, underscoring the need for long-term data, earlier observed trends changed as secondary growth occurred over the 20 years adjacent to forest fragments in Amazon (Stouffer *et al.* 2009). With secondary growth, some forest species (e.g. hummingbirds) seem capable of using secondary forests and ant followers readily traveled through second growth as did the ants. Frugivores also increased as secondary vegetation aged, possibly due to increase in fruit resources (Stouffer *et al.* 2009). Therefore, maturing patches of forests could have positive effects on understory birds. However, traditional land-use practices in Amazonia rarely allow this to happen, and these results are likely to be affected by the close proximity of the secondary forests to source populations in primary forest. This proximity increases the chance that vagrants or transient birds dispersing through secondary habitats will be detected (e.g. Terborgh and Weske 1969). Estimates of the use of secondary forests by primary forest species are considerably lower when the samples were taken over 500 m from the primary forest edge (Barlow *et al.* 2007, 2010), and we would expect them to be lower still in degraded tropical regions where source populations are isolated by an inhospitable matrix (e.g. Prugh *et al.* 2008). Finally, in the Atlantic of Brazil, bird species that tolerated secondary forest habitats were likely to be extirpated with the loss of remaining primary forests (Harris and Pimm 2004). This study suggests that even species using secondary habitats sometimes remain vulnerable to further deforestation (Harris and Pimm 2004).

4

Ecological Functions of Birds in the Tropics

Overview: Birds' ecological functions cover a wide spectrum, from creating soil to shaping primate behavior, and many species play key ecological roles, such as decomposition, predation, pollination, nutrient deposition, and seed dispersal. From an ecosystem functional perspective, birds are mobile links that are crucial for maintaining ecosystem function, memory, and resilience. The three main types of mobile links, genetic, process, and resource linkers, encompass all major avian ecosystem services. Rapid losses of tropical bird species may cause substantial reductions in certain ecosystem processes before we have time to study and understand the underlying mechanisms.

Birds are the best-studied class of organism and various studies on frugivorous, nectarivorous, and insectivorous birds have established their significance in the dynamics of diverse natural and human-dominated ecosystems (Table 4.1; Figures 4.1 and 4.2) (Stiles 1978, 1985; Redford 1992; Myers 1996; Proctor *et al.* 1996; Westcott and Graham 2000; Mols and Visser 2002; Croll *et al.* 2005; Sekercioglu 2006a; Van Bael *et al.* 2008). Birds exhibit the most diverse range of ecological functions among vertebrates. Even though mammals play comparable roles, birds have twice as many taxa, ten times more flying species, and are more resilient to extirpation (Holbrook *et al.* 2002). Many avian functions either complement those of mostly nocturnal mammals (Proctor *et al.* 1996; Clark *et al.* 2001) or have no other vertebrate equivalents, exemplified by aerial leaf litter gleaners (Gradwohl and Greenberg 1982), cavity drillers (Daily *et al.* 1993), large flying predators (Mitani *et al.* 2001), or obligate scavengers (Ruxton and Houston 2004).

Conservation of Tropical Birds, First Edition. Navjot S. Sodhi, Çağan H. Şekercioğlu, Jos Barlow and Scott K. Robinson.

Published 2011 by Blackwell Publishing Ltd.

Table 4.1 Ecological and economic contributions of avian functional groups. (After Sekercioglu 2006a)

Functional group	*Ecological process*	*Ecosystem service and economical benefits*	*Negative consequences of the loss of functional group*
Frugivores	Seed dispersal (Snow 1981; Howe and Smallwood 1982; Stiles 1985; Howe and Miriti 2000; Howe and Miriti 2004)	Removal of seeds from the parent tree (Greenberg *et al*. 1995; Avila *et al*. 1996; Sun *et al*. 1997; Wenny and Levey 1998); escape from seed predators (Janzen 1972; Howe 1993); improved germination (Murphy *et al*. 1993; Meyer and Witmer 1998); increased economical yield (Hammond *et al*. 1996; Hutchins *et al*. 1996; Narang *et al*. 2000; Yumoto 2000); increased gene flow (Howe *et al*. 1985; Hamrick *et al*. 1993; Gibson and Wheelwright 1995); recolonization and restoration of disturbed ecosystems (Robinson and Handel 1993; Tucker and Murphy 1997; Wilkinson 1997; Galindo-Gonzalez *et al*. 2000; Hjerpe *et al*. 2001)	Disruption of dispersal mutualisms (Stocker and Irvine 1983; Clark *et al*. 2001; Meehan *et al*. 2002); reduced seed removal (Cordeiro and Howe 2003); clumping of seeds under parent tree (Bleher and Böhning-Gaese 2001); increased seed predation (Howe 1993); reduced recruitment (Cordeiro and Howe 2001, 2003); reduced gene flow (Shapcott 1999; Pacheco and Simonetti 2000) and germination (Compton *et al*. 1996; Peres and van Roosmalen 1996; Meyer and Witmer 1998); reduction (Santos and Telleria 1994; Santos *et al*. 1999) or extinction (Bond 1994; Hamann and Curio 1999; Loiselle and Blake 1999; da Silva and Tabarelli 2000) of dependent species
Nectarivores	Pollination (Stiles 1978, 1985; Proctor *et al*. 1996)	Outbreeding of dependent (Keighery 1980; Ford 1985; Proctor *et al*. 1996) and/or economically important species (Nabhan and Buchmann 1997; Narang *et al*. 2000)	Pollinator limitation (Nabhan and Buchmann 1997; Murphy and Kelly 2001); inbreeding and reduced fruit yield (Feinsinger *et al*. 1982; Robertson *et al*. 1999; Cox and Elmqvist 2000; Paton 2000; Rathcke 2000; Montgomery *et al*. 2001); evolutionary consequences (Stiles 1978; Thompson 1996; Nabhan and Buchmann 1997); extinction (Bond 1994; Sakai *et al*. 2002)

Table 4.1 Cont'd

Functional group	*Ecological process*	*Ecosystem service and economical benefits*	*Negative consequences of the loss of functional group*
Scavengers	Consumption of carrion (Houston 1994)	Removal of carcasses (Pain *et al.* 2003; Prakash *et al.* 2003); leading other scavengers to carcasses (Houston 1994); nutrient recycling; sanitation (Pain *et al.* 2003; Prakash *et al.* 2003)	Slower decomposition (Houston 1994); increases in carcasses (Pain *et al.* 2003; Prakash *et al.* 2003); increases in undesirable species (Pain *et al.* 2003; Prakash *et al.* 2003); disease outbreaks (Pain *et al.* 2003; Prakash *et al.* 2003); changes in cultural practices (Parry-Jones 2001; Pain *et al.* 2003)
Insectivores	Predation on invertebrates	Control of insect populations (Crawford and Jennings 1989; Marquis and Whelan 1994; Kirk *et al.* 1996; Greenberg *et al.* 2000; Jantti *et al.* 2001; Mols and Visser 2002; Van Bael *et al.* 2003, 2008); reduced plant damage (Sipura 1999; Greenberg *et al.* 2000; Sanz 2001); alternative to pesticides (Dolbeer 1990; Naylor and Ehrlich 1997; Mourato *et al.* 2000)	Loss of natural pest control (Dolbeer 1990; Naylor and Ehrlich 1997); pest outbreaks (Crawford and Jennings 1989; Kirk *et al.* 1996; Quammen 1997); crop losses (Greenberg *et al.* 2000); trophic cascades (Terborgh *et al.* 2001)
Piscivores	Predation on fishes and invertebrates Production of guano	Controlling unwanted species (Wootton 1995); nutrient deposition around rookeries (Powell *et al.* 1991; Anderson and Polis 1999; Hawke *et al.* 1999; Palomo *et al.* 1999; Sanchez-Pinero and Polis 2000; Croll *et al.* 2005); soil formation in polar environments (Heine and Speir 1989); indicators of fish stocks (Crawford and Shelton 1978); environmental monitors (Gilbertson *et al.* 1987)	Loss of guano and associated nutrients (Oliver and Legovic 1988; Croll *et al.* 2005); impoverishment of plant communities (Oliver and Schoenberg 1989; Croll *et al.* 2005); trophic cascades (Wootton 1995; Williams *et al.* 2002; Croll *et al.* 2005); ecosystem shifts (Croll *et al.* 2005); loss of socio-economic resources (Haynes-Sutton 1987) and environmental monitors (Gilbertson *et al.* 1987)

Raptors	Predation on vertebrates	Regulation of rodent populations (Korpimaki and Norrdahl 1991; Ims and Andreassen 2000); secondary dispersal (Nogales *et al*. 2002)	Rodent pest outbreaks (Korpimaki and Norrdahl 1998); trophic cascades (Crooks and Soule 1999; Terborgh *et al*. 2001; Dunne *et al*. 2002); indirect effects (Sih *et al*. 1985)
All species	Miscellaneous	Environmental monitoring (Eriksson 1987; Bryce *et al*. 2002); indirect effects (Izhaki and Safriel 1989; Dean *et al*. 1990; Loiselle 1990; Paine *et al*. 1990; Wootton 1994; Milton *et al*. 1998; Murakami and Nakano 2002; Nogales *et al*. 2002); birdwatching tourism (Jacquemot and Filion 1987; Sekercioglu 2002b; Bouton and Frederick 2003); reduction of agricultural residue (Bird *et al*. 2000); cultural and economic uses (Diamond 1987)	Losses of socio-economic resources (Filion 1987; Sekercioglu 2002b) and environmental monitors (Peakall and Boyd 1987); unpredictable consequences (Wootton 1994)

Currently, 22.1% of bird species are considered "extinction-prone," a category that includes species that are historically extinct or extinct in the wild (1.4%), threatened (12.4%) or near threatened (8.4%) with extinction (IUCN 2010a; see Chapter 1). Ongoing reductions in the abundance (Gaston *et al.* 2003) and species richness (BirdLife International 2008b) of tropical bird species are likely to have far-reaching ecological consequences (Sekercioglu *et al.* 2004), with diverse societal impacts ranging from the spread of disease and loss of agricultural pest control to plant extinctions.

4.1 Birds as mobile links

From an ecosystem functional perspective, birds are mobile links (Gilbert 1980; Lundberg and Moberg 2003) that are crucial for maintaining ecosystem function, memory, and resilience (Nyström and Folke 2001; Figure 4.1). The three main types of mobile links, genetic, process, and resource linkers (Lundberg and Moberg 2003), encompass all major avian ecosystem services. Pollinating nectarivores and seed-dispersing frugivores are genetic linkers that carry genetic material from an individual plant to another plant or to a habitat suitable for regeneration, respectively (Figure 4.1A). Trophic process linkers are grazers, such as geese (Maron *et al.* 2006), and predatory birds, such as flycatchers and hawks, that influence the populations of plant, invertebrate, and vertebrate prey and often provide natural pest control (Holmes *et al.* 1979; Van Bael *et al.* 2008). Woodpeckers (Picidae) act both as trophic process linkers and as physical process linkers or "ecosystem engineers" (Jones *et al.* 1994). In addition to consuming insects, many woodpeckers also build nest holes used by a variety of other species (Daily *et al.* 1993; Figure 4.1D). Scavenging birds such as vultures (Figure 4.1C) are crucial process linkers that hasten the decomposition of potentially disease-carrying carcasses (Houston 1994). Piscivorous (fish-eating) birds provide good examples of resource linkers (Figure 4.1B) that transport nutrients from water to land in their droppings and often contribute significant resources to island ecosystems (Anderson and Polis 1999). These groups are not mutually exclusive. For example, many colonial waterbirds are both process linkers as predators of fish and resource linkers as transporters of nutrients from water to land in their droppings (guano).

4.2 Seed dispersal

Darwin (1859) was among the first biologists to realize that birds are "highly effective agents in the transportation of seeds." Indeed, seed dispersal may well be the most important avian ecosystem service (Figure 4.1A; see also Chapter 3). This is especially true in the tropics where avian seed dispersal may have led to the emergence of angiosperm dominance (Regal 1977; Tiffney and Mazer 1995) and is arguably key to the maintenance of extraordinary plant diversity (Janzen 1970; Connell 1971; Stiles 1985; Schupp *et al.* 2002; Terborgh *et al.* 2002). Vertebrates are the main seed vectors for angiosperms (Regal 1977; Tiffney and

Figure 4.1 Examples of the four main types of avian mobile links (Lundberg and Moberg 2003) and potential consequences of the lack of their services. Drawings are based on the cited references. (A) Genetic linkers. Loss of Philippine seed dispersers, such as Palawan Hornbills *Anthracoceros marcei*, can result in most seeds being deposited under the parent tree and being consumed by seed predators (Hamann and Curio 1999). (B) Resource linkers. Introduced foxes eliminating Aleutian seabirds, such as Tufted Puffins *Fratercula cirrhata*, can lead to reduced nutrient deposition triggering a shift from grassland to maritime tundra (Croll *et al.* 2005). (C) Trophic process linkers. Disappearance of scavenging Indian (Long-billed) Vultures *Gyps indicus*, can cause increases in the numbers of rotting carcasses and of attending mammalian scavengers (Prakash *et al.* 2003). (D) Trophic and non-trophic process linkers. Reduced numbers of Eurasian Three-toed Woodpeckers *Picoides tridactylus* in forest fragments can cause increases in spruce bark beetles (Fayt *et al.* 2005) and decreases in nesting holes used by other species (Daily *et al.* 1993). In addition to habitat loss that affects all avian functional groups, large frugivores are highly susceptible to exploitation, by-catch mortality and introduced species threaten seabirds, woodpeckers decline as a result of fragmentation, and vultures are particularly sensitive to chemicals. (Drawing by Darryl Wheye, © by Darryl Wheye/Science Art Birds.) (Please see Plate 4.1 for the color version of this figure.)

Figure 4.2 Examples of the four main types of mobile links and corresponding avian ecosystem services. From top left, clockwise: Chesnut-mandibled Toucan (*Ramphastos swainsonii*), seed dispersing genetic linker; Black-browed Albatross (*Thalassarche melanophrys*), nutrient depositing resource linker; Lappet-faced Vulture (*Torgos tracheliotos*), scavenging trophic process linker; Red-necked Woodpecker (*Campephilus rubricollis*), ecosystem engineering non-trophic process linker. (All photographs by Cagan H. Sekercioglu.) (Please see Plate 4.2 for the color version of this figure.)

Mazer 1995), particularly woody plants (Howe and Smallwood 1982; Levey *et al.* 1994; Jordano 2000). Increased seasonality in the temperate zone and consequent fluctuations in fruit and frugivore numbers make animal seed dispersal less reliable than in the tropics (Snow 1981) where vertebrate seed dispersers are especially important (Howe and Smallwood 1982; Stiles 1985) and the majority of taxa in many plant families are dispersed by birds (Hilty 1980; Snow 1981; Stiles 1985; Renner 1989; Willson and Crome 1989; Hamann and Curio 1999; Ganesh and Davidar 2001; Shanahan *et al.* 2001).

Seed dispersal is thought to benefit plants in three major ways (Howe and Smallwood 1982): (i) escape from density-dependent mortality caused by pathogens, seed predators, competitors, and herbivores, also known as the Janzen-Connell escape hypothesis; (ii) chance colonization of favorable but unpredictable sites via wide dissemination of seeds; and (iii) directed dispersal to specific sites that are particularly favorable for establishment and survival.

Seed passage through animals may also increase germination likelihood (Traveset and Verdú 2002) as a result of gut passage breaking seed dormancy (Noble 1975) or reducing seed infestation (Webber and Woodrow 2004), but most seeds do not require gut passage for germination and this mechanism is not considered to be a major advantage of dispersal (Howe and Smallwood 1982).

4.2.1 Advantages of seed dispersal

The most important contribution of seed dispersers to plant survival is by reducing the density-dependent mortality of seeds and seedlings (Janzen 1970; Connell 1971; Howe 1989; Loiselle 1990; Harms *et al.* 2000) and by enabling escape from seed predators (Janzen 1970; Sharam *et al.* 2009), competitors (Nathan and Muller-Landau 2000), herbivores (Connell 1971), and pathogens (Antonovics and Levin 1980; Packer and Clay 2000). In the Serengeti ecosystem of Tanzania, birds maintain forest by inhibiting seed predators (Sharam *et al.* 2009). In a detailed study of the seeds, saplings and adults of trees in a 2.25 ha plot at Cocha Cashu, Peru, Terborgh *et al.* (2002) showed the importance of negative density-dependent recruitment in increasing plant diversity from saplings to adults. From over 500 seeds that fell on 1 m^2 of forest floor in a given year, only four survived to become saplings (Terborgh *et al.* 2002). Over 95% of saplings originated from dispersed seeds and the probability of a seed > 75 m away from the parent tree becoming a sapling was roughly five orders of magnitude higher than a seed under the parent tree. Seed dispersal was particularly important for rare tree species, which contributed substantially to the high diversity observed and many of which would disappear if seed dispersal were reduced (Dirzo and Miranda 1991). This study confirms the importance of escape from localized enemies in greatly enhancing survival away from the parent tree.

Complementary to these results that show the importance of even modest seed dispersal distances, radio-tracking of small (13 g), relatively non-mobile, and frugivorous Ochre-bellied Flycatchers (*Mionectes oleaginous*) in Costa Rican lowland forest revealed median dispersal distances of 42–56 m for six plant species (Westcott and Graham 2000). Therefore, even small, atypical avian frugivores can provide significant seed dispersal away from the parent tree. Even though dispersal may not always provide an escape from mortality and competition (Mack *et al.* 1999) and may even be disadvantageous at times (Silander 1978), its advantages are often considerable.

In addition to its "top-down" role by enabling escape from seed predators, seed dispersal can also provide a "bottom-up" advantage to seeds by increasing the probability that seeds will colonize a site with favorable germination conditions, be they light, nutrients, temperature, humidity, or some type of required disturbance (Howe and Miriti 2004). The advantage of increased colonization potential is likely to accrue more to plants with small, abundant, and highly vagile seeds that favor open, disturbed conditions and can grow rapidly. These "weedy colonists," such as *Cecropia* spp., contrast with large-seeded and more persistent species that cannot disperse as readily, but are better competitors as a result of their greater reserves (Kennedy *et al.* 2004), and usually replace colonizing species with the passage of time. Birds can eliminate the dispersal disadvantage of large-seeded species and long-distance dispersal may be more regular than we thought (Clark *et al.* 1999). Large-seeded species with low vagility benefit most from avian seed dispersal, which seems to be crucial to maintaining the diversity of relatively stable tropical communities.

Avian dispersal is often considered "random" from the plant's perspective in that, besides the advantage of being deposited away from the parent tree, seed dispersal is not thought to be directed towards sites where plant survival probability is high. However, directed dispersal may not be rare (Wenny 2001), and may be particularly common in regenerating and arid areas (Wenny 2001). In such areas, the few available trees both attract birds and provide a favorable microclimate to seedlings. Bird droppings fertilize the soil and gut passage may increase seed germination probability and speed (Treca and Tamba 1997). Three-wattled Bellbirds (*Procnias tricarunculatus*) in Costa Rica exemplify this phenomenon. In contrast to four other native avian dispersers, bellbirds dispersed seeds >40 m from the parent tree, under song perches in canopy gaps where recruitment success was significantly higher due to a reduction in fungus-induced mortality (Wenny and Levey 1998). Similarly, in New Guinea, Dwarf Cassowaries (*Casuarius bennetti*) preferentially dispersed the seeds of *Aglaia aff. flavida* uphill from the parent tree (Mack 1995). The absence of Dwarf Cassowaries, which are heavily hunted (BirdLife International 2010a), would lead to downhill dispersal resulting in smaller and fragmented populations of this plant (Mack 1995). In some cases, however, directed dispersal may also favor the expansion of introduced species (Dean and Milton 2000). Nevertheless, directed dispersal can provide significant advantages, especially in human-dominated ecosystems where restoration is critical (Wenny 2001).

A little-mentioned but potentially crucial service of frugivorous birds is that the removal of fruit pulp can significantly reduce the risk of bacterial and fungal infections that can kill the seeds before they germinate (Howe and Vandekerckhove 1981; Jackson *et al.* 1988; and others in Witmer and Cheke 1991). In fact, this might have been the most important contribution of the extinct Dodo (*Raphus cucullatus*) to the germination of the Tambalacoque tree (*Sideroxylon grandiflorum*, previously *Calvaria major*), Sapotaceae (Witmer and Cheke 1991). The famous story of the Tambalacoque tree now being on the verge of extinction since its seeds had to pass through Dodos to germinate (Temple 1977) has been shown to be more complicated (Witmer and Cheke 1991). Unabraded seeds of this tree still germinate and there are living trees less than 300 years old, indicating that this is not an example of obligate mutualism (Witmer and Cheke 1991). Nevertheless, Dodos were doubtlessly crucial as one of the few frugivores on Mauritius who could clean and disperse the large seeds of the Tambalacoque tree and a thorough cleaning of the fruit pulp by frugivorous birds may be the key to successful germination in many plant species (Howe and Vandekerckhove 1981; Jackson *et al.* 1988).

4.2.2 Specialization, redundancy, and complementarity

Even though earlier overviews of avian seed dispersal emphasized specificity and tight coevolution (McKey 1975), starting in the 1980s (Howe and Smallwood 1982; Wheelwright and Orians 1982; Howe 1984; Estrada 1986), there has been an increasing realization that, with the possible exception of large-seeded

Box 4.1 Mistletoes and their birds

Mistletoes provide a good example of a keystone taxon highly dependent on and likely diversified by avian seed dispersers (Restrepo *et al.* 2002). Mistletoes, due to their hemiparasitic growth that buffers variation in resources, have extended phenologies, few defenses, high-quality nectar and fruits, and act as keystone resources, providing food for at least 97 vertebrate families and nesting sites for at least 50 (Watson 2001). There are marked differences between frugivores in their efficiency of the deposition of mistletoe seeds in required "safe sites" with favorable germination conditions. The viscidity of mistletoe seeds induces certain frugivores to deposit seeds in safe sites, but deters many others (Reid 1991). Therefore, mistletoes depend on a small subset of frugivorous birds, some of which, such as mistletoebirds and euphonias, need mistletoes in return. Mistletoes may require frequent seed establishment and removal of exocarps by seed dispersers, and consequently, population reductions of an avian seed disperser may limit mistletoe population size and/or distribution (Ladley and Kelly 1996). Mistletoes seem to be particularly important keystone resources in the forests of Australia (Watson 2002) and New Zealand (Ladley and Kelly 1996). Reductions in seed dispersers may have contributed to the decline of mistletoes in Australian forest fragments (Norton *et al.* 1995) and to the extinction of *Trilepidea adamsii* endemic to New Zealand (Norton 1991). On the other hand, experimental reductions in mistletoe density in two similar Australian woodland remnants resulted in a significant decrease in mistletoe-feeding and woodland-dependent bird species (Watson 2002), emphasizing the mistletoe–bird co-dependence.

species and large frugivores (Stocker and Irvine 1983; Hamann and Curio 1999), plants do not depend on a single species and diffuse coevolution is more widespread (Jordano 1987).

There are some examples of tight relationships (Reid 1991; Beehler and Dumbacher 1996; Loiselle and Blake 2002), such as mistletoe-feeding birds (Box 4.1), and specialized dispersers can provide higher quality services (Murphy *et al.* 1993). In the Costa Rican cloud forest of Monteverde, Murray (1988) found that only half of the bird species that consumed the seeds of three gap-dependent plant species dispersed the seeds in a viable condition. By combining radio-tracking data with seed passage rates, Murray calculated that these medium-sized species, namely Prong-billed Barbets (*Semnornis frantzii*), Black-faced Solitaires (*Myadestes melanops*), and Black-and-yellow Silky-flycatchers (*Phainoptila melanoxantha*), not only deposited most consumed seeds away from the parent plant, but also dispersed some seeds more than half a kilometer, thereby increasing plant reproductive success 16–36 times. Also in Costa Rica, of the five bird species consuming the seeds of *Ocotea endresiana* (Lauraceae), only Three-wattled Bellbirds dispersed the majority of seeds > 25 m away from the parent tree, and to gap sites where seedling recruitment was higher (Wenny 2000). On the Barro Colorado island of Panama, Chesnut-mandibled Toucans (*Ramphastos swainsonii*) are three to 30 times better dispersal agents of *Virola nobilis* than other birds, including the larger Crested Guans (*Penelope purpurascens*; Howe 1993).

Fruit consumption may not equal effective seed dispersal, and even legitimate and closely related seed dispersers are not necessarily equivalent. A good example of non-equivalence comes from a study of three turaco species in Rwanda (Sun *et al.* 1997). These large birds were observed to disperse a majority (>80%) of the seeds up to 304 m away from the parent tree. However, each species was best at a different aspect of seed dispersal. Ruwenzori Turacos (*Ruwenzorornis johnstoni*), which spent the shortest time in feeding trees, dispersed the highest percentage of seeds away from the parent tree, Black-billed Turacos (*Tauraco schuettii*) deposited the seeds most evenly, whereas Great Blue Turacos (*Corythaeloa cristata*), due to their large size, long flights, and extended gut retention time, dispersed seeds the farthest. Thus, these species' seed dispersal patterns were complementary, not redundant.

Clearly, declines in frugivorous birds will affect some plant taxa more than others. Most birds cannot swallow fruits >2 cm in diameter (Wheelwright 1985), although they can disperse small seeds from the large fruits they peck on (Debussche and Isenmann 1989; Pizo 2002). In fact, the relative scarcity of large seed dispersers in the Neotropics seems to have limited the evolution of large-seeded plants there (Mack 1993). In his influential review of tropical frugivorous birds and their food plants, Snow (1981) pointed out that many specialist frugivores target larger (up to 40 mm × 70 mm) and highly nutritive (up to 67% fat) fruits concentrated in the families Lauraceae, Bursaraceae, and Palmae. He suggested that these families may have coevolved with avian frugivores and therefore will be heavily impacted by the declines in the populations of their dispersers. Since frugivorous birds are often less common in the forest interior (Armesto *et al.* 2001), even some small-seeded forest understory shrubs adapted for dispersal by unspecialized frugivores (Snow 1981), as exemplified by the *Miconia* (Melastomataceae) species of Costa Rica, may be dispersal limited and more vulnerable to the extinctions of their dispersers (Loiselle and Blake 1999).

Some ecologically and economically important tree species, such as mahoganies, also depend on a few specialized avian dispersers, exemplified by the dispersal of two Papuan mahogany (Meliaceae) species by birds of paradise (Beehler and Dumbacher 1996). Indeed, the recruitments of two African mahogany species, *Entandrophragma utile* and *Khaya anthotheca*, were shown to be significantly limited by seed dispersal (Makana and Thomas 2004).

4.2.3 Large seeds

Having large seeds poses an evolutionary dilemma. On the one hand, large seed size can increase establishment success by providing more reserves to seedlings (Kennedy *et al.* 2004), by decreasing seed predation (Mack 1998; Ceballos *et al.* 2002; Jones *et al.* 2003), and by increasing survival (Tonioli *et al.* 2001), emergence rates (Tonioli *et al.* 2001), growth rates (Hegde *et al.* 1991; Sousa *et al.* 2003), and resprouting after herbivory (Green and Juniper 2004). The importance of large avian frugivores for tropical primary forests is further emphasized by the larger seeds which characterize shade-tolerant, late successional, larger tropical tree species that predominate in terms of basal area, at least in Peruvian lowland forests (Foster and Janson 1985; Silman 1996).

On the other hand, most birds cannot swallow fruits larger than a few centimeters in diameter (Levey 1987; Pizo 2002) and the dispersal efficiency of seeds declines with seed size (Levey 1987; Hegde *et al.* 1991). Therefore, large-seeded plants provide good examples (Green 1993; Corlett 1998; Hamann and Curio 1999; Kitamura *et al.* 2002) of taxa that are dependent on relatively few large frugivores (Meehan *et al.* 2002; Kitamura *et al.* 2004). The large size of these frugivores make them vulnerable to extinction (Kattan *et al.* 1994), and their demise may lead to plant recruitment bottlenecks (Peres and van Roosmalen 2002) and extinctions (da Silva and Tabarelli 2000). For example, mid- to late-successional tree species in the Philippines have highly specialized dispersal syndromes, and are mostly dispersed by hornbills and fruit pigeons, many of which are highly threatened (Hamann and Curio 1999). Consequently, the elimination of large avian seed dispersers from tropical forests may have significant long-term consequences for tree species composition and forest structure.

Large tropical frugivores have declined in many parts of the globe, and have entirely disappeared from some areas, particularly many islands in the Pacific (Pimm *et al.* 1995a; McConkey and Drake 2002; Meehan *et al.* 2002). On these islands, the extinctions of hundreds of bird species (Steadman 1995, 1997b) have already stressed ecosystems, and the introductions of rodent seed predators (*Rattus* spp.) have made seed dispersers even more important than in the past (McConkey *et al.* 2003). Introduced avian frugivores, such as silvereyes (*Zosterops* spp.), can be inefficient even for plant species with large fruits containing many small seeds, since these birds can avoid ingesting seeds (Stanley and Lill 2002). Disappearances of large frugivorous birds can also have economic consequences since many timber tree species have significantly larger seeds than non-timber species (Hammond *et al.* 1996). In addition, the elimination of many large, seed-dispersing mammals from tropical areas (Redford 1992; Laurance *et al.* 2000; Peres and van Roosmalen 2002) may mean that large tropical avian seed dispersers are becoming increasingly important (Holbrook and Smith 2000), although some research suggests that the two groups may not be able to compensate for each other (Clark *et al.* 2001).

4.2.4 Importance of long-distance dispersal

Home range size increases with body size (Jetz *et al.* 2004) and even non-migrating large avian frugivores can roam over extensive areas (Kinnaird 1998). Individual hornbills in the Central African Republic have been documented to occupy home ranges of 4472 ha (Holbrook and Smith 2000) and to make long-distance movements up to 290 km (Holbrook *et al.* 2002). Calculations of seed shadows indicate that these species can disperse roughly 80% of seeds more than 500 m from the parent tree and up to 3.5 km (Black-casqued Hornbill, *Ceratogymna atrata*) to 6.9 km (Brown-cheeked Hornbill, *Bycanistes cylindricus*). However, introductions of non-native species can turn the service of seed dispersal into a problem. In the Usambara Mountains of Tanzania, Silvery-cheeked Hornbills (*Bycanistes brevis*) are effective long-distance (up to 4 km) dispersers of the exotic *Maesopsis eminii* (Rhamnaceae) (Cordeiro *et al.* 2004), and significantly contribute to the rapid invasion of this West African species that is also dispersed by *Ceratogymna* hornbills in its native habitat (Holbrook and Smith 2000).

Nevertheless, long-distance seed dispersal by birds is mostly beneficial and modeling has also confirmed the importance of rare long-distance seed dispersal events in increasing the diversity of forest stands (Malanson and Armstrong 1996). Many "wind-dispersed" species may be dispersed by birds at larger spatial scales, with important consequences for colonizing distant areas (Vander Wall 1992). In fact, birds, including seabirds (Nogales *et al.* 2001) and owls (Grant *et al.* 1975), play major roles in the plant colonization of islands (Whittaker and Turner 1994; Willson and Traveset 2000), and endemic, bird-dispersed angiosperms on Hawaii exhibit significantly higher species richness than the taxa with other dispersal regimes (Price and Wagner 2004). Bird dispersal has led to higher speciation rates of tropical understory plants with small, fleshy fruits (Smith 2001) and higher taxonomic diversity of birds compared to mammals may have even given rise to the higher richness of bird-dispersed plant species (Fleming *et al.* 1993).

4.2.5 Reductions in seed-dispersing birds

Over a quarter of frugivorous bird species are extinction-prone (Figure 4.3), significantly above the global average (Sekercioglu *et al.* 2004). In Indonesia, frugivorous birds decline in cacao plantations with increasing distance to forest (Clough *et al.* 2009). Globally, there is a four- to fivefold decline in the proportion of frugivorous bird species among birds of tropical agricultural areas in comparison to tropical forests and agroforests (Tscharntke *et al.* 2008). Given the importance of seed dispersal for maintaining plant biodiversity, reductions in frugivorous birds can have ecological consequences. In the Gunung Palung rain forest of Indonesia, a combination of field research and modeling showed that the loss of animal seed dispersers would reduce local seedling species richness by 60% (Webb and Peart 2001). Although both birds and bats are important for tropical forest regeneration (Galindo-Gonzalez *et al.* 2000), birds are thought to disperse a wider range of plant forms (Whittaker and Turner 1994) and bat seed dispersal is unlikely to make up for the losses of avian seed dispersers.

Consequences of the reductions in avian seed dispersers can be especially dramatic on oceanic islands, which are more vulnerable to disturbance and to introduced species, and where alternative seed dispersers may be non-existent (Traveset 2002). For example, reduced species richness of avian frugivores in Madagascar with respect to South Africa has resulted in clumped tree distribution, reduced seed dispersal (8% vs. 71%), greater benefit of seed dispersal (6 times vs. 80 times), and reduced average distance to nearest conspecific (0.9 m vs. 21 m) for *Commiphora harveyi* (Bleher and Böhning-Gaese 2001). Most forest tree species in New Caledonia are thought to be bird-dispersed, and the long-term fate of large-seeded species is in balance, since their remaining principal disperser, the endemic New Caledonian Imperial-pigeon (*Ducula goliath*), is in decline (Carpenter *et al.* 2003). In the past millennium, humans and introduced vertebrates may have

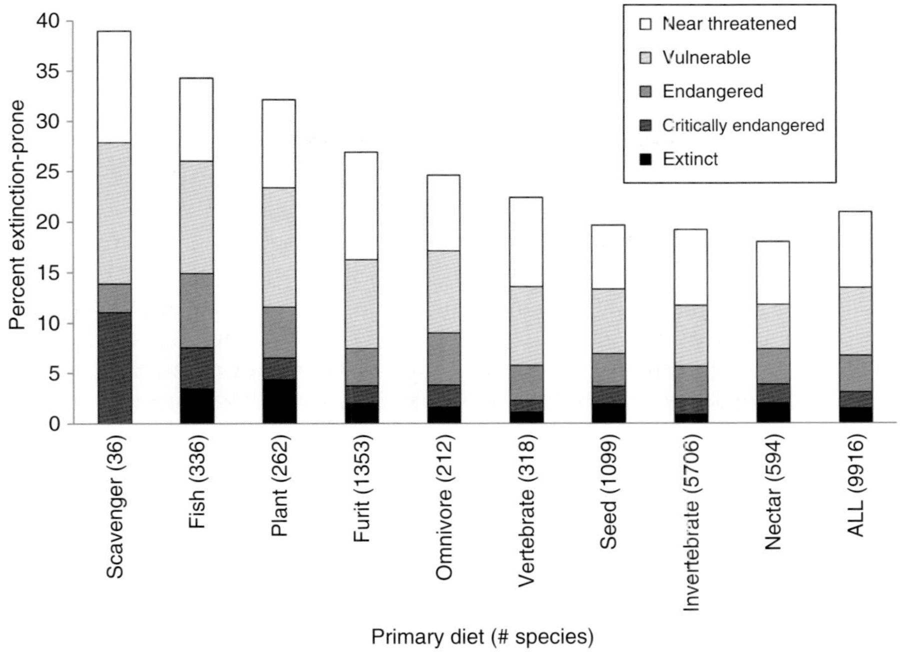

Figure 4.3 Distribution of extinction-prone species based on primary diet. Number of species in each group is in parentheses. If omnivores are reclassified based on first diet choice, percentages do not change except for scavengers (32%). (Reprinted with permission from Sekercioglu *et al.* 2004. Copyright (2004) National Academy of Sciences, USA.)

eliminated over 1000 bird species from the Pacific islands (Pimm *et al.* 1995a) and the decline in the species richness of Polynesian Columbidae is likely to have affected intra-island seed dispersal (Steadman 1997a). Such declines in avian seed dispersers of Pacific islands (Murphy and Kelly 2001; Meehan *et al.* 2002) may lead to declines or even extinctions of dependent plant species (da Silva and Tabarelli 2000), further impoverishing oceanic island ecosystems.

4.2.6 Effects of fragmentation

Since many frugivorous birds range widely to track highly variable fruit resources, forest areas below a certain size may not have enough fruiting trees to support some wide-ranging species (Price 2004), especially in the tropics. As a result, frugivorous birds, particularly large species, often decline in forest fragments (Kattan *et al.* 1994; Santos and Telleria 1994; Renjifo 1999; see Chapters 2 and 3). These declines can exacerbate the manifold effects of fragmentation (Laurance and Bierregaard 1997) and result in regional plant extinctions (da Silva and Tabarelli 2000). In central Amazonia, seedling establishment of *Heliconia acuminate* was 1.5–6 times higher in continuous forest than in 1 ha or 10 ha fragments (Bruna 2002). In Tanzania's East Usambara Mountains, Cordeiro and Howe (2001) showed that reductions in the numbers of frugivorous birds and primates in small forest fragments resulted in a threefold decrease in the recruitment of the seedlings and juveniles of 31 animal-dispersed tree species, compared to no reductions in the recruitment of wind- and gravity-dispersed species. Furthermore, recruitment was 40 times lower for ten of the animal-dispersed species that were endemic to the area. Even generalist avian frugivores can decline significantly in fragments (Cordeiro and Howe 2003), and combined with limited frugivore movement between fragments (Githiru *et al.* 2002; Hewitt and Kellman 2002), this can result in severe reductions in seed dispersal. Avian seed dispersal in forest fragments may significantly favor introduced species over native ones (Montaldo 2000), further modifying natural communities.

The increased mobility of avian seed dispersers with respect to mammals, as well as birds' higher capacity to travel through human-dominated rural landscapes (Jensch and Ellenberg 1999; Holbrook and Smith 2000; Graham 2001), can enable better gene flow between increasingly fragmented plant populations (Jordano and Godoy 2000). In fragmented ecosystems, particularly in the tropics, many specialized bird species cannot leave forest fragments (Sekercioglu 2007) and avian seed dispersal declines rapidly away from forests (Silva *et al.* 1996). In such areas, even modest efforts like planting native trees to act as stepping stones (Sekercioglu *et al.* 2007) or changing the geometry of clearings (Silva *et al.* 1996), can significantly improve seed dispersal, increase connectivity of bird and plant populations, and facilitate recolonization. These trees can also help sustain populations of some resilient native frugivores (Luck and Daily 2003), such as African Pied Hornbills (*Tockus fasciatus*) in Ivory Coast. These birds, as the only large seed dispersers crossing open areas and moving between forest fragments, transport seeds up to 3.5 km away and facilitate the regeneration of and genetic exchange between fragmented forest plant populations (Jensch and Ellenberg 1999).

4.2.7 Role of avian seed dispersal in regeneration and restoration

Avian seed dispersal also affects vegetation succession (Debussche and Isenmann 1994) and is vital for plant colonization and regeneration in naturally (Shiels and Walker 2003) and artificially (Wunderle 1997; Lwanga 2003) disturbed areas. Data from a mixed plantation of native tree species planted in a Panamanian exotic grassland suggest that birds, which generally visited large trees, may have been fundamental facilitators of seedling recruitment (Jones *et al.* 2004). In tropical secondary habitats, as few as two trees may contribute most of the genes to a founding population, and an intact seed disperser community is essential to restore genetic diversity to old-growth levels (Sezen *et al.* 2005). Avian seed dispersal can also have an important economic role in promoting natural regeneration in commercial plantations (Narang *et al.* 2000).

In Brazilian Atlantic montane forest plots ranging from 5-year-old regeneration to old growth, Tabarelli and Peres (2002) found a positive correlation between forest age and the number of woody plant species with larger seeds and those with seeds dispersed by vertebrates, indicating the long-term significance of avian seed dispersal in this critically threatened hotspot. In the Philippines, where dozens of highly threatened endemic bird species reside (BirdLife International 2010a), birds dispersed more forest seed species and individuals than bats into successional vegetation, although this was limited to 40 m from the forest edge and the distance declined with increasing seed mass (Ingle 2003). Numbers of seed-dispersing birds and consequent seed dispersal are many times higher in windbreaks (Harvey 2000) and in isolated "perch" trees (Toh *et al.* 1999; Carriere *et al.* 2002). These scattered trees also support more biodiversity (Fischer *et al.* 2010), provide increased soil moisture (Verdu and Garcia-Fayos 1996), longer water retention (Verdu and Garcia-Fayos 1996), and higher nutrient availability (Toh *et al.* 1999), and create favorable microenvironments for seed germination and establishment. Planting and maintaining windbreaks, riparian strips, and perch trees will likely increase avian seed dispersal of native plants in deforested landscapes (Lwanga 2003). Nevertheless, in some cases, dispersal of forest species may be limited (Duncan and Chapman 2002), and plant regeneration may be highly restricted due to competition with grasses (Holl *et al.* 2000; Duncan and Chapman 2002).

4.2.8 Summary

Currently, over a quarter of all frugivorous bird species are near threatened, threatened, or extinct (Figure 4.3). Avian seed dispersal is complex and varied, and changes in the populations of frugivorous birds will result in equally varied and often unpredictable changes in plant communities. The extent to which remaining species may compensate for disperser losses is unknown. Extinctions of seed-dispersing birds are likely to reduce heterogeneity (Traveset *et al.* 2001) and species richness (Tabarelli and Peres 2002) of plant communities. As is the case with bird declines in general, the effects of seed dispersal will not be uniform and will be particularly felt in certain tropical taxa, such as Lauraceae, Burseraceae, and Sapotaceae, that have large seeds with few large avian dispersers. These large frugivorous birds are significantly more threatened than average, which can have significant consequences for tropical forest communities with many shade-tolerant, late successional, and dominant tree species with large seeds. Large birds can disperse seeds dozens if not hundreds of kilometers away (Holbrook *et al.* 2002). Since it is relatively rare and difficult to observe, the importance of long-distance dispersal by birds, especially over evolutionary time scales, has been underappreciated. Long-distance dispersal is now thought to be crucial (Cain *et al.* 2000), especially over geological time scales during which some plant species have been calculated to exhibit colonization distances 20 times higher than would be possible without vertebrate seed dispersers (Cain *et al.* 2000). In this era of rapid climate change, long-distance seed dispersal by birds is becoming a necessity for more and more plant species, but this ecosystem service may be rapidly eroding in parallel with bird populations, especially of large species. As the dispersers of large seeds disappear, small-seeded, vagile species, already better colonizers that are more adapted to disturbed, rapidly changing environments (Howe and Smallwood 1982; Foster and Janson 1985), will have fewer competitors in deforested areas, and will establish themselves "by default" (Terborgh *et al.* 2002). Furthermore, avian seed dispersers can contribute to the spread of such invasive species with generalized dispersal mechanisms (Renne *et al.* 2002).

Therefore, biotic homogenization via the replacement of specialist birds with generalist birds may contribute to increases in invasive plants (see Chapter 6). Losses of frugivorous birds will have significant implications for the ecology of forests and may result in the domination of many areas by short-lived pioneer species, with long-term effects cascading through the community.

4.3 Pollination

Even though the vast majority of pollination is done by insects (Proctor *et al.* 1996), over 900 bird species (Nabhan and Buchmann 1997) pollinate about 500 of the 13,500 genera of vascular plant species (Renner 2005), concentrated in the families Bromeliaceae, Ericaceae, Fabaceae, Gesneriaceae, Heliconiaceae, Loranthaceae, Myrtaceae, Proteaceae, and about 20 others (Proctor *et al.* 1996). Flower-visiting has been recorded from approximately 2000 species (Herrera and Pellmyr 2002) in 50 bird families, in all biogeographical regions except Antarctica and most of the Palearctic. With the exception of Australia (Ford *et al.* 1979), the majority of avian pollination by far takes place in the tropics, and is mostly limited to hummingbirds, bananaquits, sunbirds, sugarbirds, honeyeaters, honeycreepers, lorikeets, and white-eyes (Proctor *et al.* 1996).

4.3.1 Quality of avian pollination

Since bird pollination requires large amounts of nectar, it is energetically expensive for plants. This has led to the evolution of floral strategies such as the bonanza-blank pattern where a small proportion of the flowers of a species may contain abundant nectar while the rest contain none, forcing hummingbirds to visit and cross-pollinate many flowers (Feinsinger 1978). Even though the greater energetic needs of birds mean increased nectar production, the same needs also force individual birds to visit up to thousands of flowers in one day (Proctor *et al.* 1996), increasing the gene flow between flowers. Individual hummingbirds have often been recorded to travel more than a kilometer during a single morning's foraging and some species, such as the near threatened Saw-billed Hermit (*Ramphodon naevius*) in Brazil, can pollinate more than 20 flower species in the course of a year (Sazima *et al.* 1995). It is an underappreciated advantage of bird pollination that birds, especially trap lining species, with their good spatial memory and multi-year lifespans, provide higher quality pollination services, particularly to self-incompatible flowers with patchy distributions (Schuchmann 1999).

4.3.2 Prevalence in ecosystems

In contrast to seed dispersal, however, birds pollinate a relatively small percentage of plant species, even in the western hemisphere where over 330 species of hummingbirds, the most specialized of avian pollinators, reside. For example, in five diverse ecosystems in Costa Rica, 6–10% of bird species feed on nectar, as opposed to 22–37% of the avifauna being frugivorous (Stiles 1985). In parallel, while 39–77% of shrub and tree species at these sites are bird dispersed, only

2.1–3.4% are bird pollinated (Stiles 1985). Most species that rely on hummingbird pollination are perennial herbs with limited nectar because the highly territorial nature of hummingbirds means that they often occupy and remain at a tree with abundant nectar, largely limiting cross-pollination between trees (Schuchmann 1999). As a result, about 1% of Costa Rican trees are bird pollinated, as opposed to 6–10% of epiphytes (such as bromeliads), with shrubs and terrestrial herbs having percentages in between (Stiles 1985). Bromeliaceae also seems to be the most important plant family for hummingbirds in the Brazilian Atlantic forest, where members of this family comprise a third of the bird-pollinated species (Buzato *et al.* 2000).

The dominant avian pollinators in Brazilian (Buzato *et al.* 2000) and Costa Rican (Stiles 1985) tropical lowland forests are the hermit hummingbirds (subfamily Phaethornineae), which are mostly replaced by non-hermit hummingbirds (Trochilineae) with increasing elevation (Stiles 1985). Endothermic hummingbirds, unlike ectothermic insects, do not need warm and sunny weather to be active, and as such, are more reliable pollinators (Schuchmann 1999), especially under the foggy, rainy, and chilly conditions that characterize many mid- to high-elevation tropical habitats. In fact, despite their small sizes and high metabolisms, hummingbirds can be surprisingly common in the high mountains of the Andes (Schuchmann 1999). The Ecuadorian Hillstar (*Oreotrochilus chimborazo*) is found as high as 5200 m (Heynen 1999). At this elevation, in order to survive through the night when the temperature often dips below freezing, this species goes into torpor, all but shutting down its metabolism, like many other hummingbirds that deal with cold weather.

Researchers in Colombia (Linhart *et al.* 1987), Mexico (Cruden 1972), and New Guinea (Stevens 1976) have shown the rising importance of bird pollination with increasing precipitation and elevation and decreasing temperatures (Stiles 1985). In fact, the 450 species in the tribe Vaccinieae, most of which are found in Neotropical cloud forests, are almost entirely pollinated by hummingbirds (Luteyn 2002). Hummingbird pollination is particularly important for many of the 800+ Neotropical species of the Ericaceae family, most of which are found in the highlands (Luteyn 2002). As with seed dispersal (Gentry 1982), bird pollination is the least important in dry, windy ecosystems, such as tropical dry forests (Stiles 1985), where hummingbird pollination is rather limited and icterids are probably the most important of avian pollinators (Stiles 1985). In fact, these birds may be the most significant of pollinators among all Neotropical songbirds (Cruden and Toledo 1977), since social passerines such as orioles, given their sizes and numbers, cannot rely on a single tree like hummingbirds and provide more effective cross-pollination as they travel from tree to tree.

4.3.3 Hermits and exhibitionists

The major taxonomical division in hummingbirds, between hermits (Phaethornithineae) and "exhibitionists" (Trochilineae), parallels important ecological divisions (Proctor *et al.* 1996) that are soon noticed even by birdwatchers visiting a Neotropical forest for the first time. As quantified in detail by Snow and

Snow (1972) in Trinidad, the drab-colored hermits are almost exclusively found in the forest understory. With their long, decurved beaks, reaching an extreme in sicklebills (genus *Eutoxeres*), these birds specialize in shade-tolerant herbs with prominent flowers, particularly in the order Scitamanieae (Stiles 1981), well exemplified by heliconias (Heliconiaceae). These understory plants produce limited nectar, which discourages territorial defense and encourages "trap lining" (Proctor *et al.* 1996). This forces birds to visit many flowers of the same species, facilitating longer pollination distances and increasing outbreeding. Hermits are among the more specialized of hummingbirds and particularly in lowland forest habitats, they are the most important avian pollinators. Their specialized bills and high-reward trap lining strategy make hermits highly effective pollinators (Schuchmann 1999).

Many species of the more colorful "exhibitionist" hummingbirds, however, are found in more open habitats, where flowers often produce more nectar. This results in increased territoriality, which reduces the pollen dispersal distance and the quality of the pollination service (Proctor *et al.* 1996). In Costa Rica, Stiles (1981) has observed three main divisions among these "typical" hummingbirds, based on increasing bill size and body mass. The largest, exemplified by Violet Sabrewings (*Campylopterus hemileucurus*), weigh up to 12 grams, possess decurved bills over 30 mm, and are the most specialized, resembling hermits in their habits. The smallest species, on the other hand, have short, sharp bills, often feed on insect-pollinated flowers and frequently steal nectar, a behavior that is highly correlated with short bill length (Proctor *et al.* 1996).

4.3.4 Bill length and nectar robbing

Avian pollinators, after all, do not aim to serve plants. These birds are in search of energy-rich nectar and if they can get to it without getting any sticky pollen on themselves, all the better (Proctor *et al.* 1996). As a result, the quality of an avian pollinator is often correlated with its bill length. Birds with longer, more decurved bills are more likely to be "legitimate" pollinators and they make it possible for plants to have deep, thick corollas inaccessible to most insects and nectar-robbers whereas shorter-billed species are more generalist feeders that visit many species. Some of the shortest-billed hummingbird species, such as the thornbills (genus *Chalcostigma*) and Fiery-tailed Awlbill (*Avocettula recurvirostris*), as well as passerine flowerpiercers (genus *Diglossa*), are mainly nectar robbers, using their sharp bills to pierce flower corollas and consume nectar without providing any pollination in return.

Among all nectarivorous bird taxa, there are frequent examples of nectar robbing, a behavior that might in fact have created the evolutionary pressure for some flowers to switch from insect to bird pollination (Sargent 1918) and may have led to the increased frequency of bird pollination in Australia (Ford 1985). Some plants probably adapted to nectarivorous birds in order to increase the probability of pollination by them (Sargent 1918). Interestingly, nectar robbing may sometimes have little to no negative effect on plant fitness, and may even result in some pollination (Graves 1982; Arizmendi *et al.* 1996; Lasso and Naranjo 2003, and other references therein).

4.3.5 Pollination syndromes

The coevolutionary relationship between nectarivores and flowers is thought to have led to increased specialization in certain taxa and to the evolution of ornithophily or "pollination syndromes," where many unrelated flower species are similar in appearance and habits (Faegri 1978; Proctor *et al.* 1996; Rodriguez-Girones and Santamaria 2004). Some of the common characteristics of ornithophilous plant species are diurnal opening, odorlessness, year-around production of abundant nectar, larger and more robust construction than "insect flowers," the presence of perching structures (where non-hummingbirds are involved), and vivid "parrot" colors (Faegri 1978; Schuchmann 1999).

It is well-known that many "bird flowers" are red and that hummingbirds will often inspect red objects carefully. It has been suggested that flowers in the red spectrum are inconspicuous to bees (Raven 1972) and that birds may be more sensitive to red than to other colors (Stiles 1981). The actual mechanism, however, is more subtle (Rodriguez-Girones and Santamaria 2004). Bees do see and visit red flowers (Chittka and Waser 1997) and neither bees nor hummingbirds have inherited color preferences (Proctor *et al.* 1996). Nevertheless, bees are not good at discriminating red flowers from a green background (Chittka and Waser 1997) and are therefore at a disadvantage compared to birds. This shortcoming, combined with optimal-foraging behavior (Possingham 1992), is likely to have led to the association of birds with red flowers (Rodriguez-Girones and Santamaria 2004). Nevertheless, such associations are often weak, the nature of plant pollination has resulted in significant generalization and dynamism (Waser *et al.* 1996), and tight linkages such as the one seen between Sword-billed Hummingbirds (*Ensifera ensifera*) and *Datura* flowers, are exceptional. As such, even between highly specialized hummingbirds and their food plants, one-to-one relationships are unknown, and the coevolution, diffuse (Schuchmann 1999).

4.3.6 Pollinator limitation

Nevertheless, bird pollination often involves fewer species that are usually more obligate than avian seed dispersers (Kelly *et al.* 2004). Some plant species mostly depend on a single (Parra *et al.* 1993) or a few (Rathcke 2000) avian pollinator species. As a result, plants are more likely to be pollinator-limited than disperser-limited (Kelly *et al.* 2004, 2010; Anderson *et al.* 2011) and a survey of pollination experiments conducted for 258 species showed that 159 species (62%) were pollinator-limited (Burd 1994). Compared to seed dispersal, pollination is more demanding due to the faster ripening rates and shorter lives of flowers (Kelly *et al.* 2004). In addition, the lack of seed dispersal does not necessarily reduce offspring production to zero, but the same cannot be said for the lack of pollination. Although most bird-pollinated plant species have more than one species that can pollinate them (Nabhan and Buchmann 1997), there are many flower species which require certain specialized birds for pollination, exemplified by the importance of Malachite Sunbirds (*Nectarinia famosa*) for the African sugarbush (*Protea roupelliae*; Hargreaves *et al.* 2004). Even species that rely on common avian pollinators, such as Bananaquits (*Coereba flaveola*), can suffer significant pollination limitation if pollinator populations decline following severe disturbances such as hurricanes (Rathcke 2000). In the Neotropics alone, thousands of plant species are thought to rely solely on hummingbirds for pollination (Schuchmann 1999). In India, 17% of 93 bird-pollinated plant species were only visited by one bird species (Subramanya and Radhamani 1993). The flowers of a Javanese mistletoe only open when visited by nectarivorous birds that trigger the flowers to explode (Docters van Leeuwen 1954). A New Zealand mistletoe might have gone extinct for having the same mechanism, but no more pollinators (Ladley and Kelly 1995). Avian pollination is particularly important in the Austral, New Zealand, and Oceanic regions, where the proportions of bird-pollinated plants are higher than in other parts of the world (Ford 1985). Most of the pre-settlement avifauna of the Pacific islands is already extinct (Steadman 1995; Pimm *et al.* 2006), contributing to significant avian pollinator limitation in the region (Montgomery *et al.* 2001; Kelly *et al.* 2010; Anderson *et al.* 2011).

Declines of pollinators in agricultural areas can result in the reduced pollination of food plants (Potts *et al.* 2010). In Indonesian cacao plantations, nectarivorous bird species have been shown to decrease with increasing distance to forest (Clough *et al.* 2009). Globally, the proportion of nectarivorous bird species is about two times higher among tropical agroforest bird species than it is among tropical forest species, but there is a substantial decline of nectarivorous species among tropical agricultural birds (Tscharntke *et al.* 2008). Collectively, these findings indicate that tropical agricultural systems could benefit from increases in bird pollination services if forest patches are retained nearby and within agricultural areas.

4.3.7 Vulnerability of island communities

As is the case with seed dispersal, pollinator limitation is often more important in island ecosystems with fewer species, tighter linkages, and higher vulnerability to disturbance and introduced species (see Chapter 6). Declines in the pollinators of island plants (Feinsinger *et al.* 1982; Robertson *et al.* 1999; Sakai *et al.* 2002; Anderson *et al.* 2011), exacerbated by numerous extinctions of island birds (Pimm *et al.* 2006), may lead to the extinctions of dependent plant species (see Figure 3.16). Island plant species do seem to be more vulnerable to the extinctions of their avian mutualists (Anderson *et al.* 2011; Sekercioglu 2011) because many island plants have lost their ability to self-pollinate and have become completely dependent on endemic pollinators (Cox and Elmqvist 2000). For example, the island of Tobago has five species of hummingbirds as opposed to 16 on the larger island of Trinidad. For the early successional hummingbird plants of Tobago, this has meant a significant reduction in pollinator visitation rates, less specialized pollination, and increased nectar production (Feinsinger *et al.* 1982), underlining the importance of pollinator diversity for the quality and energetic cost of pollination.

Pollination limitation due to reduced species richness of pollinators on islands like New Zealand (Kelly *et al.* 2010; Anderson *et al.* 2011; Sekercioglu 2011) and Madagascar (Farwig *et al.* 2004) can significantly reduce fruit sets and probably decrease the reproductive success of dioecous plant species. In Hawaii, the extinction of the competing Hawai'i O'o (*Moho nobilis*) resulted in the I'iwi (*Vestiaria coccinea*) shifting from native lobelioid flowers to ohias, possibly contributing to the native flowers' decline (Smith *et al.* 1995). In Guam (Mortensen *et al.* 2008) and New Zealand (Anderson *et al.* 2011), functional extinctions of pollinating birds have reduced pollination seed production, seedling recruitment, and plant populations.

4.3.8 Summary

Pollinators are declining worldwide (Potts *et al.* 2010). Although it is not as common as seed dispersal by birds, avian pollination has ecological, economic, evolutionary, and conservation significance, especially in certain species-rich communities, such as tropical forest understory herbs, Australian sclerophyllous plants, and Andean cloud forest shrubs. There has been little research on the economic importance of avian pollination, but birds are thought to pollinate at least 3.5% and up to 5.4% of more than 1500 species of crop or medicinal plants, three-quarters of which cannot self-pollinate (Nabhan and Buchmann 1997). Bird pollination of a number of economically important species has been demonstrated in Indomalayan (Narang *et al.* 2000) and other (Nabhan and Buchmann 1997) regions.

Reductions in avian pollinators will inevitably favor some plant species over others, as demonstrated by the Bahama swamp-bush (*Pavonia bahamensis*), which experienced significant seed set reduction as a result of avian pollinator limitation following Hurricane Lili that also created sites for plant recruitment (Rathcke 2000). Such changes in population dynamics caused by species' interactions are likely to lead to modifications in community composition in the short-term, and to have evolutionary consequences for plant lineages in the long-term (Thompson 1996).

Birds are particulary important pollinators for sparsely distributed plant species with isolated populations (Ford 1985) that suffer from increased pollen limitation (Groom 2001). Both traits increase extinction likelihood, so it would be safe to say that declines in avian pollinators can have serious consequences for many rare plant species (Mortenson *et al.* 2008; Anderson *et al.* 2011). In fact, the extinction risk of Hawaiian native plants is associated with rarity and with bird pollination (Sakai *et al.* 2002; see Chapter 3). If the extinctions of 31 species of Hawaiian Campanulaceae as a result of the disappearance of their avian pollinators (K. Wood pers. comm. in Cox and Elmqvist 2000) is any indication, hundreds of plant species may have gone extinct on Pacific islands following extensive bird extinctions (Pimm *et al.* 1995; Steadman 1997b). Introduced Polynesian rats (*Rattus exulans*) on the Easter Island may have contributed to the extinction of the *Jubaea* palm (on which islanders depended for constructing fishing boats) by causing the extinction of its psittacid pollinator as well as by consuming *Jubaea* seeds (references in Cox and Elmqvist 2000). Even though nectarivores are currently among the least threatened of bird functional groups (Figure 4.3), partially due to many hummingbird species' ability to utilize open habitats, this may change in the future (Figure 4.4) since many of these species also have small global ranges. If the expected extinctions of nectarivorous birds do materialize, not only may we lose some of the most specialized and spectacular of bird species (Figure 4.5), but we may also be faced with the disappearances of their plant mutualists, resulting in significant ecological and evolutionary repercussions.

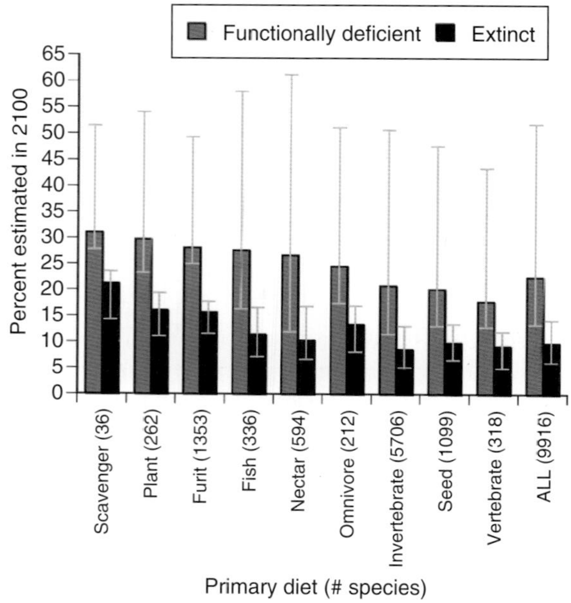

Figure 4.4 Percentages of bird species predicted to be committed to extinction and functionally deficient by the year 2100 based on an intermediate extinction scenario and no conservation intervention. For details, see Sekercioglu *et al.* (2004). Threatened and extinct species are considered functionally deficient. "Error bars", not used in a conventional sense, indicate averages of 10,000 simulations of scenarios 1 (best-case) and 3 (worst-case). (Reprinted with permission from Sekercioglu *et al.* 2004. Copyright (2004) National Academy of Sciences, USA.)

4.4 Predation and pest control – insectivores

Among all bird functional groups, insectivores have the highest species richness by far (Figure 4.3). Invertebrates comprise the primary diet choice for over half of all bird species (Figure 4.3), more than 7200 land bird species have been recorded to feed on them (Sekercioglu 2006a), and an explosive radiation of insectivorous bird species is largely responsible for the high bird diversity of tropical forest bird communities. Given this extraordinary diversity of avian insectivores and the effects of insect herbivores on plant populations (Figure 4.1D), the key question regarding bird–invertebrate interactions in tropical forests is, do birds have significant impacts on invertebrate populations?

4.4.1 Population control

Despite the relative scarcity of tropical studies compared to temperate ones (e.g. Takekawa *et al.* 1982; Holmes 1990; Marquis and Whelan 1994; Mols and Visser 2002), an increasing number of tropical studies have demonstrated the negative impacts of birds on arthropod populations in natural (Gradwohl and Greenberg 1982; Van Bael and Brawn 2005) and agricultural (Greenberg *et al.* 2000; Van Bael *et al.* 2007; Kellermann *et al.* 2008; Van Bael *et al.* 2008 and

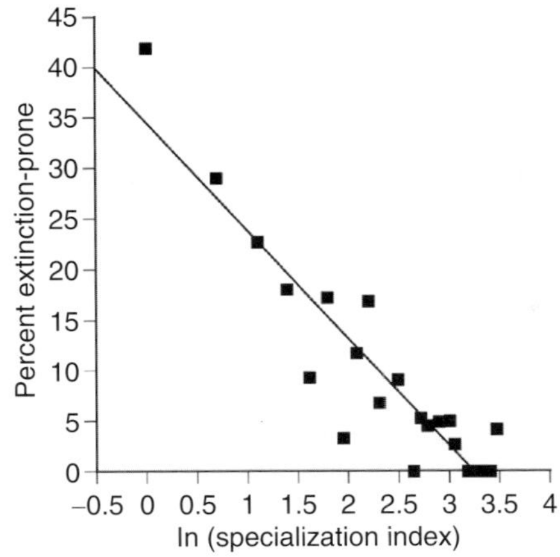

Figure 4.5 More specialization increases extinction risk; r^2 0.851. Specialization index is the product of habitats used and food types consumed. Higher numbers indicate less specialization. Species with an index of 32 or more have been pooled because of small sample sizes. (Reprinted with permission from Sekercioglu *et al.* 2004. Copyright (2004) National Academy of Sciences, USA.)

references therein) ecosystems. Earlier temperate studies largely focused on the eruptions of a few economically important lepidopteran species (Otvos 1979; Holmes 1990) and created the initial impression that birds had little control or influence over ecosystem processes (Wiens 1973). However, studies investigating other invertebrate taxa (Gradwohl and Greenberg 1982), at natural densities (Gradwohl and Greenberg 1982; Van Bael *et al.* 2003), and in tropical ecosystems (Gradwohl and Greenberg 1982; Van Bael *et al.* 2003; Perfecto *et al.* 2004; Philpott *et al.* 2004; Kellermann *et al.* 2008), provide mounting evidence that tropical forest insectivorous birds do have significant roles in controlling the populations, behavior, and evolution of their invertebrate prey. A recent meta-analysis of bird exclosure studies has shown that in tropical agroforestry systems birds reduce all arthropods, herbivores, carnivores, and, most importantly, plant damage, with increasing diversity of birds, especially migratory birds, leading to greater reductions in arthropod densities (Van Bael *et al.* 2008). Another meta-analysis of 113 experiments on vertebrate predation of arthropods has had similar findings (Mooney *et al.* 2010). Therefore, increased extinction risk of specialized insectivorous birds (Figure 4.5) and reduced representation of insectivorous bird species in tropical agroforestry and agricultural systems (Tscharntke *et al.* 2008) is likely to lead to increases in arthropod densities and damage.

The island of Guam, where the introduced brown tree snake (*Boiga irregularis*) has wiped out almost the entire insectivorous bird community (Wiles *et al.* 2003), provides an interesting test case for the roles of tropical insectivorous birds. Although anectodal (Quammen 1997) and indirect (Kerr 1993) evidence indicates spiders have responded significantly and rapidly to bird extinctions, unfortunately, there was not a long-term study of the populations of declining insectivorous birds and their prey on Guam. It would still be informative to compare these variables between Guam and nearby islands that have not been colonized by these snakes and, fortunately, such a study has just begun (Anonymous 2009).

4.4.2 Variability of avian influence

Most of the studies on bird–invertebrate interactions have taken place in the temperate zone, where seasonality increases the magnitude of population fluctuations. In the tropics, especially in forest ecosystems where many bird species are highly specialized to feed on invertebrates (Sherry 1984; del Hoyo *et al.* 2003) and where reduced seasonality may mean fewer and less severe outbreaks than in temperate systems, birds may be more significant year-around control agents, possibly contributing to the typically limited extent of tropical forest outbreaks (Van Bael *et al.* 2004). Studies increasingly show the importance of insectivorous birds, in agricultural (Greenberg *et al.* 2000; Kellermann *et al.* 2008; Koh 2008), agroforest (Perfecto *et al.* 2004; Van Bael *et al.* 2008), and forest (Gradwohl and Greenberg 1982; Van Bael *et al.* 2003) ecosystems. Global climate change is expected to increase the frequency and severity of El Niño/Southern Oscillation (ENSO) events (Timmermann *et al.* 1999) and the accompanying droughts. During these periods invertebrate outbreaks may be more likely (Van Bael *et al.* 2004) and the effects of tropical insectivorous birds on herbivores could be even greater (Mazia *et al.* 2004).

4.4.3 Counterintuitive effects

When considering bird–insect interactions, there is also the possibility of an *increase* in insect populations, as a result of birds feeding on predaceous insects and parasitoids (Hooks *et al.* 2003; Gunnarsson 2007). However, Hooks *et al.* (2003) found that excluding birds did not increase spider predation of herbivorous insects. Actually, birds alone were significantly better at controlling insects and reducing plant damage than spiders alone, with 18% of plants showing extensive defoliation with only spiders versus 0% with only birds. The argument that birds may reduce the numbers of insect parasitoids (Tscharntke 1992) by feeding on infected insects also needs to be considered with this in mind: various lepidopteran parasitoids only emerge from the pupal stage, thus not preventing defoliation by the caterpillars (Hooks *et al.* 2003). Parasites may actually lead to increased foliage consumption (Coleman 1999). Since various bird species are known to select for non-parasitized individuals (Otvos 1979) and facilitate the spread of viruses, they are often complementary to other natural enemies (Takekawa *et al.* 1982 and references therein). A recent review of vertebrate predation showed that vertebrate predation of invertebrate predators does not weaken herbivorous invertebrate suppression, but actually strengthens trophic cascades (Mooney *et al.* 2010).

Interestingly, the "classic" example of the invertebrate parasite removal service provided by oxpeckers may be an example of parasitism itself (Weeks 2000). Observations of oxpeckers on various ungulates showed that not only Red-billed Oxpeckers (*Buphagus erythrorhynchus*) did not reduce tickloads on domestic cattle (*Bos taurus*), but also they preferred to feed on blood from open wounds (Weeks 1999), prolonged their healing time (Weeks 2000), and even opened new wounds on captive black rhinos (McElligott *et al.* 2004).

4.4.4 Role of species richness

When considering the effects of insectivores on ecosystems, it may be helpful to remember that, despite extensive niche overlap among some species, each species can be considered to represent a unique combination of features in an n-dimensional hypervolume (Hutchinson 1957). Therefore, higher richness of insectivorous birds means that there will be fewer "corners" of such a hypervolume where insects can remain out of reach and cause outbreaks. Indeed, limited evidence from tropical ecosystems indicates that invertebrate control by birds may be complimentary (Philpott *et al.* 2004) and it increases with species richness of birds (Van Bael *et al.* 2008).

Due to the sampling effect (Huston 1997), higher species richness also increases the probability of having a species, such as the Rufous-capped Warbler (*Basileuterus rufifrons*), that is particularly effective in its ecological function, as Perfecto *et al.* (2004) observed in their study comparing insect predation in Mexican coffee farms with diverse and monodominant shade trees. Larvae removal rates were about 50% higher in the diverse shade control compared to the exclosure, whereas there was no difference between the control and exclosure plots in the monodominant plantation. The higher density of Rufous-capped Warblers in diverse shade plantations was thought to be the major cause of this difference.

4.4.5 Behavioral and evolutionary influences

An underappreciated impact of avian insectivores on insects (and other predators on their prey) is that, with their very presence, insectivorous birds can affect prey species' behavior and limit their movements, as well as the damage they do to plants (Holmes 1990). The highly varied morphology and foraging behavior of tropical forest insectivores (Fitzpatrick 1981; Stiles 1985) result in significant selection pressures on tropical insects, contributing to the astounding diversity and elaboration of their camouflage (Powell 1979). Holmes (1990) convincingly argued that the evolutionary pressure on invertebrates applied by avian predation has manifested itself in the form of elaborate mimicry, aposematism, nonmimetic polymorphisms, and anti-predator behavior, as well as changes in invertebrate morphology, sex ratios, life styles, and feeding behavior. All these adaptations have significant ecological consequences for the food plants. By limiting their movements to avoid bird predation, many insects will also take longer to develop, increasing their exposure to parasitoids, disease, and predators (Holmes 1990).

4.4.6 Consequences for plants

In many instances, insectivorous birds do have significant behavioral, ecological, and evolutionary effects on their invertebrate prey. An equally important question is, do these behavioral changes and population reductions of insect herbivores have secondary, cascading effects (Schmitz *et al.* 2000) on the food plants (Murakami and Nakano 2000)? In some cases, bird-induced reductions in insect herbivores may not translate to reductions in plant damage (Bock *et al.* 1992). This is more likely in systems where plants have significant anti-herbivore defenses or where the herbivore community is highly diverse, both of which result in the attenuation of trophic cascades (Schmitz *et al.* 2000). It is also critical to note the plant response variable measured, since the change in plant damage may be higher than the more meaningful measures of changes in biomass and/or reproductive output (Schmitz *et al.* 2000; Halaj and Wise 2001; Lichtenberg and Lichtenberg 2002).

Nevertheless, reviews of terrestrial trophic cascades have found that removals of predators often result in increases both in herbivores and in plant damage (Schmitz *et al.* 2000; Halaj and Wise 2001; Van Bael *et al.* 2008), and that the effects of vertebrate carnivores are greater than those of invertebrate carnivores (Schmitz *et al.* 2000; Mooney *et al.* 2010). A detailed study by Mols and Visser (2002) showed that Great Tits (*Parus major*) reduced the numbers of caterpillars and the resulting fruit damage in apple orchards. The authors' review of the literature revealed that such reductions in plant damage caused by avian insectivory were not uncommon. There is also some indirect evidence for the importance of insectivorous birds for plant populations. By releasing volatile compounds, plants may be attracting insectivorous birds to defend against insect herbivory, exemplified by Willow Warblers (*Phylloscopus trochilus*) in Finland that preferred sawfly-damaged branches of mountain birch (*Betula pubescens czerepanovii*) to control branches (Mantyla *et al.* 2004). Invertebrate predators and parasitoids use volatile compounds to detect prey, and birds may be using olfaction and/or ultraviolet vision for the same purpose (Mantyla *et al.* 2004). This would make them more effective control agents than if they foraged randomly.

4.4.7 Economic value of insectivorous birds

Avian control of insect herbivores and consequent reductions in plant damage can have important economic value (Takekawa *et al.* 1982; Marquis and Whelan 1994). Birds can reduce the intensity of spruce budworm (*Choristoneura fumiferana*) outbreaks and mitigate damage on spruce plantations (Crawford and Jennings 1989) at magnitudes comparable to the most effective insecticides (Takekawa *et al.* 1982). In northern Washington state (USA), avian control of spruce budworm was calculated to be worth at least \$1473/km^2/year (Takekawa and Garton 1984). Increasing insectivorous bird numbers via nest boxes is a widespread forest management tool in Europe (Takekawa *et al.* 1982), resulting in the high mortality of leaf-eating caterpillars and consequent declines in damage to economically important species such as white oaks (*Quercus alba*; Marquis and Whelan 1994) and Pyrenean oaks (*Quercus pyrenaica;* Sanz 2001). Insectivorous birds have also been documented to significantly reduce insect pest damage in agricultural systems (Kirk *et al.* 1996; Greenberg *et al.* 2000; Mols and Visser 2002). The last study is particularly noteworthy since the authors found that the damage reduction translated to a significant increase in the yield of domestic apples (*Malus domestica*), from 4.7 kg to 7.8 kg of apples per tree, underlying the potential financial importance of insectivorous birds for agriculture. There are few such studies in the tropics (Perfecto *et al.* 2004; Koh 2008), but recently, the economic value of the reduction of the coffee berry borer by birds was calculated to be \$310 ha^{-1} in a Jamaican coffee plantation (Johnson *et al.* 2010).

4.4.8 Summary

Comprising by far the most diverse avian functional group, insectivorous birds are ubiquitous, abundant, and essential components of most terrestrial ecosystems, and they reach their peak diversity in tropical forests. Not only do these birds often have considerable influences on the behavior, evolution, ecology, and population sizes of their invertebrate prey, they can also modify the population dynamics and even the evolution of plants through indirect effects. Furthermore, as invertebrate pests develop resistance to chemicals that often eliminate invertebrate predators, as increasing numbers of farmers switch to organic agriculture, and as pesticide use is curbed by public attitudes, environmental regulations, and consumer trends (Naylor and Ehrlich 1997; Mourato *et al.* 2000; Mols and Visser 2002), insectivorous birds will have growing significance as providers of natural pest control, components of integrated pest management, and indicators of healthy agroecosystems. Therefore, it is rather disconcerting that 12–51% of all bird species feeding on invertebrates are expected to be functionally deficient by 2100 (Figure 4.4). Insectivorous birds include far more extinction-prone species than any other group (Figure 4.3) and widespread declines in tropical forest insectivorous birds (Thiollay 1997a; Sekercioglu 2002a; Sodhi *et al.* 2004b), 26% of which are extinction-prone, should be a cause for concern. Extreme specializations of many insectivorous birds, especially in the tropics (Sherry 1984; del Hoyo *et al.* 2003), make it unlikely that other taxa can replace these birds' essential ecological services. Increased specialization in bird species increases chances of extinction (Figure 4.5; Sekercioglu *et al.* 2004; Sekercioglu 2007) and declines and extinctions of specialized species will have ecological consequences for the world's ecosystems.

4.5 Predation and pest control – raptors

Although birds are usually not considered important top predators (Pearson 1966; Paine and Schindler 2002), accumulating evidence suggests the contrary for some ecosystems, especially when indirect effects are considered (Parrish *et al.* 2001; Roemer *et al.* 2002). Compared to most predators, raptors are highly mobile, which can both increase their influence on prey populations, as when predatory birds arrive en masse to take advantage of lemming populations booms in northern Alaska (Pitelka *et al.* 1955), but also decrease it, as when bird predators leave in response to declining vole populations in central California while mammalian "carnivores stay on the job" (Pearson 1966). As with insectivorous birds, raptors can also detect areas of high prey densities, sometimes by detecting rodent scent marks that are only visible under ultraviolet light (Viitala *et al.* 1995), and thus have significantly more impact than if they hunted randomly.

4.5.1 Ecological redundancy?

In some parts of the world, such as African savanna woodlands or Neotropical humid forests, many raptor species of similar size and seemingly overlapping diets co-exist, creating the impression that some of these species may be functionally "redundant." However, many raptor species are highly specialized and respond differently to ecological changes. The assumption that species in similar trophic positions are functionally equivalent is likely to be erroneous (Chalcraft and Resetarits 2003) and an impression of ecological redundancy may often be an artifact of limited knowledge of complicated systems that exhibit significant spatio-temporal fluctuations (Jaksic *et al.* 1996). In the semidesert of Chile, for example, although some raptor species initially seemed to be ecologically "redundant," based on diet similarity, after the first three years of research, guild structure shifted significantly, emphasizing the variable nature of raptors' contributions to ecosystem function (Jaksic *et al.* 1996). Additionally, while in some years transient species seemed redundant, in other years they had unique trophic roles. The authors concluded that short-term ecological data on this guild would have provided misguided decisions of conservation triage.

However, it is not known how raptor diversity in tropical forests provides temporally variable ecosystem services, due to the scarcity of detailed studies. A perusal of the literature on the effects of avian predation shows a significant bias towards temperate, low diversity ecosystems with open vegetation structure, particularly the deserts and the tundra. Not only the large majority of raptor species are found in tropical ecosystems, but also the influence of predation on individual species is likely to increase as one moves towards the equator and average prey population size decreases with increasing species richness. Unfortunately, due to the difficulty of studying predation (Mitani *et al.* 2001), especially in closed habitats, there are few detailed studies on the role of avian predation in shaping tropical ecosystems (Groom 1992; Robinson 1994; Mitani *et al.* 2001; Boinski *et al.* 2003) and some of the conclusions are anectodal and speculative. Nevertheless, available evidence hints at the significance of direct and indirect effects of tropical forest raptors. For example, Robinson (1985) found that the dense clustering of Yellow-rumped Cacique (*Cacicus sela*) nests in southeastern Peru were partly driven by nest defense against Black Caracaras (*Daptrius ater*) and the cacique population fluctuated almost tenfold, mostly as a result of Great Black-hawk (*Buteogallus urubitinga*) nest predation. Raptor attacks on parrots in the same area seem to affect parrots' foraging patterns (C. Munn, pers. comm. in Robinson 1994) and may also affect psittacid social dynamics (Munn 1986).

Predation by large forest raptors, such as eagles, may be a major source of mortality for arboreal mammals, such as sloths (Boinski *et al.* 2003) and monkeys (Mitchell *et al.* 1991; Mitani *et al.* 2001). Mitchell *et al.* (1991) found that raptor predation was the major source of mortality for the squirrel monkey species *Saimiri boliviensis* and *S. oerstedii*, and primates formed 82–88% of the prey remains under Crowned Hawk-eagle (*Stephanoaetus coronatus*) nests in Kibale National Park, Uganda (Skorupa 1989; Struhsaker and Leakey 1990; Mitani *et al.* 2001). In fact, avian predation is likely to be the main factor driving larger group size in arboreal primates (Terborgh and Janson 1986; Shultz *et al.* 2004). Increased group size increases the likelihood of noticing a raptor and reduces an individual's probability of being captured, whereas competition for resources tends to reduce group size. In addition, increased risk of avian predation results in more vigilance behavior and may reduce foraging activity in some primates

(Boinski *et al.* 2003). Given the potential importance of raptors in tropical forest ecosystems, tropical avian predator–prey dynamics is a research frontier offering the possibility of novel and exciting findings.

4.5.2 Indirect effects

Avian nest predators can significantly reduce the breeding success of other bird species, especially in fragmented ecosystems where avian nest predation often increases (Patten and Bolger 2003). Increased nest predation in northeastern US forest fragments has had a substantial role in the decline of many Neotropical migratory songbirds (Wilcove 1985; see also Chapter 9). High nest predation rates are also likely to have contributed to the extinctions of various understory bird species from Barro Colorado Island, Panama (Karr 1990), although in this system increased nest predation is likely exacerbated by *decreases* in the numbers of birds of prey such as Harpy Eagles (*Harpia harpyja*) that feed on potential nest predators like white-faced capuchin monkeys (*Cebus capucinus*) and coatimundis (*Nasua nasua*). Even though evidence for the direct effects of avian predators on prey populations is limited, data suggest that indirect effects can be equally or more important (Brown *et al.* 1988) and birds of prey can have significant indirect, counterintuitive, or even positive effects on their prey species.

4.5.3 Trophic cascades

A trophic cascade is a classic example of an indirect effect where the loss of a predator such as the Harpy Eagle can result in cascading population changes in lower trophic levels, including increases in herbivory (Wootton 1995; Hamback *et al.* 2004), ecological release of mesopredators (Crooks and Soulé 1999; Terborgh *et al.* 2001), and consequent declines in the abundance and diversity of plants, nesting birds, and other species.

Even though it has been argued that terrestrial systems are unlikely to experience trophic cascades due to spatial heterogeneity, prey variability, and food web complexity (Finke and Denno 2004), increasing evidence suggests that this may not be the case (Crooks and Soulé 1999; Post *et al.* 1999; Terborgh *et al.* 2001; Hamback *et al.* 2004). Trophic cascades may be dampened in strongly seasonal systems (Norrdahl *et al.* 2002), and with few exceptions (Terborgh *et al.* 2001), most of the studies on trophic cascades have taken place in temperate systems that are strongly seasonal. This is disconcerting since most species of birds of prey live in less seasonal tropical forest ecosystems where the consequences of the losses of raptor species may be greater, especially since higher predator diversity also reduces the effects of predator cascades (Finke and Denno 2004).

4.5.4 The landscape of fear

Birds of prey may affect prey populations by their very presence. By establishing a "landscape of fear" (Laundre *et al.* 2001), avian predators can have indirect effects that may be more important than these birds' direct impacts on prey populations. As is the case with insectivores and invertebrates, prey species' perceived risk of predation can significantly affect prey behavior (Sodhi *et al.* 1990), stabilize predator–prey dynamics (Ives and Dobson 1987), and lead to greater species richness via competitive coexistence (Brown *et al.* 1988). The fear of being hunted can limit the population size of a prey species by limiting its foraging behavior (Brown and Kotler 2004) and reducing its access to food (Power 1984). For example, Brown *et al.* (1988) found that three species of heteromyid rodents in Arizona spent less time in more open habitats, reduced foraging time, and left food patches sooner under the risk of owl predation. The reduction in foraging time due to the fear of predation resulted in a substantial (47% to 91%) reduction in energy intake. These rodents responded rapidly to changes in owl predation risk, and constantly adjusted their foraging levels accordingly. Moonlight increases the risk of owl predation (Kotler *et al.* 1988) and many nocturnal rodents reduce their foraging activities under moonlight (Brown and Kotler 2004), limiting their intake of plants. In one of the few large-scale, experimental exclusions of vertebrate predators, including raptors, Lagos *et al.* (1995) discovered that individuals of the herbivorous rodent *Octodon degus* in northern Chile had smaller home ranges and had more runways between shrubs when predators were absent, indicating the importance of perceived predation risk in altering the behavior and ecological impact of prey populations. Field experiments in Israel with a trained Barn Owl (*Tyto alba*) revealed that two gerbil species *Gerbillus allenbyi* and *G. pyramidum* not only showed reduced activity and spent more time in bushes in the presence of owls, but their activity increased rapidly following the removal of owls (Abramsky *et al.* 1996). Abramsky *et al.* (2002) also revealed that the presence of avian predators can significantly reduce seed consumption by small rodents. The presence of the avian predator Common Fiscal Shrike (*Lanius collaris*) results in fewer nest visits by the prey species East African Stonechat (*Saxicola torquata axillaris*), which results in reduced growth of nestling stonechats (Scheuerlein and Gwinner 2006).

On the other hand, raptors' roles in indirectly "defending" the nests of other bird species from more generalist predators are also well documented (Paine *et al.* 1990; Norrdahl *et al.* 1995; Blanco and Tella 1997; Bogliani *et al.* 1999; Haemig 2001; Ueta 2001; Quinn *et al.* 2003; Halme *et al.* 2004). Collectively, these studies of indirect effects emphasize the role avian predators can have on prey and non-prey species, not only as a result of direct mortality, but also via more subtle, but equally or more important behavioral, ecological, and evolutionary influences induced by these birds' very presence. Reductions in avian predators can cause their prey to perceive less predation risk, potentially leading to significant increases in foraging activity, body condition, and population size. Thus, declines in avian predators are likely to have cascading effects in ecosystems as prey species no longer experience mortality from birds, and the behavior, population dynamics, and evolution of prey species change accordingly.

4.5.5 Summary

Although raptors as a group have a lower percentage of extinction-prone species than most other functional groups (Figure 4.3), large raptor species are more sensitive to disturbance and are more threatened than average. Furthermore, the expected functional extinctions of 13–22% of avian vertebrate predator species

(Figure 4.4) may lead to trophic cascades in some ecosystems, particulary in the tropics where most of these extinctions are expected to take place. Declines in the largest tropical forest raptors, such as Crowned Eagles (*Harpyhaliaetus coronatus*, Endangered), Harpy Eagles (Near Threatened), New Guinea Eagles (*Harpyopsis novaeguinae*, Vulnerable), and Philippine Eagles (*Pithecophaga jefferyi*, Critically Endangered), may have significant impacts on the numbers (Mitani *et al.* 2001) and behavior (Cordeiro 1992) of their prey, with further changes possible at lower trophic levels (Terborgh *et al.* 2001). Birds of prey often feed on many species and are well-connected hubs. Human-caused extinctions usually select against such large top predators (Ebenman *et al.* 2004) and food webs are very vulnerable to the selective losses of hubs (Allesina and Bodini 2004). Consequently, as populations of raptors, particularly large, tropical species decline and disappear, not only we are deprived of the thrill of observing some of the most majestic, inspirational, and symbolic creatures in existence, but we may also have to deal with the ecological and economic consequences of eliminating the drivers of crucial ecosystem processes.

4.6 Scavenging

Since most scavenging birds are highly specialized to rapidly dispose of the bodies of large animals, these birds are crucial in the recycling of nutrients, leading other scavengers, including people (Mundy *et al.* 1992; Eaton 2003), to dead animals (Houston 1979, 1994), consuming the majority of carcasses mammalian scavengers never find (Houston 1974), keeping energy flows higher in food webs (Putnam 1983 in DeVault *et al.* 2003), and limiting the spread of diseases to human communities that would be caused by slowly decomposing carcasses (Figure 4.1C). Vultures are the only known obligate vertebrate scavengers and energetics necessitate obligate terrestrial vertebrate scavengers to be large, soaring fliers (Ruxton and Houston 2004). Vultures may well be the most accomplished fliers in existence, patrolling the heights of the Andes and the Himalayas, effortlessly soaring to thousands of meters, and covering hundreds of kilometers in a day's work. A collision at 11,600 m over Ivory Coast between a plane and a White-backed Vulture (*Gyps africanus*) is by far the highest altitude at which any bird has been recorded (Laybourne 1974).

4.6.1 Ecological significance

Even though we tend to think of large mammalian predators as the dominant meat eaters, in many ecosystems, such as African savannas (Houston 1979), Indian woodlands (Houston 1983), and Neotropical forests (Houston 1986), vultures are (or were) the major carnivores due to their efficiency in finding and consuming dead animals. Houston (1983) observed that 86% of the experimental carcasses he put out in the Tanzanian savanna were discovered only by vultures. He estimated that in the Serengeti, vultures consume at least 370 kg of meat per km^2 $year^{-1}$, as much as all mammalian carnivores combined. In fact, vulture community richness can even give us an idea of the importance of predation in some communities, since when predators are the major source of mortality, large avian scavenger species are reduced or absent (Houston 1986). Avian scavengers rely on finding carcasses before predators can get to them, which is often by following vultures. In the Serengeti, lions locate at least 11% of the carcasses they scavenge by watching vultures (Schaller 1968), and this likely applies to many other species of mammals.

Vultures, especially rapidly declining *Gpys* species, are highly effective in quickly discovering and disposing of carcasses (Houston 1983). Since Old World vultures cannot smell, they are mostly absent from tropical forests (Houston 1994). New World vultures of the genus *Cathartes* can smell, and members of this genus and the other vultures that follow them to carcasses are the major scavengers in Neotropical forest ecosystems (Houston 1986). For example, Greater Yellow-headed Vultures (*Cathartes melambrotus*) found 63% of experimental carcasses in Colombia, whereas mammals found only 5% (Gomez *et al.* 1994). Houston (1986, 1988) observed that 116 out of 120 domestic chicken carcasses he placed in Panamanian and Venezuelan forests were discovered and consumed by Turkey Vultures (*Cathartes aura*) in 3 days. Interestingly, he also observed that in forests with reduced mammal abundance, Turkey Vultures missed many of the carcasses which were mainly consumed by invertebrates (Houston 1987). Not only does this indicate a change in scavenging dynamics as a result of mammalian reductions and reduced specialization of vultures on mammalian carcasses, but also suggests that *Cathartes* vulture numbers can be used to make rapid assessments of Neotropical mammal numbers, serving as an index of mammalian defaunation (Dirzo and Miranda 1990).

4.6.2 Sanitary services

Besides their ecological significance, vultures are particularly important in many tropical developing countries where sanitary waste and carcass disposal programs may be limited or non-existent (Prakash *et al.* 2003), and where vultures contribute to human and ecosystem health by getting rid of refuse (Pomeroy 1975), feces (Negro *et al.* 2002), and dead animals (Prakash *et al.* 2003). Sixty-six percent of the Indian population practices open defecation (Jha 2003) and coprophagy by vultures can improve community hygiene and may reduce disease. The alarming reduction in the vultures of the Indian subcontinent may result in an increase in exposed feces, possibly augmenting pest populations and increasing the incidence of water-borne diseases. Vultures have an impressive ability to resist and possibly detoxify bacterial toxins in rotting flesh. Extremely acidic secretions of the vulture stomach, with pH values as low as 1, kill all but the most resistant spores and this is thought to significantly reduce the bacterial sources of infection from the carcasses vultures feed on (Houston and Cooper 1975). Understanding the physiological mechanisms of vulture resistance to infectious agents may even lead to medical discoveries.

4.6.3 Worldwide declines

Indeed, avian scavengers worldwide comprise the most threatened avian functional group, with about 40% of the species being threatened or near threatened with extinction (Figure 4.3). Sadly, this is not surprising. The same factors that make vultures efficient scavengers (Ruxton and Houston 2004), namely large body size, long travel distances, and the consumption of many carcasses, also make them vulnerable to accidental and deliberate poisoning, persecution, collisions with powerlines, habitat loss, disturbance, and even ritualistic killings inspired by their majesty (Mundy *et al.* 1992; Houston 1994; Snyder and Snyder 2000). Therefore, even non-threatened (Least Concern) species such as Bearded Vultures (*Gypaetus barbatus*) have been experiencing population reductions (Brown 1991). Due to recent and catastrophic population declines (Liberatori and Penteriani 2001; Donazar *et al.* 2002), the formerly Least Concern Egyptian Vulture (*Neophron percnopterus*) was downgraded to Endangered in 2007 (BirdLife International 2010b). Avian scavengers are also highly susceptible to poisoning as a result of swallowing lead bullets while consuming carcasses unclaimed by hunters (Clark and Scheuhammer 2003). Like birds in general (Figure 4.5; Sekercioglu *et al.* 2004), more specialized Neotropical vulture species are more prone to extinction, and some of the largest species in the New World have become extinct in the Pleistocene (Hertel 1994). African vultures are exposed to many anthropogenic threats (Schüz and König 1983; Mundy *et al.* 1992), including habitat destruction, reductions in food availability, inadvertent killing during "problem animal" control campaigns (poisons, gin traps, etc.), electrocution on electricity pylons, nest disturbance, harvesting for traditional medicine, and even smoking vulture brains due to the supersitition that it will help predict outcomes for World Cup gambling (Smith 2009). Since 1970, the numbers of West African vultures, with the exception of Hooded Vultures (*Necrosyrstes monachus*), declined by an average of 95% and White-headed (*Trigonoceps occipitalis*) and Lappet-faced Vultures (*Torgos tracheliotos*) have virtually disappeared from the region (Thiollay and Rondeau 2004).

4.6.4 Disappearance of South Asian vultures

Vulture declines have been particularly severe in South Asia (Prakash 1999; Oaks *et al.* 2004; see Figure 1.21). Once some of the most common raptors in the world and with large urban populations (Galushin 1971), White-rumped (*Gyps bengalensis*), Indian (*Gyps indicus*), and Slender-billed Vultures (*Gyps tenuirostris*; the latter two split from Long-billed Vulture (*Gyps indicus*)) of the Indian subcontinent have declined rapidly, from Least Concern in 1994 (Collar *et al.* 1994) to Critically Endangered by 2000 (BirdLife International 2000). Prakash (1999) was the first to report on declines of 96–97% in the Keoladeo National Park between 1988 and 1999 (also see Figure 1.22), and in a carcass dump at Uttar Pradesh, Long-billed Vultures declined from 6000 in 1991–92 to five in 1994 (Eaton 2003). Although infectious disease initially seemed most consistent with the observed symptoms (Cunningham *et al.* 2003), further studies have confirmed renal failure caused by the cattle anti-inflammatory drug diclofenac as the underlying cause (Green *et al.* 2004; Oaks *et al.* 2004). The combination of this sudden population crash and a potential increase in carcass numbers, combined with infectious agents and high human population density, may cause increases in incidences of anthrax, bubonic plague, and rabies (Pain *et al.* 2003), but these crucial interactions have not been studied.

4.6.5 Costs and consequences

Even though no one has estimated the potential cost of the loss of decomposition services provided by vultures, increased disease transmission and consequent health spending is likely. Observations in India, where most people do not consume cattle and where most animal carcasses are simply left outside for vultures, indicate substantial increases in the numbers of rotting carcasses, especially around human habitations (Prakash *et al.* 2003). Between 1992 and 2001, Prakash observed a 20-fold increase in the numbers of feral dogs at an Indian garbage dump (Prakash *et al.* 2003). Every year, more than 55,000 people worldwide die from rabies; 24,000 of these take place in Africa and 20,000 rabies deaths take place in India (World Health Organization 2010). Wordwide, 15 million people require post-exposure treatment of rabies every year (Bourhy *et al.* 2010). Increased numbers of cattle carcasses also increase the chances of the spreading of livestock diseases such as anthrax (Prakash *et al.* 2003). It may be no coincidence that the 1994 outbreak of bubonic plague in western India occurred soon after the start of the crash of vulture populations. Although the ecological factors behind this outbreak were not systematically studied (Gratz 1999), it is thought to have been initiated by an unusually hot summer killing many cattle, the carcasses of which led to an explosion in rat numbers (Kaplan 1997). The plague infected at least 876 people, killed 54, resulted in quarantines and evacuations, and the resulting media attention led to trade and tourism boycotts (Gratz 1999). The whole episode cost India over US$2 billion (Kaplan 1997).

Even though feral dogs seem to be replacing vultures in parts of India, their scavenging is less efficient and less sanitary than the vultures'. While vultures can remove all soft tissues extremely rapidly (Houston and Cooper 1975), dogs eat only the choice bits, leaving much rotting flesh behind. Not only does this provide a breeding ground for pathogenic bacteria, it also allows resistant spores to form (Houston and Cooper 1975). Dogs' stomachs are about four times less acidic (Lui *et al.* 1986) than those of vultures (Houston and Cooper 1975), increasing the likelihoods of survival and transmission of pathogenic bacteria such as *Brucella abortus* or *Bacillus anthracis* found in carcasses fed on by dogs. In addition, increasing numbers of feral dogs and rats mean more vectors of human diseases such as rabies and bubonic plague, as well as wildlife and livestock diseases caused by canine distemper virus, canine parvovirus, *Leptospira* bacteria, and other pathogens (Pain *et al.* 2003; Butler *et al.* 2004). Furthermore, subsidized by a carcass surplus, the increased populations of these opportunistic mammalian scavenger-predators are likely to put a higher predation pressure on vulnerable wildlife (Pain *et al.* 2003).

4.6.6 Socioeconomic effects

The disappearance of vultures from India has also had social, economic, and even religious consequences. Their loss has meant the loss of income for the impoverished "bone collectors" who rely on the efficient and relatively hygienic cleaning of vultures (Pain *et al.* 2003). Since the Parsis, who comprise a sect of Zoroasthrianism, believe that death pollutes the sacred elements of earth, fire, and water, they cannot use any of these elements to dispose of their dead. For centuries the Parsis have been leaving their dead to the elements, to be cleaned by avian scavengers, a practice with roots going back 8000 years, to the Neolithic site of Çatalhöyük, Turkey (Eaton 2003), where drawings of Cinerous Vultures (*Aegypius monachus*) circling headless bodies have been found. On *dakhmas* or "Towers of Silence" constructed for this purpose, vultures would normally take about half an hour to clean a corpse and three corpses would be left each day. After the crash of vulture populations, however, hardly any vultures visit and smaller avian scavengers are not effective. The corpses can no longer be disposed of according to the Parsi religious doctrine, which has resulted in a spiritual crisis (Parry-Jones 2001).

4.6.7 Summary

Scavengers, especially the obligate scavengers consisting of the Old and New World vultures (Houston 1979), provide one of the most important yet under-appreciated and little-studied ecosystem services of any avian group due to the difficulty of and human aversion towards studying rotting substances (DeVault *et al.* 2003). Although there are some studies quantifying carrion consumption by avian scavengers (Houston 1988; DeVault *et al.* 2003, and references therein), despite an extensive literature search, we were unable to find a published study that compared carcass decomposition rates between two areas with intact and reduced avian scavenger communities. Such a "before and after" study would have been especially valuable in quantifying the effects of the Indian subcontinent vulture population crash (Prakash *et al.* 2003). As such, this and many other unconducted studies on the ecological roles of avian scavengers represent significant and urgent research opportunities for avian ecologists. If the declines in vulture populations continue, it may soon be too late to find an intact "before" community in most parts of the world.

Compared to other avian functional groups, the species richness of the obligate scavenger guild is tiny, comprising only a few dozen species whose food consumption is predominantly based on scavenging. As such, even the declines or extinctions of a small number of species can result in significant reductions in avian scavenging, especially when one considers that in any one part of the world there are at most nine species of vultures. Their scavenging nature requires that these birds represent the epitome of animal flight, and ecologically and evolutionarily, vultures are in a unique and highly threatened class of their own. From prehistoric Africans likely following vultures to obtain carcasses to Andean and Californian natives revering condors to Neolithic Anatolians and present-day Parsis leaving their dead on *dakhmas*, vultures' unique status in ecosystems has always been paralleled in their special place in the human cultural landscape. It is now upon us to make sure that these majestic birds stay around to continue playing their crucial roles in the biosphere and in the human psyche for the millennia to come.

4.7 Nutrient deposition

An underappreciated role of birds is transporting nutrients from one habitat to another (Figure 4.1B). This is particularly important in the case of seabirds transferring marine productivity to terrestrial ecosystems, especially in coastal areas and unproductive island systems (Sanchez-Pinero and Polis 2000). Seabird droppings are enriched in important plant nutrients such as calcium, magnesium, nitrogen, phosphorous, and potassium (Gillham 1956). Murphy (1981) estimated that seabirds around the world transfer 10^4 to 10^5 tons of phosphorous from sea to land every year, and this guano also provides an important source of fertilizer and income to many people living near seabird colonies. Ironically, the very currents, such as Benguela, California, and Humboldt, that facilitate spectacular marine productivity, also create temperature inversions that result in low productivity deserts on nearby landmasses. Marine birds, by providing allochthonous inputs in the form of guano and carcasses, help offset this imbalance and are crucial conduits between some marine and terrestrial areas (Sanchez-Pinero and Polis 2000; Croll *et al.* 2005). Avian enrichment of soils with nutrient-rich guano can have indirect but significant and cascading effects on the populations of plants (Stapp *et al.* 1999; Croll *et al.* 2005), invertebrates (Polis and Hurd 1996), rodents (Stapp and Polis 2003), and even large mammals (Iason *et al.* 1986; Wolfe *et al.* 2004).

4.7.1 Community and ecosystem-level effects

Besides affecting productivity and abundance of organisms, an increased influence of seabird colonies can also change community composition. On the Mediterranean island of Bagaud, the presence of Yellow-legged Gull (*Larus michahellis*) colonies significantly changed beetle assemblages, with a marked shift from phytophagous species to polyphagous tenebrionid species (Orgeas *et al.* 2003), which seem to benefit from bird colonies more than other beetle families (Sanchez-Pinero and Polis 2000). Seabird colonies can also have dramatic impacts on the productivity, diversity, and composition of nearby plant communities (Anderson and Polis 1999), sometimes at the expense of native taxa (Vidal *et al.* 2000, 2003). Moderate disturbance by seabirds may maximize diversity (Vidal *et al.* 2003), with increased disturbance resulting in an increase of alien species at the expense of native species (Vidal *et al.* 2003). On the other hand, some plant species, such as those found in the genus *Lepidium* (Brassicaceae) endemic to New Zealand, have so adapted to the nutrient enrichment and disturbance regimes associated with seabird (and seal) colonies that some are extinct or threatened with extinction as a result of seabird and seal declines (Norton *et al.* 1997).

Although many studies on the effects of seabird allochotonous input have focused on specific taxa, these inputs often influence entire ecosystems. Simultaneous increases in invertebrates and lizards as a result of the addition of nutrients by seabirds have been documented on islands off New Zealand (Markwell and Daugherty 2002). Seabird-derived nutrients were found in the tissues of plants, invertebrates, and lizards, and the addition of this resource led to a significant increase in organismal abundance across the board. Harding *et al.* (2004) showed that 28–38% of the nitrogen in the biota of streams near globally Vulnerable Westland Petrel (*Procellaria westlandica*) breeding colonies in New Zealand were marine-derived. The authors also emphasized the potential consequences for nitrogen cycling and ecosystem productivity of the reduction in marine-derived nutrients following extensive reductions in seabird colonies on the New Zealand mainland.

4.7.2 Life history consequences

The effects of seabird nutrient input on the life histories of island species can be substantial, even affecting the survival of small mammals (Wolfe *et al.* 2004) and the reproductive success of deer (Iason *et al.* 1986). Iason *et al.* (1986) found that local nutrient addition by Herring Gulls (*Larus argentatus*) on the Isle of Rhum, Scotland, resulted in an enrichment of soil nitrogen and phosphorus, followed by an increase in the nitrogen content of vegetation, which led to a rise in the lifetime reproductive success of red deer (*Cervus elaphus*) females feeding on this vegetation. The populations of the endangered marsupial dibbler (*Parantechinus apicalis*) on two islands off western Australia provide a particularly fascinating example of nutrient subsidies (Wolfe *et al.* 2004). Males of some dasyurid species such as dibblers experience substantial to complete stress-related post-breeding mortality which is reduced by improved body condition (Wolfe *et al.* 2004). On the island with greater nesting seabird density, plant-available soil nutrients were 5–18 times more enriched, there were more invertebrates, and the body conditions of insectivorous dibblers were significantly better. This led to significantly higher post-breeding survivorship of males, indicating the crucial importance of seabird-derived resources for these animals.

4.7.3 Excessive nutrient inputs

Although avian nutrient inputs benefit many species, communities, and ecosystems, there can be too much of a good thing. Excessive inputs of avian guano can inhibit plant growth (Gillham 1960). Abundance and species richness of invasive plant species may increase in areas of greatest intensity of bird colonies (Vidal *et al.* 1998, 2003), since these plants often have high disturbance tolerance and colonizing capacity. Human presence can increase the establishment success of invasive plants near bird colonies, as exemplified by the increased number of alien plants in the King Penguin (*Aptenodytes patagonicus*) colonies closer to the scientific research station on the Posession Island, in the Crozet archipelago (Vidal *et al.* 2003). Interactions with people can also result in population explosions of opportunistic species, such as Yellow-legged Gulls, whose breeding colonies can do significant damage to fragile and rare plants and animals (Vidal *et al.* 1998). As also exemplified by invasive plant species dispersed by birds (Cronk and Fuller 1995) and introduced pigs catalyzing the elimination of Channel island foxes by Golden Eagles (*Aquila chrysaetos*; Roemer *et al.* 2002), human activity and the species that benefit from it can significantly modify ecological interactions and turn positive synergisms into negative ones (Lundberg and Moberg 2003).

4.8 Ecosystem engineers and other ecological actors

Birds have a plethora of other roles in ecosystems that cannot be pigeon-holed into the main categories above. For example, grazing birds, such as geese and ducks, can have significant impacts on the vegetation of some areas, particularly in wetlands and coastal areas where anatids are often concentrated. Although intensive grazing can lead to the degradation of some areas, such as the salt marshes on the shores of Hudson Bay damaged by Snow Geese (*Chen caerulescens*) (Kerbes *et al.* 1990), ducks and geese can also reduce agricultural residues in an environmentally-compatible manner, as opposed to open-field burning that has been restricted by legislation (Bird *et al.* 2000).

4.8.1 Ecosystem engineers

Perhaps the least appreciated ecological contribution of birds are as ecosystem engineers (Jones *et al.* 1994). This is partly because avian engineering does not have the very visible effects of more prominent engineers such as beavers or trees, but nevertheless some birds are ecosystem engineers (Figure 4.1D). Another reason for the relative lack of awareness is that ecosystem engineering itself has received little recognition until recently (Jones *et al.* 1994). By definition: "Ecosystem engineers are organisms that directly or indirectly modulate the availability of resources (other than themselves) to other species, by causing physical state changes in biotic or abiotic materials. In so doing they modify, maintain, and/or create habitats" (Jones *et al.* 1994).

Given that birds have limited capacity to transform their surroundings physically (as opposed to corals, earthworms, or prairie dogs, for example), some of the best examples of avian engineering come from bird nests. Even small bird nests often house beetles, moths, and other invertebrates (Collias and Collias 1984). At the other extreme, colonial Sociable Weavers (*Philateirus socius*) in southern Africa construct the largest nest of any bird species. In addition to providing a dwelling to many other organisms, such as snakes, Pygmy Falcons (*Polihierax semitorquatus*), and countless invertebrates, these massive structures can even bring down trees (C.H. Sekercioglu, personal observation). Even though not as extreme, there are various other examples of large avian nests, particularly of raptors, weavers, and oropendolas, that have effects that go beyond the original nest builder. Trogons engineer in tropical forests (Valdivia-Hoeflich *et al.* 2005) whereas burrow-nesting seabird colonies can change soil fertility and lead to massive erosion (Furness 1991).

The best examples of nest construction resulting in ecosystem engineering come from woodpeckers (Figure 4.1D). Their unique behavior of drilling nest holes is arguably a more important contribution to ecosystems than the insectivorous habits they share with many other species, although woodpeckers' superior ability to extract invertebrates certainly benefits many trees. Because they drill nesting cavities which are later used by other, secondary cavity-nesting species, woodpeckers provide novel resources to other species by changing the physical structure of their environment and, therefore, are ecosystem engineers par excellence. Since cavity nesting bird species often have higher nesting success (Knutson *et al.* 2004) and, consequently, larger clutch sizes (Jetz *et al.* 2008a), woodpeckers are important components of many avian communities. Therefore, it is encouraging that woodpeckers (Picidae) comprise the only avian family that contains significantly fewer threatened species than expected (Bennett and Owens 1997), which may be a consequence of woodpeckers' ability to extensively engineer their habitats. By drilling trees, some woodpeckers also provide sap for other species, a behavior also seen in Akiapolaau (*Hemignathus munroi*), a globally endangered Hawaiian honeycreeper, whose decline may have averse affects on the species that benefit from this sap (Pejchar and Jeffrey 2004).

4.8.2 Seed consumers

Although birds may well be the most important seed dispersers, especially in the tropics, some granivorous birds, such as finches, and parrots, can be significant seed predators too. Avian seed predation may increase in tropical forest fragments since many tropical granivorous birds are more common in forest fragments and outside forests than in extensive forest. Granivorous birds can be quite common in some coffee plantations (Komar 2006) and granivorous bird species' richness increases in Indonesian cacao plantations with increasing distance to forest (Clough *et al.* 2009). The global pattern is in agreement with these field studies: among tropical agricultural bird species, the proportion of granivorous bird species is four or five times higher than their proportions in tropical forest or agro-forest bird communities, respectively (Tscharntke *et al.* 2008).

In the forest fragments of southeast Brazil, where rodent seed predators have declined and granivorous birds have increased, birds have become important, if not the main seed predators of *Croton priscus* (Euphorbiaceae; Pizo 1997). In fact, granivorous birds are the most important avian pests of agriculture, although damage estimates are often exaggerated and not collected in a scientific manner (Weatherhead *et al.* 1982).

The most notorious example of an avian seed predator is the Afrotropical bird Red-billed Quelea (*Quelea quelea*). It is thought to be the world's most numerous bird with 1–3 billion individuals (Elliott and Lenton 1989) and the predominant avian pest in Africa. Nevertheless, detailed studies indicate that although local damage may be high, the impact on food production Africa-wide is negligible, with losses to cereal crops amounting to less than 1% of the production (Elliott and Lenton 1989). This is in the region of losses caused by bird pests in other parts of the world (Weatherhead *et al.* 1982; Elliott and Lenton 1989). Also considering the important ecological roles played by Red-billed Queleas as predators of insects, including pest species, as providers of nutrients which also fertilize fields and orchards, and as important food sources for many birds, mammals, and people (Elliott and Lenton 1989), the extensive environmental damage and non-target deaths caused by explosives, fire bombs, and especially aerially sprayed fenthion (Meinzingen *et al.* 1989) cannot be justified. Fenthion has especially severe effects on aquatic species found in water bodies near quelea roosting sites and on predatory and scavenging birds (McWilliam and Cheke 2004). Birds of prey can reduce quelea populations significantly (Bruggers and Elliott 1989), but many of them die after spraying operations (Meinzingen *et al.* 1989). Furthermore, many Africans collect and consume queleas killed by avicides and are thus routinely exposed to dangerous chemicals (Jaeger and Elliott 1989).

4.9 Birds and humanity

If we fail in our endeavor to reach a sustainable compromise with global biodiversity, including thousands of bird species, not only will we fail morally in our roles as stewards of other species and will be confined to an aesthetically impoverished planet, but we will also be faced with more concrete consequences. Important avian guilds are in rapid decline and consequent reductions in ecosystem processes are to be expected. The societal importance of ecological services is usually appreciated only after their loss. Historically extinct birds are the seemingly distant rumble before an imminent flash flood of bird declines (Gaston *et al.* 2003) and population losses (Hughes *et al.* 1997), likely followed by concomitant decreases in ecological interactions and evolutionary processes (Thompson 1996). The ecological and evolutionary consequences of the reductions in bird species and populations are hard enough to estimate, let aside put a price on. However, it is almost certain that there will be financial losses as a result of the reductions in ecological services provided by birds. In areas that suffer heavy avian losses, birdwatching tourism income will decline considerably (see Chapter 10). Investments in understanding and preventing declines in the populations of birds and other organisms will pay off only while there is still time to act. Even putting ecological and economic consequences aside, one does not have to be a birdwatcher or an ornithologist to feel a profound sense of loss from the disappearance of hundreds, if not thousands, of bird species.

Note

This chapter is based on Sekercioglu, C.H. 2006. Ecological significance of bird populations. Handbook of the Birds of the World 11: 15–51, and has been updated with many important papers.

5

Fire and the Conservation of Tropical Birds

Overview: Fire is a natural ecological process, but humans are thought to have altered fire-regimes in 70% of tropical habitats. Changes in the grazing regimes and the loss of traditional burning practices can lead to woody encroachment in savanna systems, threatening many birds that specialize in open areas. In addition, selective logging, habitat fragmentation and increased drought severity mean that many fire-sensitive ecosystems such as humid tropical forests are now burning at unprecedented rates, driving biodiversity loss. The consequences of fires for bird conservation are highly context specific, depending on a multitude of complex factors including variability in life history strategies, the history of fire in each ecosystem, and the spatial and temporal scale of the burn events.

5.1 Introduction

Fire appeared in the geological record over 400 million years ago following the evolution of terrestrial plants (Scott and Glasspool 2006; Bowman *et al.* 2009) and about 250 million years before the evolution of the avian ancestral dinosaur *Archaeopteryx* in the Late Jurassic period (Padian and Chiappe 1998). As an ecological process, fire has had a profound influence on the distribution of the world's major biomes (Bond and Keeley 2005). Its effects are most keenly felt in the tropics where most of the present day global fire activity is concentrated, with a particular focus on regions that are more seasonal (Giglio *et al.* 2006) and have high net primary productivity (Bowman *et al.* 2009).

Conservation of Tropical Birds, First Edition. Navjot S. Sodhi, Çağan H. Şekercioğlu, Jos Barlow and Scott K. Robinson.

Published 2011 by Blackwell Publishing Ltd.

Our understanding of wildfires is centred on the concept of a fire regime, which was developed by Gill (1975) to describe the predictable aspects of fire in Australian ecosystems. It has recently been modified to encompass: (i) fuel consumption and fire spread patterns; (ii) intensity (energy release); (iii) severity (the impact on the ecosystem); (iv) frequency; and (v) seasonality (Bond and Keeley 2005). This concept has become an integral part of fire ecology across the world. For example, entering "fire regime" into Thompson ISI Web of Science returns just two articles from the 1980s, 149 from the 1990s, and 618 from 2000–2009.

Fire regimes are considered intact when they are within their range of natural variability (Shlisky *et al.* 2009), and altered if any of the individual components are affected by human activity. Degraded fire regimes have been defined as those where conditions are "outside their range of natural variation, but are considered restorable." Very degraded fire regimes are "far outside their natural range or variability, and may not be restorable" (Shlisky *et al.* 2009). Following these definitions, humans are thought to have altered fire-regimes in over 60% of the world's terrestrial habitats (Shlisky *et al.* 2009). The situation is particularly acute in the tropics, where the fire regimes are considered to be altered in more than 70% across the region, with the only exceptions being fire-independent ecosystems such as deserts, where there are insufficient fuel-loads, as well as the wettest and most remote tropical forests.

What has caused the widespread alteration of fire regimes in the tropics? The causes are complex, but are unlikely to surprise those familiar with other aspects of tropical conservation biology. A report by The Nature Conservancy (Myers 2006) lists seven different causes of altered or undesirable fire regimes, including ignitions from agricultural clearing and site preparation by farmers, other activities related to land clearing or land use that increase fuels and their susceptibility to ignition, arson, changes in grazing practices, invasive plants changing fuel characteristics, landscape fragmentation, and climate change affecting seasonality and vegetation productivity (Myers 2006; see also Chapters 2 and 8). These factors rarely act alone, and most altered fire regimes are likely to be a result of the complex interactions between two or more of these human influences. For example, abnormally severe drought in the Amazon only results in fires in regions where there is agricultural activity to provide the ignition source (Aragão *et al.* 2007).

5.1.1 Fire and birds

Given the importance of fire as an evolutionary and ecological force in tropical biomes, it is unsurprising that shifts in fire-regimes can have a significant impact on the present-day conservation of tropical birds. Indeed, the global distribution of fire activity (Bowman *et al.* 2009) often overlaps with regions that contain a high species richness of threatened birds (Orme *et al.* 2005), most notably in Madagascar and Southeast Asia. Historical precedent gives support to the idea that altered fire regimes can lead to the extinction of bird species. For example, fossil evidence suggests that human-induced changes in the fire-regime may have been responsible for the widespread ecosystem change that drove the rapid extinction of specialized feeders such as the large flightless Demon Ducks

(*Genyornis newtoni*) from Australia 45–50,000 years ago, shortly after human colonization (Miller *et al.* 2005).

This chapter examines the present-day relevance of fires for the conservation of tropical birds, with implications for terrestrial biodiversity in general. We briefly examine direct effects of fire, before focussing on the influence of altered fire regimes in fire-dependent and fire-sensitive ecosystems (Hardesty *et al.* 2005). We do not discuss fire-independent environments (such as deserts), where fire is almost always absent due to the lack of fuel.

5.2 Direct effects of fire

Although fire is a dramatic event, the direct impacts of the flames and smoke are likely to be rather insignificant compared to the indirect effects mediated through changes in vegetation structure and composition. However, there are some direct consequences of fire that have been observed in the tropics that are worthy of mention here. For example, flightless or nesting birds may be at particular risk from fire. Peres (1999) reports finding the beaks of Cuvier's Toucan (*Ramphastos cuvieri*) and Lettered Aracari (*Pteroglossus inscriptus*) on the forest floor following a wildfire in the central Brazilian Amazon; these birds had presumably succumbed to smoke asphyxiation during the night. Teixeira and Negret (1984) describe the responses of the vulnerable Dwarf Tinamou (*Taoniscus nanus*), which favors grasslands in the Cerrado regions of Brazil but is a very poor flyer, and is often unable to escape fast-moving fires after becoming dizzy from smoke asphyxiation. Conversely, fire can also attract birds, and raptors such as the White-Tailed Hawks (*Buteo albicaudatus*) are often attracted to feed on invertebrates and vertebrates fleeing grassland fires (Tewes 1984; A. Lees, pers. comm.).

5.3 Indirect effects of fire and altered fire regimes

The most significant effects of fire on birds are likely to occur long after the last flames have been extinguished, following the fire-induced changes in vegetation structure and composition. These indirect effects can have profound impacts on birds, and may be driving some species towards extinction. Critically endangered bird species are those considered to be at extremely high risk of extinction in the wild. Of 190 critically endangered bird species, 145 are well known enough to have had their threats identified (BirdLife International 2008b). Of these, changes in the fire regime are a direct threat for nine species, eight of which are found within the tropics (the only exception being the Orange-bellied Parrot *Neophema chrysogaster*, which is restricted to southern Australia and Tasmania; Table 5.1).

These figures are likely to underestimate the potential importance of fire in endangering the world's most threatened birds, as the effects of altered fire regimes may occur over long time scales, and are likely to have gone undetected in the tropics. Furthermore, the vast majority of research on fire regimes has focussed on vegetation responses to fire in temperate systems. Of a total of 810

Table 5.1 Critically endangered bird species (those considered to be at extremely high risk of extinction in the wild) known to be threatened by changes in the fire regime. Population estimates are based on the number of mature individuals. (Data derived from BirdLife International 2008a)

Common name	*Latin binomial*	*Country(ies)*	*Population estimate*	*Habitat*	*Fire-related threat*
Nihoa Finch	*Telespiza ultima*	USA (Hawaii)	2100–3550	Low shrubs and grasses	Risk of large uncontrolled wildfire on island
Niceforo's Wren	*Thryothorus nicefori*	Colombia	<50	Dry forest habitat	Seasonal burning for farming
Pale-headed Brush-finch	*Atlapetes pallidiceps*	Ecuador	250–999	Early successional scrub	No clear fire-related threat identified
Royal Cinclodes	*Cinclodes aricomae*	Peru, Bolivia	160–166	Montane scrub and woodland	Uncontrolled use of fire
Lear's Macaw[a]	*Anodorhynchus leari*	Brazil	250–500	Arid caatinga with licurí palms	Large wildfires could wipe out food supply
Pernambuco Pygmy-owl	*Glaucidium mooreorum*	Brazil	<50	Tropical forests	Wildfires leading to forest degradation
Alagoas Antwren	*Myrmotherula snowi*	Brazil	50–249	Semi-humid tropical forest	High threat of fire in fragmented forests
Alagoas Foliage-gleaner	*Philydor novaesi*	Brazil	50–249	Upland tropical forest	High threat of fire in fragmented forests
Orange-bellied Parrot	*Neophema chrysogaster*	Australia (Tasmania)	150	Eucapyptus-moorland mosaic	Change in fire regime in breeding habitat

[a] Recently downlisted to Endangered due to increasing population size.

articles including the term "fire regime" published (Thompson ISI Web of Knowledge) up to the time of writing (May 2010), only 67 shared the word "tropical." Links between fire regimes and avian research are particularly limited, and less than 3% (20) of articles shared the term "fire regime" with "bird", "birds" or "avian." Although these quickly returned search results are inevitably imprecise (excluding some relevant articles and including many inappropriate

Figure 5.1 Fire in two contrasting tropical reserves. (a) An understorey wildfire burns a private forest reserve in the northeast Brazilian Amazon. (b) A fast-moving grassland fire burns the Chapada das Mesas reserve in the Brazilian Cerrado. (Photographs by: (a) Jos Barlow; (b) Alex Lees.) (Please see Plate 5.1 for the color version of this figure.)

articles), they nevertheless provide an indication of the lack of knowledge regarding the importance of fire for tropical bird conservation.

5.4 Altered fire regimes in fire-dependent ecosystems

Fire-dependent ecosystems are those "where fire is essential and the species have evolved adaptations to respond positively to fire" (Myers 2006). Tropical examples of fire-dependent systems include the Brazilian Cerrado in South America, African Savannas, and the tropical dry forests that exist across the tropics including Western Madagascar (Moat and Smith 2007), the Western Ghats (Kodandapani *et al.* 2008), and Central America (Otterstrom *et al.* 2006). Fire is highly variable within these systems, ranging from low intensity ground fires which burn fine grassy fuels (Figure 5.1) to high intensity stand-replacing canopy fires that reset succession (Myers 2006). Given these differences, we examine savanna and tropical dry forests separately.

5.4.1 Savanna and savanna woodlands

A long dry season and a high incidence of lightning means that fires are especially frequent in tropical savannas. For example, 25–50% of Madagascar's grasslands are thought to burn annually (Kull and Laris 2009), and the fire regime is one of the most important factors controlling the relative cover of grass and woody vegetation in savannas (Van Langevelde *et al.* 2003). Fires are also known to be an important determinant of bird community structure in savanna systems, mediated through these changes in the vegetation (Hannah *et al.* 2007). Threats to the tropical bird communities in these systems are mostly due to changes in the fire regime brought about by changes in agricultural practices.

Many of the best studies on the consequences of changing fire regimes in savannas and savanna woodlands come from the use of long-term records in Australia and South Africa. In Queensland, Australia, Woinarski and Cattarall (2004) made use of historical bird records collected at a farm named Coomooboolaroo from 1873 to 1933. They resurveyed the 454 km^2 property in 1999, and found significant changes in the bird community. Of 192 species, 18 had disappeared from the study area altogether and 68 had declined in abundance, while 13 species increased and 11 had apparently colonized the area (see Figure 3.7). Many of these changes are consistent with changes in the structure and composition of vegetation that would have resulted from the abandonment of traditional Aboriginal burning practice: the species that had declined or been extirpated included those with a preference for dense grassy habitats that would have been favored by frequent burning (Woinarski and Cattarall 2004).

The results presented by Woinarski and Cattarall (2004) are likely to be representative of changes taking place across most of the northern Australian savannas, which cover around 20% of the entire continent. Carefully reconstructed records of bird occurrence have revealed widespread declines in the abundance and distribution of resident granivorous birds since the arrival of European pastoralists in the 19th century and the resulting displacement of Aborigines (Franklin 1999). Twelve out of the 49 bird species studied have suffered a significant decline in their range, including the endangered Golden-shouldered Parrot (*Psephotus chrysopterygius*) and the colorful Paradise Parrot (*Psephotus pulcherrimu*) (Franklin 1999), which sadly is now classified as extinct (BirdLife International 2008b). In a similar region of Northern Australia, species such as the Carpentarian Grasswren (*Amytornis dorotheae*) have recently been upgraded to vulnerable on the Red List, with populations being extirpated by unprecedented large fires (BirdLife International 2010c). As the region has been largely untouched by intensive agriculture the underlying assumption is that the declines and extinction of birds are the result of changes in the fire regime due to the replacement of Aboriginal land management with European pastoralism.

Historical records from South African savannas paint a similar picture of change, with a high proportion of grassland species becoming locally extinct over time (e.g. Krook *et al.* 2007). A recent assessment of historical bird species population trends from across southern Africa during the second half of the 20th century suggests that scrub encroachment on savannas is one of the main drivers of changes in the bird populations, jeopardizing the regional persistence of savanna specialists such as Temminck's Courser (*Cursorius temminckii*) (Sirami *et al.* 2009).

Long-term studies are vital for detecting change over ecologically meaningful time periods. However, they do not provide mechanistic links between the change in the fire regime, changes in vegetation, and the resulting change in the bird community. Shorter-term ecological studies have examined these interactions in more detail.

One of the most striking results of these short-term studies is the resilience of the savanna bird community to single fire events that occur within the bounds of natural variability. For example, no bird species were absent from burned areas two months after a fire event in South Africa (Mills 2004), and O'Reilly *et al.* (2006) found no change in bird species richness 22 months after fire in Kenyan savannas. Detailed studies in temperate ecosystems help support this view, and show the importance of fire for bird conservation in these ecosystems. Bird species richness (or the number of species of conservation concern) is generally higher following fires in a range of fire-dependent North American ecosystems, including the conifer-dominated forests of the western United States (Hutto 1995), oak savannas (Davis *et al.* 2000), and southeastern pine savannas (Brawn *et al.* 2001).

Although many bird species are apparently resilient to single fire events, many other studies demonstrate the negative impact that long-term changes in fire regimes can have. We first examine the importance of landscape heterogeneity, and then assess how the interaction between grazing and wildfire can influence bird conservation in savanna systems.

5.4.2 Fires and landscape heterogeneity in fire-dependent systems

Birds respond to changes in both the spatial and temporal scale of fires (Brawn *et al.* 2001). In particular, fires that are limited in their spatial extent can increase the landscape heterogeneity that is essential for many savanna birds, with frequent small fires providing a mixed range of habitats for species that require different habitat types for different parts of their life history. For example, the Australian Partridge Pigeon (*Geophaps smithii*) requires unburned habitat for nesting and shelter, while burned areas provide the bare ground it requires to feed on fallen seeds (BirdLife International 2010d). The species still persists where present-day fire management reflects Aboriginal management, with small early dry-season fires ensuring the maintenance of habitat heterogeneity. In another Australian example, burning of small patches enhances the temporal availability of seeds across the landscape by delaying the seed production of the grass *Schizachyrium* spp., benefiting granivorous species such as the endangered Golden-shouldered Parrot (*Psephotus chrysopterygius*) (Crowley and Garnett 1999). This parrot is endemic to southern and central Cape York Peninsula, Queensland, Australia, and is directly threatened by changes to the burn regime and the expansion of woodland onto its preferred grassland habitat (BirdLife International 2010e).

5.4.3 Interaction with grazing

There are many ecological parallels between fire and herbivory. Both consume plant material and exert a strong influence on vegetation composition and structure (Bond and Keeley 2005). Moreover, the two may interact with each other to alter vegetation structure, with cascading effects on savanna fauna. Changes in herbivory through livestock management or changes in the numbers and distribution of wild herbivores can alter fire regimes in savanna systems (Van Langevelde *et al.* 2003). Grazers reduce the grass biomass (which is the main fuel load in savannas), which in turn reduces the intensity and frequency of fires. This allows more trees and woody vegetation to survive each fire, leading to an increase in the dominance of woody vegetation. This woody encroachment has been identified as one of the principle threats to the populations of birds dependent on savanna grasslands in South Africa (Skowno and Bond 2003; Bond and Parr 2010).

Increasing the frequency with which a landscape burns can also affect herbivore behavior, with knock-on consequences for birds. For example, Archibald *et al.* (2005) reported results from an intriguing study on a dynamic grassland system in South Africa. Intense and repeated grazing by white rhinos promoted short lawns with stoloniferous grasses, while burning promoted tall bunch grasses. The management of the fire–grazer interaction in this system is important for the management of the full suite of grasslands birds as the two grassland types hold distinct bird communities – grazing lawns were required by three species (African Pipit *Anthus cinnamomeus*, Crowned Lapwing *Vanellus coronatus*, and Sabota Lark *Mirafra sabota*), while seven species were most frequently associated with the tall grasses favored by fire (Krook *et al.* 2007). If fires are too frequent, they can inhibit the development and maintenance of the short grazing lawns, as herbivores are attracted to the lush regrowth available in the wider landscape allowing the grazing lawns to recover their biomass and become susceptible to a future burn (Archibald *et al.* 2005).

5.4.4 Tropical dry forests

Tropical dry forests are extremely threatened – for example, the dry forests of Madagascar had been reduced to less than 3% of their original extent (Smith *et al.* 1997). Trees typical of dry forests are more tolerant to fire than those found in humid tropical forests (Otterstrom *et al.* 2006), presumably because the long history of fire within these ecosystems has either selected for fire-adapted species or the species within these biomes have evolved fire-adapted traits. Although the number of studies on bird responses to fire in tropical dry forests is limited, evidence suggests that the consequences are much less severe than in fire-sensitive humid tropical forests. For example, recently burned Madagascan dry forests can have a higher conservation value for birds than dry forests with no recent fire history, although some endemic species remain restricted to the unburned forest (Pons and Wendenburg 2005).

5.5 Altered fire regimes in fire-sensitive ecosystems

Humid forests are the most widespread fire-sensitive ecosystem in the tropics. The major expanses of tropical forests include Mesoamerica, the Amazon, South American Atlantic forests, West and Central Africa, parts of Madagascar, the Western Ghats in India, much of Southeast Asia, and large portions of Oceania including northeast Australia. Their humid nature means these fire-sensitive ecosystems are not very flammable in their natural state, and are primarily composed of species that have evolved in the absence of fire. For example, fires in Amazonian forests were unlikely to have occurred more than once or twice per millennium (Sanford *et al.* 1985; Turcq *et al.* 1998) and the flora and fauna shows little adaptation to these episodic disturbance events (Uhl and Kaufmann 1990; Barlow *et al.* 2003).

The stable and non-flammable status of humid tropical forests has changed rapidly in the past 30 years. Wildfires are estimated to have burned a total of 20 million hectares in Southeast Asia and Latin America during the 1997–1998 El Niño event (Cochrane 2003), and some Amazonian forests currently exhibit fire-return rates of once or twice per decade (Cochrane and Laurance 2002). This increase in fire activity in humid tropical forests has been driven by local anthropogenic activity (Cardoso *et al.* 2003; Aragão and Shimabukuro 2010), and has been enhanced by recent severe droughts resulting from El Niño Southern Oscillation events and Atlantic sea temperature anomalies (Alencar *et al.* 2004; Aragão *et al.* 2007; Cochrane *et al.* 1999; Malhi *et al.* 2008). Large-scale forest clearance leads to reduced precipitation (Oyama and Nobre 2003), and both edge creation and selective logging increase forest flammability by allowing the understory to dry out (Uhl and Buschbacher 1985; Nepstad *et al.* 1999; Cochrane and Laurance 2002; Alencar *et al.* 2004). In addition, many deforested areas are subsequently replaced by secondary forests (e.g. Lucas *et al.* 2000) which can become rapidly desiccated and flammable during dry periods (Ray *et al.* 2005). The increased physical access from improving roads and other infrastructure encourages the spread of human populations who routinely use fires to clear agricultural plots (e.g. Nepstad *et al.* 2001; Cardoso *et al.* 2003). This frontier expansion increases the susceptibility of forests to fires as well as providing the ignition sources that are responsible for initiating most fires.

The impact of wildfires on the understory tropical forest bird community is strongly linked to the severity of the fire, which in turn can be related to the number of times a forest has burned. When an undisturbed or lightly disturbed primary forest burns for the first time, the fire is generally restricted to the leaf litter, with flame heights rarely exceeding 10–30 cm under normal fuel loads and humidity conditions (Holdsworth and Uhl 1997; Cochrane and Schulze 1999) (Figure 5.1a).

Despite their rather innocuous appearance, these low intensity surface fires can kill most of the understory plants and up to half of the trees ≥ 10 cm in diameter at breast height in Amazonian forests (Barlow and Peres 2004a). Unsurprisingly, such drastic changes have severe impacts on the composition of forest birds, which track post-fire changes in the vegetation structure in both the Brazilian Amazon (Barlow and Peres 2004b) and in Southeast Asia (Slik and Van Balen 2006). The high rate of tree mortality means the forest canopy becomes broken and much more open, leading to a loss of the species adapted to foraging in the relatively dark and humid understory, including many insectivores and species typical of mixed species flocks (Barlow *et al.* 2002; Slik and Van Balen 2006). In the Brazilian Amazon, these include endemic species with restricted ranges and high conservation concern, such as the Pale-faced (*Phlegopsis borbae*) and Harlequin Antbirds (*Rhegmatorhina berlepshi*) (Barlow *et al.* 2002). These are often replaced by birds of low conservation concern, such as those typical of forest edges, or young secondary forests.

In contrast to savanna systems, birds typical of humid tropical forests recover slowly after low-intensity surface fires. In the first few years after a wildfire, bird community composition can become increasingly dissimilar to that found in unburned forest (Barlow and Peres 2004a), with an increase in the abundance of species associated with or tolerant of second growth, such as the Moustached Wren (*Thryothorus genibarbis*) and the Blackish Antbird (*Cercomacra nigrescens*) in the Brazilian Amazon. This increasing dissimilarity reflects patterns of avifaunal recovery following selective logging (Mason 1996) and presumably occurs because of ongoing tree mortality and the high number of tree falls (of live and dead stems) that continue to open up the canopy, and the changes in the structure and composition of the regenerating understory vegetation which resembles secondary forests (e.g. Cochrane *et al.* 1999; Barlow and Peres 2004a, 2008).

While low-intensity surface fires are clearly a cause of great concern for the conservation of understory specialists, their impacts on tropical forest birds are overshadowed by recurrent fires that occur when a forest has already burned. Initial low-intensity wildfires (shown in Figure 5.1a) can set in motion a positive feedback process that renders forests increasingly likely to burn again, due to the more open canopy allowing the understory to dry out and the additional fuel on the forest floor (Cochrane *et al.* 1999). These recurrent fires are generally much hotter and more severe than the initial low-intensity fires, and kill many of the remaining trees (Cochrane and Schulze 1999) and cause a marked shift in the composition of the regenerating vegetation (Barlow and Peres 2004a, 2008).

The pronounced impact of recurrent fires on vegetation has clear consequences for the understory bird fauna. Barlow and Peres (2004b) show a strong relationship between surrogates of fire severity in once- and twice-burned Amazonian forests and the species richness of disturbance-sensitive and disturbance-tolerant birds (Figure 5.2), and forests that had succumbed to recurrent fires had virtually no species in common with unburned forests three years after the most recent fires. Adeney *et al.* (2006) demonstrate how repeated burns can have a strong influence on the subsequent composition of birds (at the level of genera and family) in Sumatra. In contrast, Slik and Van Balen (2006) did not find any obvious additional effect of repeated burning on the avifauna in forests that burned in 1983 and 1998. This lack of effect may be due to the length of the interval (15 years) between the first and second burn, as concurrent analyses of the vegetation structure indicated that once- and twice-burned forests were more or less equally affected by the 1998 fires (Slik and Eichhorn 2003).

Overall, the spread of fires into humid tropical forests should be seen as one of the greatest threats to the conservation of forest birds. A comparison of the effects of selective logging and forest fragmentation in the Amazon show that low-intensity fires alter avifaunal composition of forests to a similar extent as the most destructive forms of selective logging (with additional post-felling silvicultural practices) and the isolation of forest fragments of just one hectare, while recurrent fires were by far the most severe form of forest degradation examined (Barlow *et al.* 2006), extirpating almost all bird species typical of the unburned control plots (Figure 5.3). Furthermore, these results are from burned plots

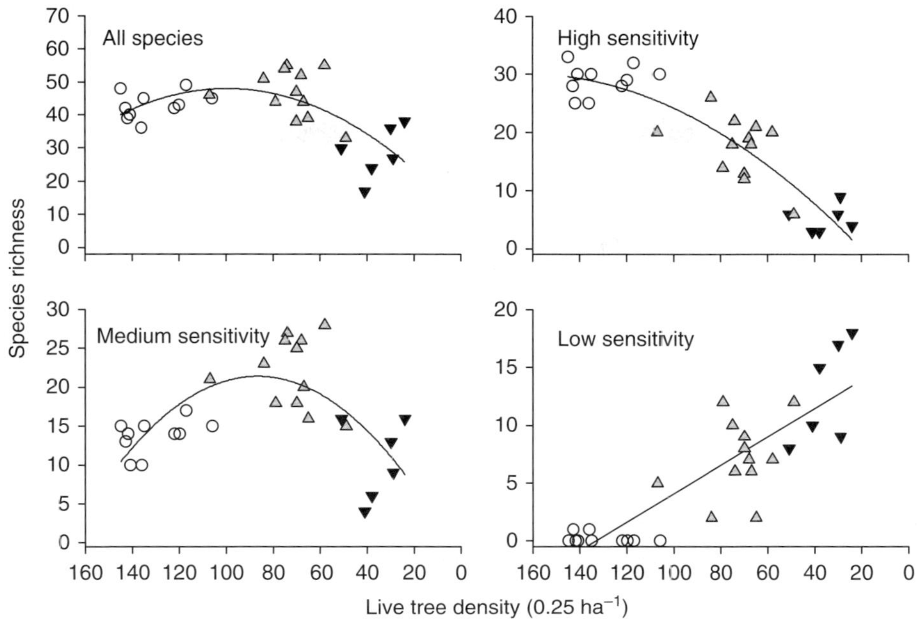

Figure 5.2 Relationship between burn severity (measured as live tree density) and bird species richness, split by response guild sensitivity to disturbance (following Stotz *et al.* 1996). Open, gray, and solid symbols represent unburned, once-burned, and twice-burned forest plots, respectively. The figure demonstrates how increasing burn severity in humid tropical forests brings about a loss of the most-disturbance-sensitive species, and favours species adapted to disturbance. (Adapted from Barlow and Peres 2004a.)

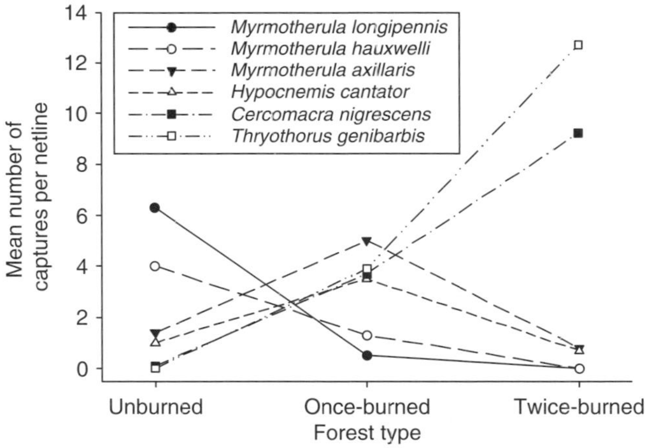

Figure 5.3 Three contrasting responses to fire exhibited by the most abundant arboreal-gleaning insectivorous species captured in the forest understorey in the Central Brazilian Amazon. The figure demonstrates how disturbance-tolerant species replace disturbance-intolerant species in once- and twice-burned forest, and shows the complete turnover in species composition that occurs when a forest burns for a second time. (Adapted from Barlow and Peres 2004a.)

located in continuous forest, and within a few kilometers of extensive areas of unburned forest. Equivalent surface fires in forest landscapes that are logged, fragmented, with a small proportion of remaining primary forest, or far from potential sources of recolonization, can all be expected to have greater and longer-term impacts.

5.6 Interaction between fires and other threats to tropical birds

While the previous sections have examined the role of fires in isolation from other ongoing disturbance events, it is important to remember that humans are changing the world in many ways, and species are likely to be threatened by a combination of effects acting synergistically (Laurance and Useche 2009; see Figures 1.8 and 3.12). Fire is particularly important in this context, as it interacts with many of the ongoing human activities that threaten the conservation of tropical birds. Although the impacts of these interactions remain poorly studied, we outline some of the most important of these below.

5.6.1 Fire and agriculture

Agriculture provides a plentiful source of ignition events in tropical regions, where many people depend on fire to clear land. It has been estimated that between 300 and 500 million people depend on slash and burn agriculture on some way (Brady 1996), with much of this farming activity taking place at the interface between human-dominated landscapes and the remaining areas of natural habitat. In addition, fire is frequently used as a management tool in many large-scale agricultural land uses, including cattle ranching and sugar cane. The use of fire in agriculture is particularly threatening to small forest remnants that persist in highly fragmented landscapes, as forest edges are drier and are more likely to burn than the forest interior (Alencar *et al.* 2004). For example, the practice of burning sugar cane plantations threatens to burn the forest fragments that contain the remaining populations of the Alagoas Antwren (*Myrmotherula snowi*) and the Alagoas Foliage-gleaner (*Philydor novaesi*) in the northeast of Brazil (BirdLife International 2008a; see Barnett *et al.* 2005).

5.6.2 Fire and invasive species

Invasive plants can be critically important in changing fire regimes. A process known as the grass–fire cycle shows how invasive grasses (often combined with other human pressure) can cause dramatic changes in the fire regime, and convert woody ecosystems to grassland (D'Antonio and Vitousek 1992; Brooks *et al.* 2004). The cycle begins when exotic grasses are introduced to new regions to promote grazing. Many of these alien grasses spread into the surrounding landscape, changing the fuel dynamics and increasing the likelihood that the landscape will burn. The invasive grasses are often the main beneficiaries of an initial fire, and increase in extent, further enhancing landscape flammability and the size and intensity of any subsequent fires.

One of the most sobering examples of this grass–fire cycle occurring is in Hawaii, where the invasion of C_4 (carbon-fixation) perennial grasses (*Schizachyrium condensatum* and *Melinis minutiflora*) have led to the severe degradation of seasonal submontane woodlands. Prior to the invasion by grasses, there were 27 fires in 48 years, with each fire affecting an average area of 4 ha. After the invasion of C_4 grasses there were 58 fires in just 20 years, and the average area affected by each fire increased to 205 ha (D'Antonio and Vitousek 1992).

Introduced grasses can also increase biomass (and therefore fuel loads) in tropical savannas, leading to hotter and possibly more frequent fires. This has been documented with the introduction and subsequent invasion of African grasses in Venezuelan and Australian savanna systems (Rossiter *et al.* 2003). Finally, in some circumstances invasive plants such as succulent cacti and some trees can actually suppress fire (Brooks *et al.* 2004).

The impact of biotic invasions on tropical birds focuses mostly on the effects of animals and parasites and disease (see Chapter 6), and there is surprisingly little information on the potential consequences of the grass–fire cycle for tropical bird conservation. However, a temperate example suggests the problem is deserving of further study: alien annual grasses such as *Bromum tectorum* have increased fire frequency so much in western North America that native steppe species can no longer recover, with negative effects for native bird species such as the Greater Sage Grouse (*Centrocercus urophasianus*) (Knick *et al.* 2003).

5.6.3 Fire and selective logging

Logging is identified as a threat to 43% of critically endangered species, with the vast majority of these found in tropical forests (see Chapter 1). Selective logging also has important interaction with fire in humid tropical forests: by opening up the canopy and adding fuel to the understory, timber harvesting can push normally fire-resistant forests over the flammability threshold, substantially increasing the chance that they will burn (Uhl and Buschbacher 1985; Cochrane *et al.* 1999). In these cases, any fire that follows selective logging is likely to have a much greater additional effect on the bird community than the logging itself (Barlow *et al.* 2006).

5.6.4 Fire and climate change

Many of the ongoing and predicted changes in climate increase the risk of fires spreading into humid tropical forests (see also Chapter 8). For example, climate change is likely to be accompanied by increased air temperature and dry season length over large regions of the Amazon forest (Cox *et al.* 2004), and a likely increased frequency of severe seasonal droughts initiated by El Niño/Southern Oscillation (ENSO) events and Atlantic sea surface temperature (SST) anomalies (Li *et al.* 2006). These severe droughts are known to both augment and further desiccate the available fuel load, forcing forests over the flammability threshold (Ray *et al.* 2005) with potentially disastrous consequences. Widespread fires linked to these severe drought events have affected large areas of continuous forest, including many reserves and sustainable use forests (Kinnaird and O'Brien 1998; Barlow and Peres 2004a).

5.7 Managing fire for bird conservation

The conservation of many tropical birds will depend on developing cost-effective solutions for fire management that promote desirable fire regimes in fire-dependent habitats, and prevent damaging fires from occurring in fire-sensitive ecosystems. Conservation does not take place in a vacuum, and *Integrated Fire Management* has been developed as a way of "addressing the problems and issues posed by both damaging and beneficial fires within the context of the natural environments and socioeconomic systems in which they occur" and "evaluating and balancing the relative risks posed by fire with the beneficial or necessary ecological and economic roles that it may play in a given conservation area, landscape or region" (Myers 2006). Despite their context dependency, general management strategies can be separated according to whether the ecosystem is fire-sensitive or fire-dependent.

5.7.1 Fire management in fire-sensitive systems

Fire management in temperate zones often focuses on the suppression of large wildfires. However, this is very expensive and would be beyond the logistical and economic means of most tropical countries where fires take place across vast spatial scales. Moreover, the effectiveness of fire-fighting techniques has yet to be validated even where equipment and technology are readily available (Johnson *et al.* 2001; Minnich 2001). The only realistic method of fire control in fire-sensitive humid tropics is to prevent their occurrence in the first place, either by removing or controlling ignition sources, which are almost always anthropogenic, or reducing the flammability of the landscape.

Removing or controlling ignition sources is complex and controversial in most tropical landscapes. For example, while controlling or preventing the use of fire in slash and burn agriculture may be appealing to help prevent the spread of wildfires in tropical forests, it can be incredibly damaging to the livelihoods of rural populations and will be very difficult to enforce where there are no viable alternatives to the use of fire. Furthermore, because fires often spread from neighboring landholdings (Nepstad *et al.* 2001), and can burn for tens of kilometers where forests are contiguous, people investing in fire prevention often fail to accrue the benefits (Nepstad *et al.* 2001). Controlling the use of fire through legislation may also have unintended consequences. For example, one of the perverse outcomes of increasing the control of fire in Brazil's Cerrado has been to encourage the conversion of native savanna to exotic grasslands that no longer require burning (Almeida *et al.* in press).

One way of reducing fire may be to encourage a shift from fire-dependent agriculture to mechanized agriculture and perennial crops that reduce the prevalence of fire in the landscape (Aragão and Shimabukuro 2010). Furthermore, investment by farmers in fire-sensitive agriculture (e.g. including fences and perennial crops) increases the cost of accidental fires for farmers, and therefore their motivation to reduce fire risks using firebreaks and the timing of controlled burns (Nepstad *et al.* 2001). However, while this approach may reduce reliance on fire in aging frontiers that are already degraded and impoverished, it may not restrict the use of fire in more active and recently inhabited forest frontiers where fire-based agricultural practices continue to predominate.

Given the social and ecological complexity of reducing the use of fire, the complementary approach of reducing forest flammability may prove essential. In both temperate and tropical forests, fire risk is closely associated with fuel humidity (Johnson *et al.* 2001) and fuel loads (Minnich 2001). As large-scale weather events (such as extreme droughts) are beyond our control, management should concentrate on minimizing the lowered understory humidity and increased understory fuel loads attributable to anthropogenic forms of disturbance such as selective logging and forest fragmentation.

If forests are destined to be logged, substantial reductions in forest flammability can be gained through the adoption of reduced-impact logging practices (Putz *et al.* 2001) that reduce both canopy openness and fuel loads on the forest floor and could help break the strong association between logging and fires (Siegert *et al.* 2001). Even where the adoption of these methods does not prevent the occurrence of fires, they will at least reduce fire severity, and hence their impacts on biodiversity, forest recovery times, and the chances of a recurrent-burn regime from becoming established. However, any logging activity in fire-prone seasonal tropical forests must be considered carefully, avoiding activity in forests close to ignition sources and preventing human incursions along the logging roads.

Fire can also be reduced by changing landscape configuration. As discussed in the previous section, there is a strong interaction between fire and edge effects. This is particularly detrimental in landscapes created through ad hoc development programs, which are often characterized by small irregularly shaped forest patches with high edge to interior ratios that are juxtaposed next to pastures, where maintenance fires are frequently set to rejuvenate the grasses and clear scrub (Uhl and Buschbacher 1985; Cochrane and Laurance 2002; Alencar *et al.* 2004). Management should focus on reducing the interface between fire-dependent agriculture and the remaining forest fragments. Options include reforestation, the strategic placement of forest reserves, and careful economic and ecological zoning to minimize or remove the zone of contact between fire-dependent agriculture and forest reserves.

5.7.2 Fire management in fire-dependent ecosystems

In the same way that fire is a cheap and labor-efficient way for farmers to manage lands, it can also be a cost effective management strategy for managing habitats for bird conservation in fire-adapted ecosystems. Fire management in fire-dependent ecosystems is similar to the management of any disturbance event (Brawn *et al.* 2001), and is largely dependent on selecting the appropriate frequency, scale and seasonality of the fire event. This can be achieved by using prescribed fires, which involves actively burning some areas in order to reach some desired or predicted outcome, or by manipulating unplanned wildfires to obtain beneficial outcomes (Myers 2006). Prescribed fires are particularly useful on small scales, where it is possible to closely manage the quantity and distribution of fuel (e.g. removing fuels, creating fire breaks), the timing of the fire (time of day or time of year), type of wind, whether to burn with or against the wind, and the spatial scale of the burn.

At a landscape scale, many conservation strategies aim to replicate traditional burning practices based on frequent small fires (e.g. Yibarbuk *et al.* 2001), creating a fine-scale mosaic of patches with a range of fire-histories and preventing the severe fire regimes that result following excessive fuel accumulation. However, in practice it can be extremely difficult to find the appropriate fire regime (Woinarski *et al.* 2004), and care must be taken to recognize that species of conservation concern may have different requirements. In tropical regions of Australia, too frequent fires can prejudice many species adapted to scrubbier vegetation, while fire prevention would likely contribute to woody encroachment and the loss of grassland specialist birds (Woinarski *et al.* 2004).

One of the simplest ways of managing the scale and severity of fire is to alter the season in which timed burns take place. Burning when the vegetation is more humid (in the early dry season, or even the wet season) has been advocated as a way of reducing the size of individual fires and increasing landscape heterogeneity. A study in Australia shows how wet season and dry season burns can favor different bird assemblages 4 years after the burning treatments. However, dry season fires were necessary to create habitat suitable for nectarivorous and granivorous species (Valentine *et al.* 2007), showing the importance of varying the fire management strategy within regions. The small fire paradigm has also recently been challenged in temperate systems, where severe large-scale fires in ponderosa pine forests in the United States may be beneficial for some species, including the Black-backed Woodpecker (*Picoides arcticus*) (Hutto 2008). Selecting an appropriate fire regime for conservation clearly depends on the habitat preferences of the species that are of most conservation concern, as well as the social-economic realities of the landscape in question.

5.8 Birds and post-fire regeneration

The majority of this chapter has examined how the impact of fire on vegetation structure and composition can have important consequences for the conservation of tropical birds. However, birds make an important contribution to the maintenance of key ecosystem functions (see Chapter 4) and can help shape ecosystem responses and recovery to fire. Birds are likely to play an important role in post-fire regeneration, dispersing seeds, and could also control the population of insect herbivores (Sekercioglu 2006b). Again, while tropical examples are limited, work from temperate Mediterranean ecosystems demonstrates the importance of bird-dependent seed rain in burned and managed forests, which allows seeds from sources in unburned areas to reach the areas affected by fire (Rost *et al.* 2009) This is likely to be especially important in humid tropical forests, where many plants depend on vertebrates to disperse their seeds (Howe and Smallwood 1982), and where dispersal by birds has been shown to be important on abandoned and degraded agricultural lands (Wunderle 1997; Duncan and Chapman 1999).

5.9 Conclusion

This chapter has shown how altered fire regimes can affect bird communities, mostly acting through changes in vegetation structure and composition. The consequences of fires for bird conservation are highly context specific, depending on a multitude of complex factors including the life history strategy of the bird species, the history of fire in that ecosystem, and the spatial and temporal scale of the burn events. Furthermore, a fire regime can be changed in many ways, and these changes could have a myriad of responses depending on the ecosystem and fauna under consideration.

We still lack a detailed understanding of how the effects of fire change across different spatial and temporal scales, and the potential interactions with other ongoing climatic and anthropogenic changes. However, it is clear that the effective conservation of many tropical birds will be dependent on developing a better understanding of the consequences of fire, and effective ways of managing its effects in a socially and ecologically complex world.

6

Biotic Invasions and Tropical Birds

Overview: Worldwide, 597 bird species are threatened or have gone extinct owing to the negative effects of invasive species. Cats and rodents are the main culprits in these losses. Thus management or eradication of invasive species threatening birds is often needed. A number of bird species are deliberately released through the pet trade, for agricultural purposes, or to support recreational, subsistence or commercial hunting. Some of these species establish invasive populations. However, the successful management of invasive bird species needs strong biological knowledge. Sadly, avian invasion ecology remains a poorly developed field in the tropics.

Biotic invaders remain an ubiquitous threat to tropical birds. In this chapter, we first highlight the effects of invasive species on tropical birds. Because some bird species themselves can become invaders, we briefly discuss the underpinnings of avian invasion ecology. We also spotlight some of the key tropical invasive birds and ways to manage them. In this chapter, invasives are defined as those species that have spread from the point of introduction into an area they have not previously occupied and become abundant, threatening the native biota or human enterprise (Kolar and Lodge 2001). Invasive species are interchangeably referred to as introduced species, non-indigenous species, exotics, or aliens.

6.1 Effects of invasive species on tropical birds

Globally, 597 threatened and extinct bird species are or were affected by invasive species, the majority by mammals (Figure 6.1; Butchart 2008). Domestic (feral) cat (*Felis catus*) and rodents impact (or impacted) 52% of these species. Common

Conservation of Tropical Birds, First Edition. Navjot S. Sodhi, Çağan H. Şekercioğlu, Jos Barlow and Scott K. Robinson.

Published 2011 by Blackwell Publishing Ltd.

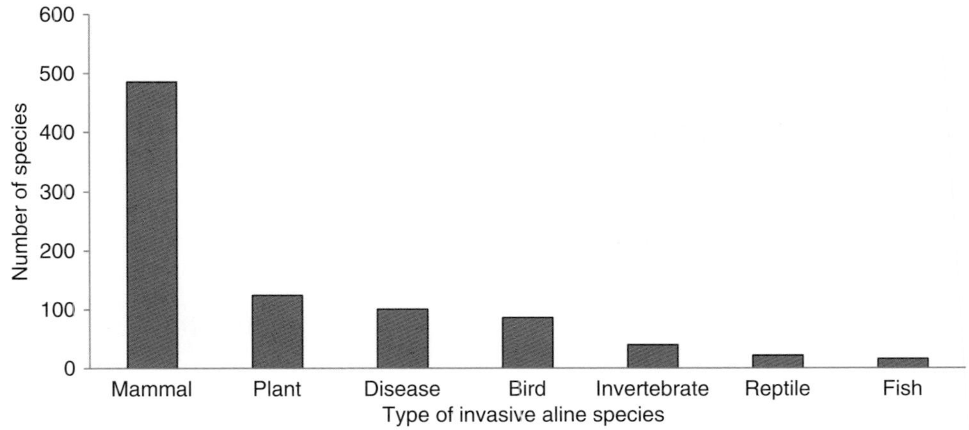

Figure 6.1 Number of bird species impacted by different types of invasive alien species. (After Butchart 2008.)

Myna (*Acridotheres tristis*), brown tree snake (*Boiga irregularis*), avian malaria (*Plasmodium relictum*), and avian pox (*Poxvirus avium*) are key invaders impacting tropical birds (Butchart 2008; see below). The main effects of invasive species include direct mortality, reduction in reproductive success, degradation of habitats, and competition (Figure 6.2; Butchart 2008). Native birds of islands remain particularly vulnerable to invaders (Blackburn *et al.* 2004; Karels *et al.* 2008; Galetti *et al.* 2009; also see below). Islands may have either received more exotic species introductions or their simple ecosystems may be more permeable to introduced species establishment than comparable mainlands (Cassey 2003). Around the globe, except in Oceania, island birds with small ranges remain particularly vulnerable to invasive species (Figure 6.3; Clavero *et al.* 2009). Due to their high habitat specialization, restricted range species remain vulnerable to environmental perturbations (Sodhi *et al.* 2004b; see Chapter 3).

We discuss some of the prominent cases of invasive species impacting tropical birds below. The point to note is that invasive species may synergistically act with habitat degradation, climate change, and hunting to threaten tropical birds (Brook *et al.* 2008; see Chapters 1 and 3).

The brown tree snake provides a quintessential example of how introduced predators can harm native tropical birds. The snake was introduced on the island of Guam in the South Pacific shortly after the Second World War probably as a stowaway in ship cargo (Fritts and Leasman-Tanner 2001). Tree snakes have been likely directly responsible for the loss of 12 of 18 native bird species there (Wiles *et al.* 2003). Most of the remaining bird species in Guam are declining (Figure 6.4; Wiles *et al.* 2003). While the evidence for the direct link between invasion and extinction is still inconclusive for some of Guam's extirpated species, alternative explanations for this biotic catastrophe, like disease, habitat alteration, and overhunting, have all been discounted. Despite an annual expenditure of US$44.6 million, tree snakes on Guam are still not under control (Simberloff 2001), largely because of their ability to penetrate artificial snake barriers such as fences.

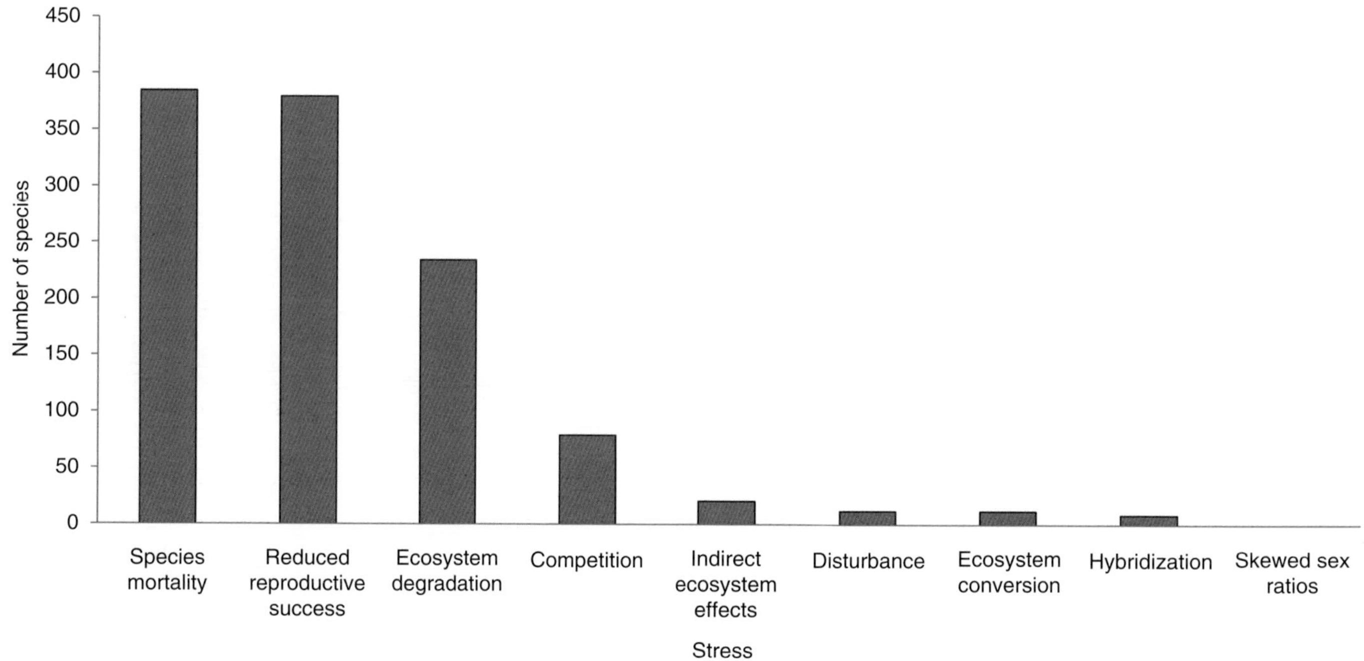

Figure 6.2 Number of bird species impacted in different ways by invasive alien species. (After Butchart 2008.)

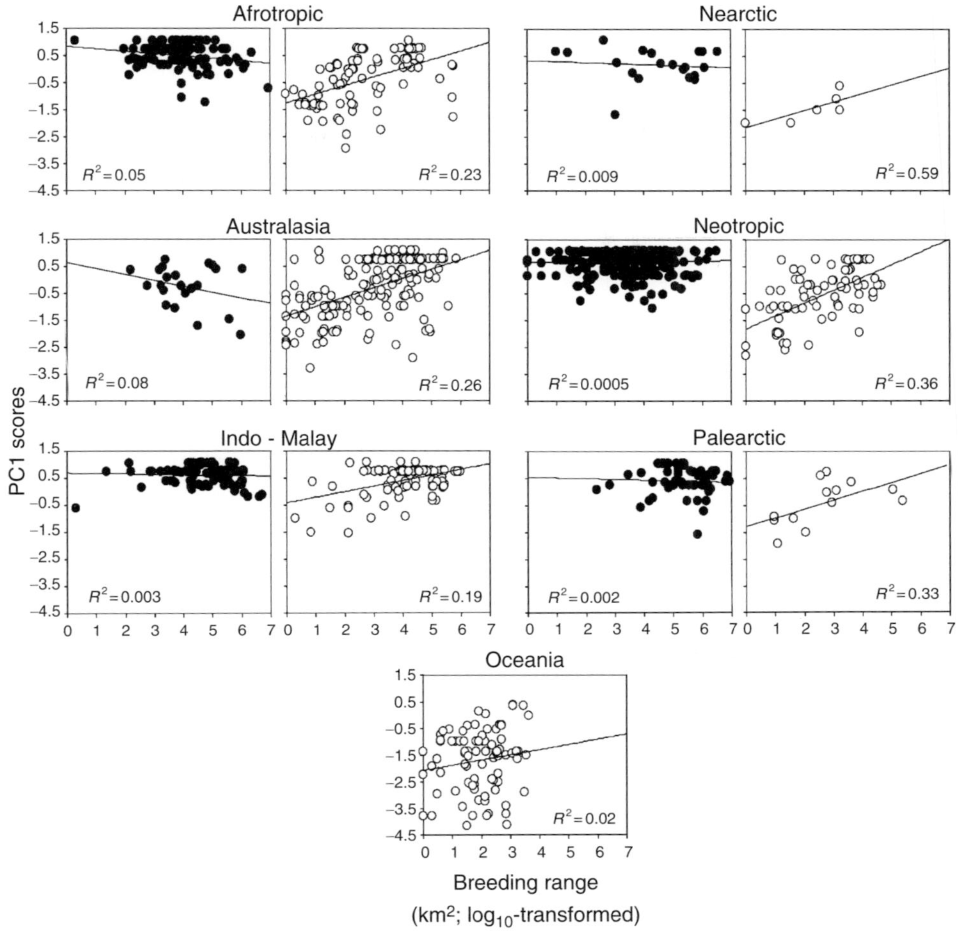

Figure 6.3 Island birds (open circles) and those with small ranges are generally impacted by invasive species. Continental birds are indicated by filled circles. PC1 = Principal Components Scores on Axis 1. (After Clavero *et al.* 2009.)

As mentioned, invasive mammals remain the main threat to native bird species. On many islands in French Polynesia, monarch birds (*Pomarea* spp.) are rarely found on islands invaded by the black rat (*Rattus rattus*) (Seitre and Seitre 1992), suggesting that they avoid this potential nest predator (Thibault *et al.* 2002). Alternatively, rats may have wiped out monarchs– they are implicated in monarch extinctions and decline (Table 6.1; Thibault and Meyer 2001). Although population declines in monarchs are caused primarily by nest predation, they are exacerbated by competition with another invader — the Red-vented Bulbul (*Pycnonotus cafer*) (Thibault *et al.* 2002). Indicating heavy mortality caused by rats in other areas, depredation by introduced rats (e.g. Norway rat, *R. norvegicus*) was the only cause of mortality in 53 radiotagged Bicknell's Thrushes (*Catharus bicknelli*) in the Dominican Republic (Townsend *et al.* 2009).

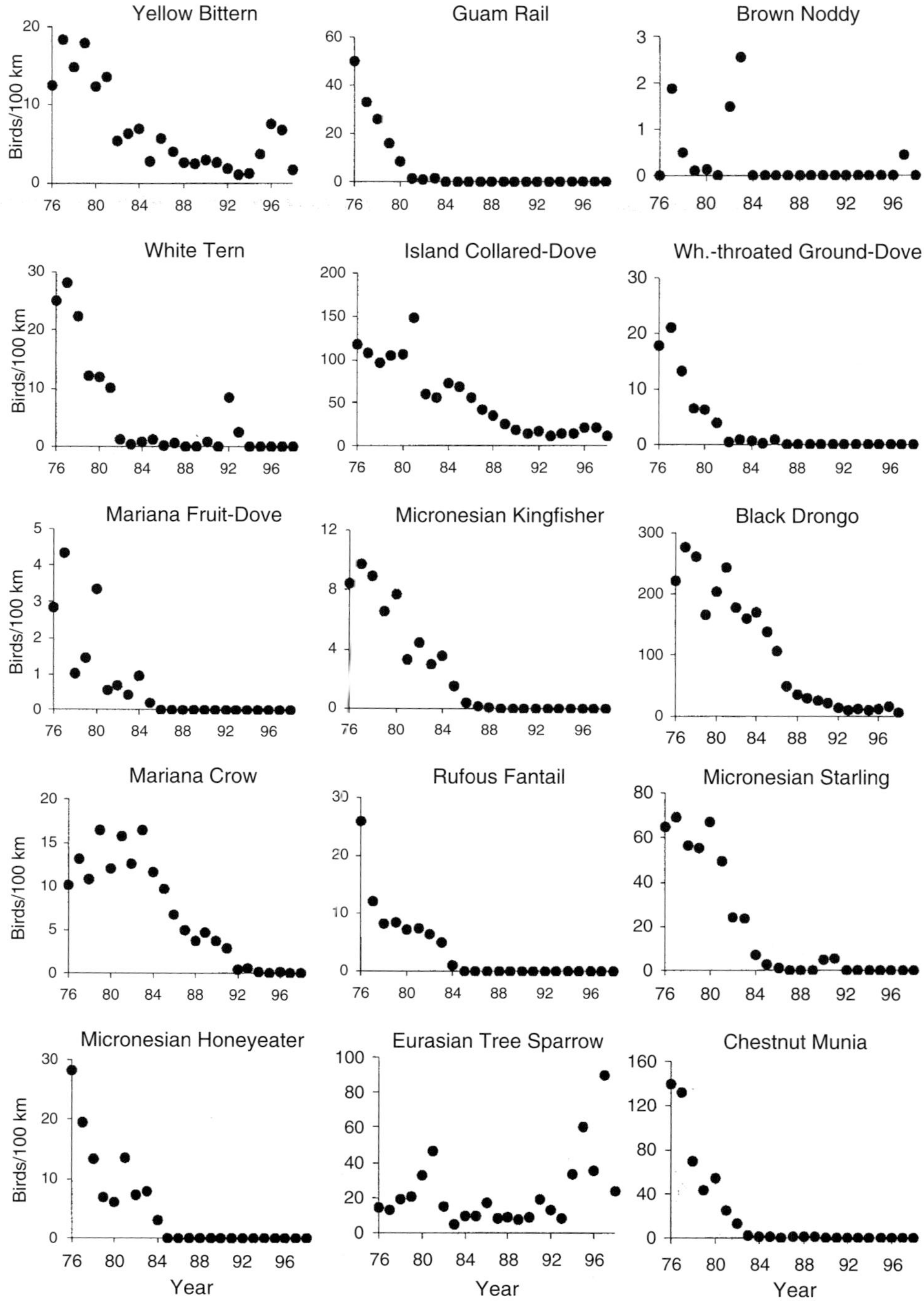

Figure 6.4 Population trends for Guam birds as indicated by roadside surveys, 1976–1998. Wh. = White. (After Wiles *et al*. 2003.)

Table 6.1 Conservation status and main causes of extinction and current threats to the monarchs (*Pomarea* spp.) in French Polynesia. (After Thibault and Meyer 2001)

Species name	*Conservation status*	*Sub-species name*	*Causes of extinction or decline*
Society Islands			
P. nigra	CR		Black rat, introduced birds, invasive plants
P. pomarea	EX		†, unknown (black rat, intense secondarization of the vegetation?)
Marquesas Islands			
P. iphis	VU	*P. i. fluxa*	†, unknown (black rat, overgrazing by feral sheep or disease?)
		P. i. iphis	Decrease of forest by overgrazing
P. mendozae	EN	*P. m. mendozae*	†, black rat, decrease of native forest (invasive plants)?
		P. m. nukuhivensis	†, black rat, decrease of native forest (invasive plants, cattle)?
		P. m. mira	†, black rat, decrease of native forest (invasive plants)?
		P. m. motanensis	Overgrazing by feral sheep
P. whitneyi	VU		Black rat, decrease of native forest (fire, goats)

CR = critical; EX = extinct; VU = vulnerable; EN = endangered; † = extinct form.

Additionally, predation by introduced rats remains the primary cause of seabird extinction and endangerment (Jones *et al.* 2008b). Breeding seabirds on islands are particularly vulnerable to introduced mammalian predators possibly due to lack of behavioral, morphological, and life-history defenses (Keitt *et al.* 2002). For example, 90% of the diet of feral cats on Natividad Island (Mexico) comprised shearwaters (Keitt *et al.* 2002). It was estimated that annual growth rate of Black-vented Shearwater (*Puffinus opisthomelas*) declined by 5% for every 20 cats per 150,000 birds on this island (Keitt *et al.* 2002). Similarly on Reunion Island (Western Indian Ocean), as much as 65% of the cat diet can comprise seabirds, but cats also prey on introduced predatory rats thus alleviating some of the rat predation on seabirds (Faulquier *et al.* 2009). In addition, cat predation threatens the survival of the endemic Fuerteventura Chat (*Saxicola dacotiae*) on subtropical Canary Islands (Medina and Nogales 2009). Like cats, introduced house mouse (*Mus musculus*) can cause heavy mortality in seabird chicks and thus pose a considerable risk to their populations in the tropics (Wanless *et al.* 2007).

In addition to invasive predators, introduced parasites and disease can devastate tropical bird populations. The mosquito (*Culex quinquefasciatus*) that carries the avian malaria parasite (*Plasmodium relictum*) was inadvertently introduced to Hawaii in 1826, with the parasite arriving thereafter possibly in an invasive bird (Mack *et al.* 2000; Woodworth *et al.* 2005). Avian poxvirus was also introduced in Hawaii around the same time, most likely through an introduced bird species (Atkinson and LaPointe 2009). Avian poxvirus is transmitted through contact or mouthparts of blood-sucking arthropods and causes lesions on the body. Since the 1800s, avian malaria and pox have been largely responsible for the decline and extinction of numerous endemic bird species on the Hawaiian Islands (Van Riper *et al.* 1986; Atkinson *et al.* 2005). Since 1837, 13 Hawaiian bird species have gone extinct because of disease and possibly other factors such as habitat loss (Table 6.2; Smith *et al.* 2006). Without having evolved in the presence of the disease, Hawaiian bird populations were generally unable to cope with the novel pathogens. However, since the establishment of avian malaria over 100 years ago, some native thrushes (*Myadestes* spp.) are now showing resistance to the disease, although many other species such as endangered honeycreepers (e.g. Hawaii Creeper, *Oreomystis mana*) still are vulnerable (Atkinson *et al.* 2005). Further, birds infected with avian poxvirus remain vulnerable to malaria exacerbating the effect of the former (Atkinson and LaPointe 2009). Both avian malaria and pox occur at low prevalence at higher elevations (Atkinson *et al.* 2005). However, with global warming, mosquitoes may increase their upper elevation range thus making the high-elevation avian refugia vulnerable to malaria and pox (Figure 6.5; Atkinson and LaPointe 2009; see Chapter 8). The most feasible method of reducing transmission of malaria seems be to reduce or eliminate vector mosquito populations through chemical treatments and elimination of larval habitats (Atkinson *et al.* 2005).

Other introduced threats to Hawaiian birds are also a cause for concern. Ectoparasites, especially chewing lice infections, are increasing there (Freed *et al.* 2008a). Chewing lice and competition with introduced Japanese White-eyes (*Zosterops japonica*) are the main causes of population crashes in some Hawaiian birds such as the Hawaii Akepa (*Loxops coccineus*) (Freed *et al.* 2008b). Therefore, increasing abundance of invasive bird species in Hawaii may further endanger native bird populations by spreading diseases and increasing competition (Figure 6.6; Foster *et al.* 2004).

Disease (malaria and avian pox) and parasites are also endangering the birds of the "evolutionary heritage site" – Galapagos archipelago (Wikelski *et al.* 2004). There was 33% annual increase in avian poxvirus infection from 2000 to 2004 in Darwin's finches on Santa Cruz Island (Kleindorfer and Dudaniec 2006). In the males of the Small Ground Finch (*Geospiza fuliginosa*), mating success was lower (17% paired) in pox-infected individuals than in uninfected ones (77%) (Kleindorfer and Dudaniec 2006). This indicates fitness (reproductive success) costs associated with poxvirus infection. Similarly, in this species, parasitism of nestlings by the introduced ectoparasitic fly *Philornis downsi* elevated mortality (Dudaniec *et al.* 2006). *Philornis downsi* also caused up to 37% and 41% mortality in the nestlings the of Medium Ground Finch (*G. fortis*) and Darwin's Medium Tree Finch (*Camarhynchus pauper*), respectively (Figure 6.7; Huber 2008; O'Connor *et al.* 2010). Thus, this fly has a potential to depress populations of native birds of Galapagos Islands. Further, the increased poultry production there

Table 6.2 Known bird species extinctions purportedly facilitated by disease. (After Smith *et al.* 2006)

Species	*Distribution*	*Date*[a]	*Causal factors*
Akialoa ellisiana	Hawaiian Islands	1837c	Habitat loss; disease from introduced species[b]
Akialoa stejnegeri	Hawaiian Islands	1969r	Habitat loss; disease from introduced species
Akialoa lanaiensis	Hawaiian Islands	1892c	Habitat loss; disease from introduced species
Akialoa obscura	Hawaiian Islands	1940r	Habitat loss; disease from introduced species; hunting
Corvus hawaiiensis[c]	Hawaiian Islands	2002r	Habitat loss; disease from introduced species; hunting
Drepanis pacifica	Hawaiian Islands	1899r	Habitat loss; disease from introduced species; hunting
Hemignathus obscures	Hawaiian Islands	1940r	Habitat loss; disease from introduced species
Moho apicalis	Hawaiian Islands	1837c	Habitat loss; disease from introduced species
Moho bishop	Hawaiian Islands	1915e	Habitat loss; disease from introduced species
Moho braccatus	Hawaiian Islands	1985r	Habitat loss; predation and disease from introduced species; natural disaster
Moho nobilis	Hawaiian Islands	1934r	Habitat loss; disease from introduced species
Myadestes myadestinus	Hawaiian Islands	1989r	Habitat loss; competition and disease from introduced species; natural disaster
Paroreomyza flammea	Hawaiian Islands	1963r	Habitat loss; disease from introduced species
Fregilupus varius	Mascarene Islands	1860e	Habitat loss; disease from introduced species; natural disaster
Heteralocha acutirostris	New Zealand	1907r	Habitat loss; disease from introduced species; hunting
Coturnix novaezelandiae	New Zealand	1875e	Habitat loss; predation and disease from introduced species
Ectopistes migratorius	North America	1900e	Habitat loss; disease from introduced species; hunting; intrinsic factors
Psephotus pulcherrimus	Australia	1927r	Habitat loss; predation and disease from introduced species; hunting; natural disaster; changes in native species dynamics

[a] Date of extinction defined as last reported in the wild (r), last collected (c), extinct (e) (IUCN 2004).
[b] Referred to as parasites and pathogens in the IUCN database.
[c] Denotes species extinct in the wild.

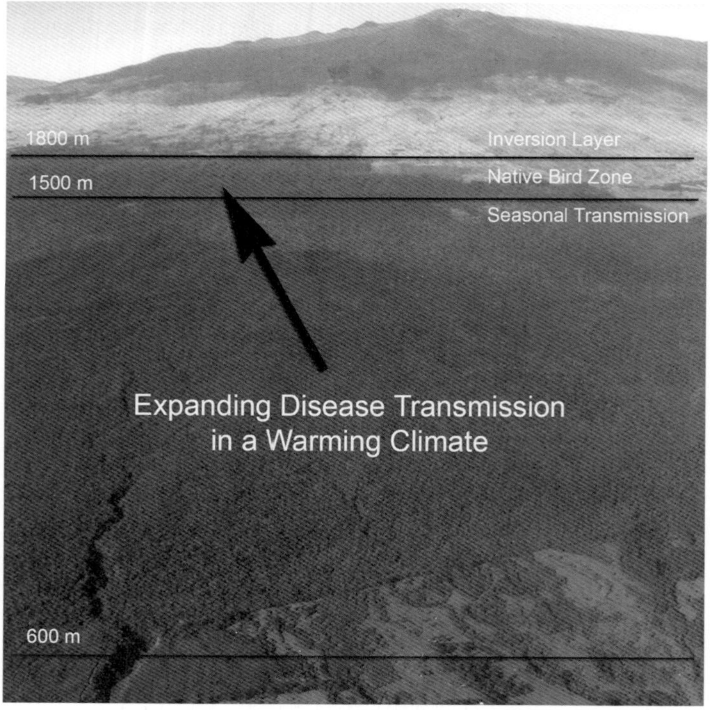

Figure 6.5 Aerial photograph of the windward, eastern slope of Mauna Kea Volcano on the island of Hawaii. In a warming climate, high elevation populations of Hawaii's most endangered birds may be threatened by expanding disease transmission. (After Atkinson and LaPointe 2009; courtesy of United States Geological Survey.)

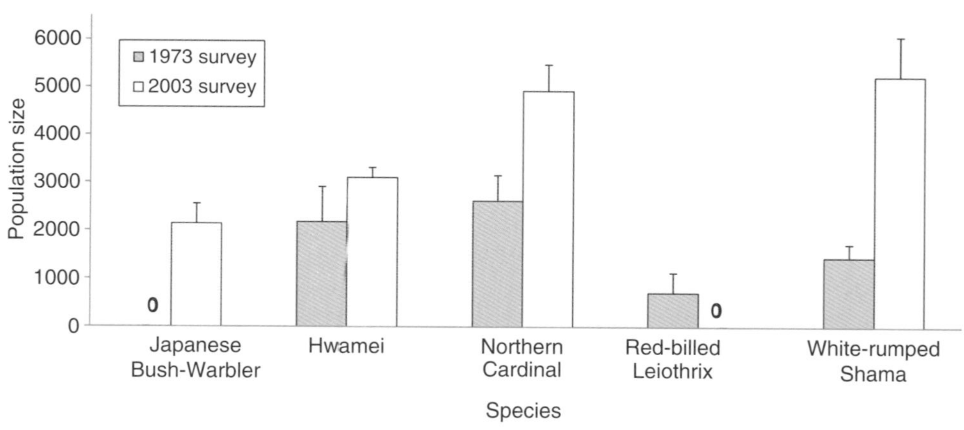

Figure 6.6 Estimated population sizes of introduced birds species in Kaua'i (Hawaii). Error bars represent standard errors. (After Foster *et al*. 2004.)

may increase the risk of disease transfer to native birds that may have little resistance to introduced pathogens (Wikelski *et al*. 2004). Therefore, sound poultry biosecurity measures (e.g. immunization of poultry) should be enacted in the Galapagos archipelago to protect native avifauna (Gottdenker *et al*. 2005).

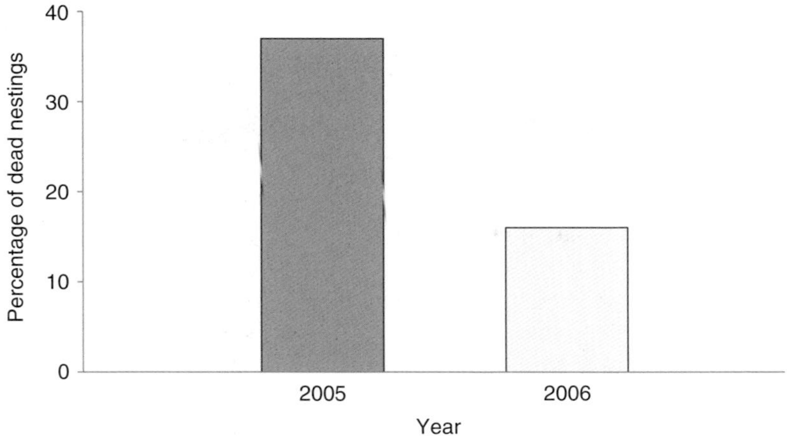

Figure 6.7 Percentage of dead nestlings of the Medium Ground Finch in nests with parasites. Data were collected from the Galápagos Islands (Ecuador). (Data derived from Huber 2008.)

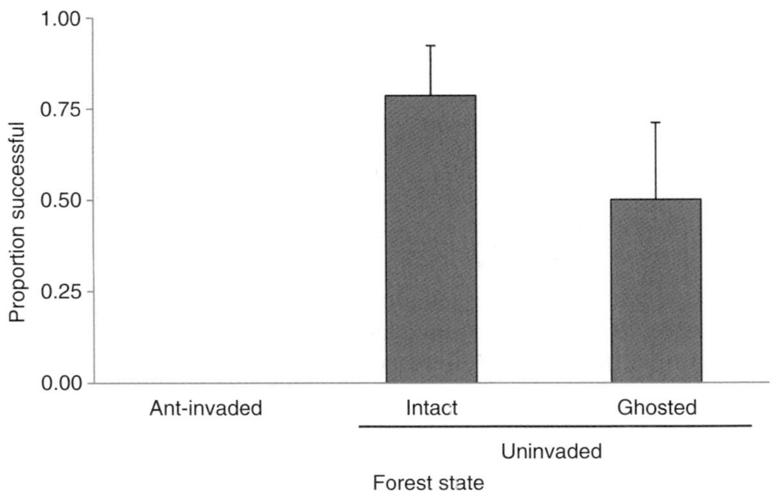

Figure 6.8 The proportion of the Island Thrush nests that successfully fledged nestlings in ant-invaded versus uninvaded forests. All these, included ghosted were primary forests. Error bars represent standard errors. (After Davis *et al*. 2008.)

Responding to invaders, some native species alter their behavior. For example, on Christmas Island (Indian Ocean), the Island Thrush (*Turdus poliocephalus*) and Emerald Dove (*Chalcophaps indica*) avoided foraging on ground in the forests that were invaded by introduced yellow crazy ants (*Anoplolepis gracilipes*) (Davis *et al*. 2008). Although nest predation was not observed, none of the Island Thrush nests were successful in an ant-invaded forest (Figure 6.8; Davis *et al.* 2008). Paradoxically, the foraging success of native Christmas Island White-eye (*Zosterops natalis*) was higher (6% versus 2–4%) in an ant-invaded forest than in forests not invaded by ants. White-eyes fed on honey-secreting scale insects that increased in abundance in ant-invaded forest due to their mutualism with the ants (Davis *et al.* 2008). Ants receive energy-rich carbohydrates from scale insects in return for protecting them from their predators (Ness and Bronstein 2004).

Similarly in Mauritius, introduced honey bees (*Apis* sp.) may also be interfering in the mutualistic interactions between endemic plants and white-eyes (Hansen *et al.* 2002). White-eyes seemed to avoid times and plants frequented by honey bees, and this could impact pollination because honey bees are less efficient pollinators of native plants in Mauritius (Hansen *et al.* 2002).

To alleviate the effects of invasive species on tropical birds: (i) international trade needs to be better managed; (ii) potential introduction pathways (see below) of invasive species spread and establishment need to be severed; and (iii) invasive species should be better controlled or eradicated in tropical areas (Butchart 2008). Eradication of some of the invasive species (e.g. rats and cats) has been successful in some tropical locations with positive effects on tropical birds (Nelson *et al.* 2002; Rodriguez *et al.* 2006). For example, cat eradication likely increased the Sooty Tern (*Onychoprion fuscatus*) population by an average of 21% over 3 years (2005–07) on Ascension Island (South Atlantic Ocean) (Hughes *et al.* 2008).

However, the control of invasive species should be carefully planned considering the "newly formed" trophic relationships in the community, because removal of some species can disrupt the ecosystems (Caut *et al.* 2008). Some invasive species may have a positive role in the ecosystem. For instance, only 16% of breeding pairs of endangered Mauritius Fody (*Foudia rubra*) nest on native trees, the rest on introduced *Pinus* spp. or *Cryptomeria japonica* trees, where nest predation is lower (Cristinacce *et al.* 2009). Additionally, some invasive species can serve as prey for native birds. For example, a House Wren (*Troglodytes aedon*) has been observed to feed on juvenile house gecko (*Hemidactylus frenatus*) in Costa Rica (Barquero and Hilije 2005). Similarly, due to a decline in its main prey (guanacos, *Lama guanicoe* and rheas, *Rhea* sp.), threatened Andean Condors (*Vultur gryphus*) now survive because they have shifted their diet to exotic herbivores such as sheep (*Ovis aries*) and goat (*Capra* sp.) (Lambertucci *et al.* 2009). Similarly, carcasses of invasive goats and rabbits (*Oryctolagus* sp.) provide food for scavenging birds on Canary Islands (Gangoso *et al.* 2006). However, the benefits of invasive species may also incur costs. The Lesser Short-toed Larks (*Calandrella rufescens*) in the Canary Islands feeding on goat farms where food is abundant become susceptible to avian poxvirus (Carrete *et al.* 2009). Additionally, sometimes invasive species (e.g. Puerto Rican frog, *Eleutherodactylus coqui*) serve as prey for endemic birds, but also do the same for the introduced predators such as rats and mongoose (Herpestidae), thus bolstering their populations (Beard and Pitt 2005). Thus, there may be a net negative effect of the "introduced prey."

6.2 Avian invasions

In addition to being impacted by invasive species, some bird species can themselves become invaders. Introduced bird species are deliberately released through the pet trade, for agricultural purposes, or to support recreational, subsistence or commercial hunting. Introduced bird species may also be a byproduct of commerce, e.g. stowaways in ships. Some bird invasions can also be a result of range expansion facilitated by changing human footprint (e.g. human-mediated land-use changes; Blackburn *et al.* 2008; Chiron *et al.* 2009). For example, the expansion of Common Mynas in Peninsular Malaysia and Singapore has been facilitated most likely by heavy deforestation and accompanying habitat modifications (Sodhi and Sharp 2006). Vegetation (i.e. nesting and roosting sites) and food source (e.g. garbage) can facilitate the spread and establishment of some exotic bird species in urban centers (see below).

Probably the very first bird introduction was about 8000 years ago when Red Junglefowls (*Gallus gallus*) were transported to China (West and Zhou 1989). Worldwide, 1400 attempts to introduce 400 bird species have been recorded (Duncan *et al.* 2003). Approximately 70% of bird introductions have been to islands, although islands make up only about 3% of Earth's ice-free land area (Sodhi 2010). Just over half of global bird introductions have been to Pacific Islands (notably to Hawaii) and to Australasia. Most of bird introductions took place in the 18th and 19th centuries, during a period of major European colonization (Long 1981; Lever 1987). Because of this, the large proportion of invasive bird species originated from temperate regions (Figure 6.9; Blackburn *et al.* 2009). Birds were introduced by settlers to various countries predominantly for aesthetics, hunting, and biocontrol. Two-thirds of the species chosen for introduction have been from six of 145 bird families: Anatidae (ducks), Columbidae (pigeons and doves), Fringillidae (finches), Passeridae (sparrows), Phasianidae (pheasants), and Psittacidae (parrots). This over-representation of certain families points to the reasons for these introductions. For example, ducks and pheasants were introduced for hunting, whereas finches, sparrows, and parrots were introduced as pets (Lockwood 1999; Sodhi 2010).

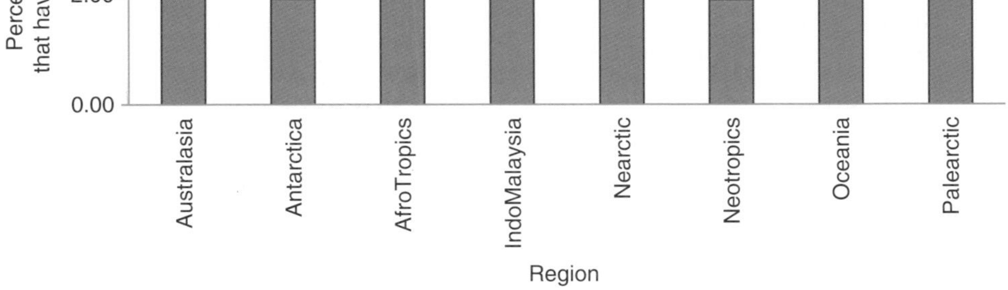

Figure 6.9 Percentage of bird species in the region that have established exotic populations elsewhere. (Data derived from Blackburn *et al.* 2009.)

Hong Kong epitomizes the present and future of avian invasions in degraded subtropical landscapes. The natural vegetation of Hong Kong was cleared by humans before the 18th century. Nineteen bird species introduced by humans have colonized Hong Kong (Leven and Corlett 2004). There are no signs that the successful colonization by avian invaders will cease in Hong Kong (Figure 6.10; Blackburn *et al.* 2009). Usually there is a "lag time" between introductions and naturalization, and these lag times can vary with taxa and region (Daehler 2009) and they may be responsible for continued establishment of introduced birds in Hong Kong. Due to high levels of habitat conversion in the tropics (see Chapter 1), these areas may offer conducive habitats for future bird invaders (Peh 2010).

6.3 Pathology of avian invasions

Every bird introduction does not always result in an established population. A case in point is the introduction of more than half a million common quail (*Coturnis coturnix*) to more than 30 states in North America between 1875 and 1958. Incredibly, this experiment failed (Sodhi 2010). Based on analysis of invasions in Britain, Williamson and Brown (1986) proposed a rule of tens. That is, 10% of imported species make it into the wild, 10% of these species establish self-sustaining populations, and only 10% of established species became pests (i.e. causing ecological and/or economic harms). However evidence for tens rule has not always been forthcoming – e.g. Jeschke (2008) found that globally 50% of the bird introductions have resulted in the establishment, and 34% of these became invasives (i.e. harmful).

In order to establish exotic populations, a bird species has to cross a number of hurdles in "invasion pathways" (Figure 6.11; Blackburn *et al.* 2009). Invasion pathways can be broken into four phases: import, introduction, establishment,

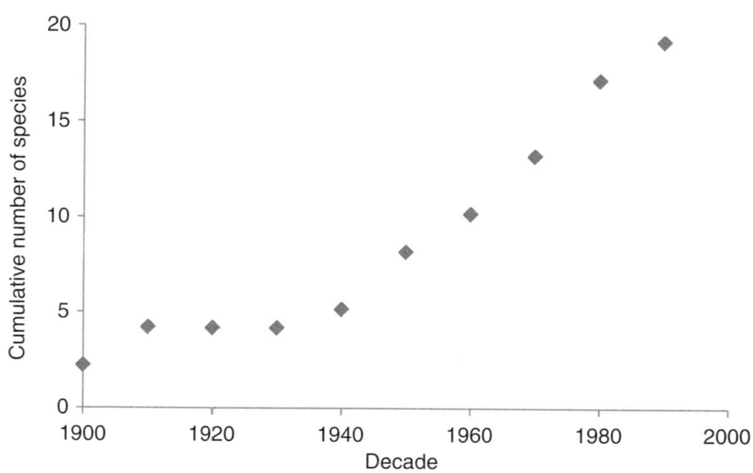

Figure 6.10 Cumulative number of exotic bird species that have established in Hong Kong (China). (Data derived from Leven and Corlett 2004.)

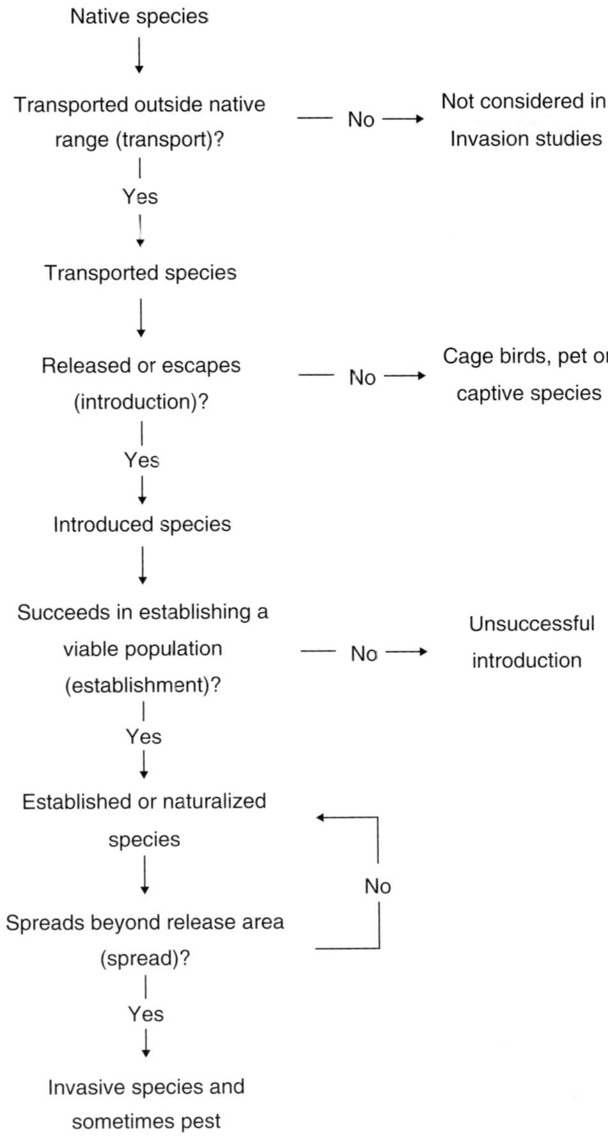

Figure 6.11 Schematic representation of the invasion pathway. (After Blackburn *et al.* 2009.)

and becoming a pest (Williamson 1996). Characteristics needed to pass through each of these phases may vary. While the same factor may affect all invasion stages, usually, socioeconomic factors are important during early (see above), while biogeographical, ecological and evolutionary factors becoming more important during later stages (Williamson 2006). For example, the economic isolation of Eastern European countries during the cold war protected it from bird introductions (Chiron *et al.* 2010). This suggests that better policies need to be realized to reduce the risk of avian invasions into these countries now due to increase in trade and human movement.

Basically, a successful avian invader should: be capable of population growth even in small populations, have broader ecological niches, and be capable of behavioral flexibility (Blackburn *et al.* 2009). Supporting the last criteria, Martin and Fitzgerald (2005) showed that House Sparrows (*Passer domesticus*) from a recently invaded (28 years ago) population in Panama were more likely to approach and consume novel food items than those from a long-established population (150 years) in USA. Establishment success in birds is also influenced by larger brains relative to body size, which may facilitate innovation propensity and thus help in coping with novel environments (Sol *et al.* 2005; Vall-llosera and Sol 2009).

Breeding early and throughout the year may result in high reproductive output in some invasive birds. For example, the successful invader, the Feral Pigeon (*Columba livia*), can breed at the age of one and throughout the year (Sodhi and Sharp 2006). The latter may be because they feed on seeds, which are generally available throughout the year. Established invasive bird species may also have large clutch sizes resulting in high productivity (Cassey 2002).

The success of an invasive bird species depends upon "propagule pressure" or the size of the "founding" population (Williamson 1996). Larger introduced populations can have higher success than smaller populations. Smaller populations have a greater risk of extirpation because of factors such as loss of genetic heterogeneity or allee effects. Allee effects are factors (e.g. lack of mates) that cause lower growth rate in small populations (Courchamp *et al.* 2008). In support of the propagule pressure hypothesis, the median individuals for successful Galliform introductions in Hawaii were 179 compared to 14 for the failing species (Moulton *et al.* 2001). However, in addition to the propagule pressure, the condition or composition (e.g. age and sex) of the founding population and subsequent population augmentation through immigration can influence successful establishment (Drake and Lodge 2006; Blackburn *et al.* 2009).

Species traits such as large range sizes, large body size, non-migratory behavior, bi-parental care, and lack of sexual dichromatism also facilitate the establishment of introduced birds. Large range sizes usually correlate with behavioral flexibility thus allowing the invaders to cope and thrive in the unpredictable novel environments (Cassey 2002; Croci *et al.* 2007; Hayes and Barry 2008). Because of a positive relationship between adult survival and body size (Pimm *et al.* 1988), larger avian invaders are expected to survive better than smaller ones (Jeschke and Strayer 2008). Migratory invasive species may find it difficult to locate mates in invaded areas (Lever 1987). Biparental care may increase reproductive input of avian invaders through mechanisms such as better nest protection (McLain *et al.* 1999). High energetic investment in sexual traits may make sexually dichromatic invading birds less able to cope with the novel environments (McLain *et al.* 1999).

Two other hypotheses have been proposed regarding the success of invasive species. According to the "enemy release hypothesis", the escape from the "usual" predators and parasites assists in the flourishing of introduced species in newly colonized areas (Torchin *et al.* 2003). "Climate-matching hypothesis" suggests that introduced populations have a higher chance of becoming established if the climate between natural and exotic locations matches closely (Cassey 2003; Williamson 2006; Hayes and Barry 2008). The results from the Rose-ringed Parakeet (*Psittacula krameri*) seem to support both these hypotheses (Shwartz *et al.* 2009). These parakeets suffer higher nest predation in their native India than in the invaded UK. Similarly, their reproductive success is higher in the warmer Mediterranean than in the colder UK, both being invaded locations. However, Ishtiaq *et al.* (2006) found mixed support for the enemy release hypothesis when they compared the blood parasite (*Plasmodium* and *Haemoproteus*) prevalence in native and exotic populations of the Common Myna. Only exotic populations on two oceanic islands (Fiji and Hawaii) had lower blood parasite prevalence than native populations. There was also evidence that the introduced populations may be infected by new parasites suggesting that mynas were confronting "new enemies" in exotic locations (Ishtiaq *et al.* 2006). Other factors may also influence invasion success. Some introduced bird species may not be able to establish populations because they are subjected to heavy hunting, or environmental catastrophes like cyclones may wipe their established populations (Simberloff 1992c).

Extrinsic factors such as competition may also affect invasion success (Blackburn *et al.* 2009). Thirty-one passeriform bird species have been introduced on the island of Saint Helena (South Atlantic Ocean) (Brooke *et al.* 1995). The first bird introduction took place there in 1776, and the last in 1929. None of the 14 bird species released in 1929 was successful, suggesting that competition from already existing avian invaders may have prevented successful colonizations (Brooke *et al.* 1995).

The characteristics possessed by invasive species have been postulated to be the reverse of species threatened with extinction (Bradshaw *et al.* 2008). However data from birds do not support this "two-sides-of-the-same-coin" hypothesis (Blackburn and Jeschke 2009).

6.4 Tropical avian invaders

Three bird species included in the 100 worst invasive species are the European Starling (*Sturnus vulgaris*), Red-vented Bulbul, and Common Myna (www.issg.org). Generally, invasive species can have four major effects: (i) extinction of native biota; (ii) alteration of the abiotic environment such as fire regimes; (iii) damage to agriculture as pests; and (iv) harm to humans, native flora, and fauna through the introduction of virulent diseases (Mack *et al.* 2000). Approximately 20% of 680 reported animal extinctions were likely caused by invasive species (Clavero and García-Berthou 2005). In addition, there usually is a positive correlation between the dominance of invasive species and native species decline (Didham *et al.* 2005). Although invasive species may not always be exclusively responsible for negative consequences to biodiversity (Didham *et al.* 2005), there are a number of examples showing how invasive species generally create ecological havoc in the tropics (see above).

Invasive bird species can depress the reproductive success of native birds through predation and competition. For example, in Kenya, House Crows (*Corvus splendens*) have been observed raiding the nests of weaver and other small bird species (Ryall 1992). Invasive birds may impact humans, since many of these are found in human-altered habitats. Some species such as munias and sparrows, parrots, crows, geese, and cranes can damage crops, vegetables, and fruits. Invasive "pest" birds have been reported as reducing the yield of rice (*Oryzis* sp.) by 50% in Thailand (Yap and Sodhi 2004). Communally roosting invasive bird species such as crows and mynas foul gardens, pedestrian footpaths, buildings, and vehicles beneath or near their roosts. Excessive noise from roosts, especially in early morning, can cause annoyance to people. Invasive species have also sometimes been found to carry the bird flu virus (H5N1); for example, the House Crow in Hong Kong (Sodhi 2010).

However, the field of avian invasion ecology is still in its infancy in the tropics (Peh 2010), so it is likely that more evidence for their negative effects in the tropics will emerge over time. Furthermore, processes such as habitat modification and loss (e.g. logging) may promote the establishment of non-native species, thus acting synergistically to threaten native biodiversity (Brook *et al.* 2008; see Chapter 1). To exemplify the multi-faced effects of tropical avian invaders, we briefly discuss the two worst bird invaders in the region – Common Myna and Red-vented Bulbul.

6.4.1 Common Myna

A native of Afghanistan and India, the Common Myna (Figure 6.12a) arrived, for example, in Singapore from the north in the 1930s, via South Vietnam, Thailand, and the Malayan peninsula (Sodhi and Sharp 2006). The bird can now be found

(a) (b)

Figure 6.12 (a) Common Myna. (b) Red-vented Bulbul. (Photographs by J. M. Garg.) (Please see Plate 6.12 for the color version of this figure.)

over a range from Afghanistan to southwestern China and Japan, and from South Africa to Hawaii, and even in Australia and South Africa (Sodhi and Sharp 2006). The spread of Common Mynas has been due largely to range expansion concomitant with anthropogenic land-use changes (see above), although they were deliberately introduced into Australia in 1863, into Hawaii in 1865, into South Africa in about 1900, and into Hong Kong in about 1952 (Leven and Corlett 2004; Sodhi and Sharp 2006). Most of the deliberate releases have been to control insect pests of crops (Sodhi and Sharp 2006). As a case in point, the Common Myna has also spread westwards from Afghanistan into Uzbekistan, where its reported population for Tashkent city alone is now around five million, largely due to deliberate introduction in the 1960s and 1970s as a biocontrol for insect pests. In Hong Kong, its numbers seem to have been held in check by the native Crested Myna (*Acridotheres cristatellus*) (Leven and Corlett 2004).

Common Mynas can prey on the eggs and nestlings of other birds and may aggressively defend territories and thus impacts native birds (www.issg.org). The Common Myna nests in tree hollows and may compete for these resources with native hollow-nesting species (e.g. native parrots in Australia; Pell and Tidemann 1997). Common Mynas disturb the nesting threatened Seychelles Magpie Robin (*Copsychus schellarum*) on Fregate Island in the Seychelles (Komdeur 1996). These nest disturbances had an adverse effect on the breeding success of the robins. Similarly, it is thought to be reducing, through competition and possibly clutch destruction, the reproductive success of the endangered Tahiti Flycatcher (*Pomarea nigra*) (Blanvillain *et al.* 2003). The Common Myna is also implicated in the decline of locally threatened hole-nesting Oriental Magpie Robin (*Copsychus saularis*) in Singapore (Huong and Sodhi 1997), and may also disturb the locally rare nesting Oriental Pied Hornbills (*Anthracoceros albirostris*) (Peh 2010). Mynas have been the main predator of Sooty Tern (*Onychoprion fuscata*) eggs on Ascension Island, responsible for 25% of nest failures (Hughes *et al.* 2008). This myna may carry bird malaria, a threat to native birds where it has been introduced (Ishtiaq *et al.* 2006; see above). It may also be a reservoir of other diseases and parasites and may mob native raptors and mammals, thus chasing them from the area (www.issg.org).

In Hawaii, the Common Myna could cause "invasion meltdown" (impacts of one invader exacerbated by interactions with another invader; Simberloff and Von Holle 1999) because it is one of the main agents helping in dispersal of the invasive weed Lantana (*Lantana camara*) (Pimentel *et al.* 2005).

In southern Asia, the Common Myna is not always considered to be a pest (Sodhi and Sharp 2006). It is to a large extent a commensal of humans and is frequently abundant around towns where it roosts communally. These roosts can cause annoyance to people through fecal droppings and noise (Yap *et al.* 2002). Their fecal dropping can facilitate the growth of fungus (e.g. *Histoplasma capsulatum*) that can cause human respiratory ailments (Yap *et al.* 2002). Yap *et al.* (2002) conducted food and tree canopy manipulation to determine the relative importance of these two factors in roost selection by Common Mynas in Singapore. The canopy of two roosts was pruned, while a 1 km^2 area within another two roosts was maintained garbage free (main food source of mynas). Yap *et al.* (2002) found that although both canopy and food reduction resulted in decrease in the number of roosting mynas, canopy reduction had a stronger effect. They recommended that existing roosting trees should be pruned, while large dense-canopy trees should not be planted. Stringent control of garbage leakage can further reduce the number of mynas.

The species can be beneficial to farmers by eating many insect pests (Parasharya *et al.* 1994), but they often cause damage to orchard fruits, oil palms, and crops such as wheat and rice (Sodhi and Sharp 2006). Because of their fruit-eating habits the species is regarded by many as being a potential pest in areas where it has been introduced (Heather and Robertson 1996).

6.4.2 Red-vented Bulbul

Originally found in south Asia (India, Sri Lanka, Myanmar, and southern China), the Red-vented Bulbul (Figure 6.12b) has now established exotic populations in the Pacific Islands including Hawaii, Fiji, and Samoa, and Middle Eastern countries such as Dubai and New Zealand (Watling 1978; Long 1981). Most of its releases are likely through accidental or deliberate release of caged birds but ships could have assisted their spread (www.issg.org).

Similar to the Common Myna, the Red-vented Bulbul is aggressive and may outcompete native species. These bulbuls have been suspected to suppress the reproductive success of endangered Tahiti Flycatchers – territories containing bulbuls had lower nest success (Blanvillain *et al.* 2003). The decline may be due interspecific aggression and/or food completion. Red-vented Bulbuls are also observed to be aggressive towards nesting monarchs (*Pomaren* spp.) in Polynesian Islands, thus contributing to their population decline (Thibault *et al.* 2002; also see above).

Like Common Myna, this species may also cause invasion meltdown as it disperses the seeds of invasive plants such as *Lantana camara* and *Miconia calvescens* (Berger 1975; Medeiros *et al.* 1997). Similarly, congeneric and invasive Red-whiskered Bulbul (*P. jocosus*) disperses the seeds of a number of introduced plants on Reunion Island (Mandon-Dalger *et al.* 2004). Red-vented Bulbuls are also considered to be agricultural pests because they prey on fruits, flowers, beans, tomatoes, and peas (Cummings *et al.* 1994; Bhatt and Kumar 2001). Along with Japanese White-eyes, they can damage up to 70% of plants of Hawaiian orchids (Orchidaceae) (Cummings *et al.* 1994). Therefore both mynas and bulbuls affect native birds and humans.

6.4.3 Avian hybridization

Some invasive bird species in tropics may hybridize with native species resulting in the production of sterile offspring and thus waste reproductive effort. The population of endemic and endangered Hawaiian duck (*Anas wyvilliana*) was initially threatened by habitat loss and degradation, but now its major threat is hybridization with the introduced Mallard Duck (*A. platyrhynchos*) (Rhymer and Simberloff 1996). Similarly in Singapore, the Red Junglefowl is threatened by hybridization with feral chicken (Lim 1992). There are also reports of interbreeding between native Yellow-vented Bulbuls (*Pycnonotus goiavier*) and introduced Sooty-headed Bulbuls (*P. aurigaster*) in Singapore (Peh 2010).

Figure 6.13 Birds being sold for prayer release in Thailand. (After Karesh *et al.* 2005.)

6.4.4 Prayer bird release

In addition to the reasons for bird release discussed above, in Asia, some people release captive animals, mostly birds, during religious ceremonies. This practice is called "prayer animal release" and is performed in countries such as Cambodia, China, Korea, Malaysia, Singapore, and Thailand (Figure 6.13; Severinghaus and Chi 1999; Karesh *et al.* 2005). Prayer animals constitute both native and exotic species and are obtained from pet shops. For example, in Thailand, millions of exotic Yellow-breasted Buntings (*Emberiza aureola*) are caught and released every year for religious ceremonies (Sodhi and Sharp 2006). Similarly in Malaysia and Singapore, thousands of exotic Scaly-breasted Munias (*Lonchura punctulata*) are released every year by Chinese temple-goers (Sodhi and Sharp 2006). Exotic birds released through prayers may be the main factor in the establishment of successful breeding populations of 28 bird species (30% of 93 released bird species) in Taiwan (Shieh *et al.* 2006). Less educated people are the main participants in prayer animal releases (Severinghaus and Chi 1999). Education programs particularly targeting this group can alleviate the release of birds (and other animals) during religious ceremonies. Commendably in Singapore, the non-governmental Nature Society is working with Buddhist temples to educate the public on the ill effects of releasing animals into the wild (Sodhi and Sharp 2006).

6.4.5 Some benefits of avian invaders in the tropics

Rarely, invasive bird species can carry on the ecosystem processes disrupted due to extinction of native species. The Hawaiian Islands have lost many of its native seed dispersers and pollinators. A vine (*Freycinetia arborea*) that was originally

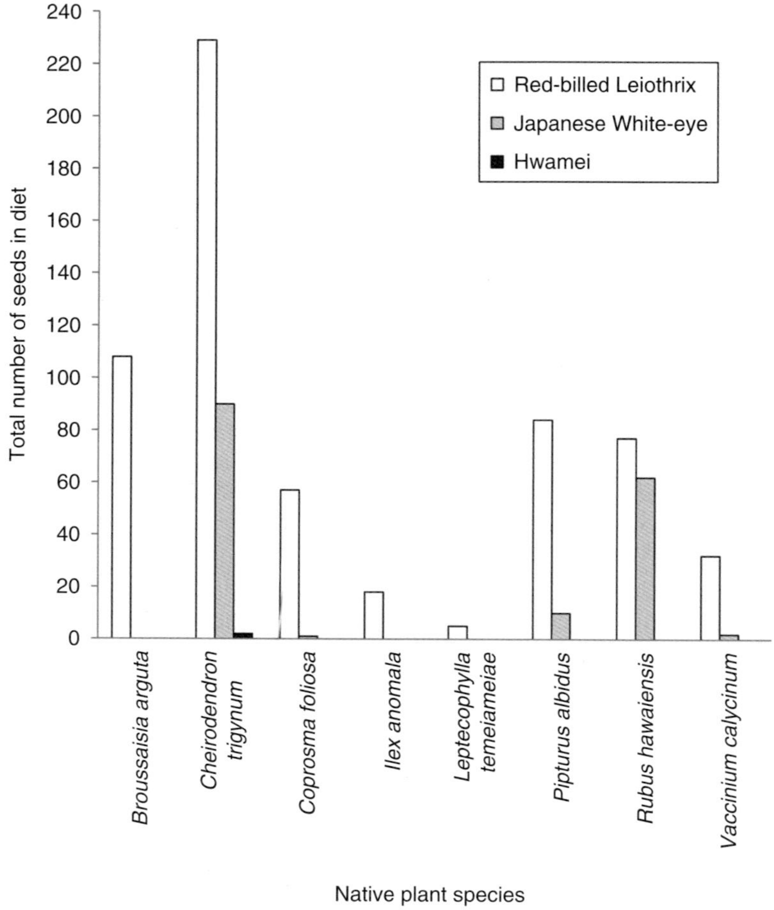

Figure 6.14 Seeds of native plant species in the diet of invasive bird species on the island of Maui (Hawaii). (Data derived from Foster and Robinson 2007.)

pollinated by an extinct bird now survives due to pollination services provided by an introduced Japanese White-eye (Cox and Elmqvist 2000). Most common understory plants in the rainforests of Hawaii now depend on seed dispersal by introduced bird species (Figure 6.14; Foster and Robinson 2007). Some of these introduced birds have dispersed the seeds of native plants even in exotic-dominated forests and thus have facilitated their reestablishment (Foster and Robinson 2007). However, a recent study by Chimera and Drake (2010) in Hawaii found that invasive birds were rarely dispersing the seeds of native trees. Nonetheless, in Hong Kong, the introduced Greater Necklaced Laughingthrush (*Garrulax pectoralis*) is now the largest-gaped fruit-eating bird capable of swallowing and dispersing certain large fruits in shrublands and secondary forests. Until this bird arrived, there was no other animal around to disperse these native fruits (Leven and Corlett 2004). Additionally, the Red-whiskered Bulbul, a pest in many other places but because it is a fruit-eater, possibly now is one of the most important seed dispersal agents in Hong Kong. In addition to helping forest regeneration through seed dispersal, bulbuls can remove insect pests in agricultural areas (Leven and Corlett 2004). Other insectivorous invasive bird species may also be providing the latter benefit. Some invasive bird species such as crows can be useful scavengers in urban areas and may keep rats populations in check (Brook *et al.* 2003b).

6.5 Management of avian invaders

As mentioned, some of the invasive bird species become crop pests, while other species are undesirable because of their communal roosting behavior or fouling of buildings and other property. Yet others compete with native birds. Therefore a need for the control and management of these birds may arise with the establishment of populations in pest proportions. Attempts have been made around the world to disperse or eradicate invasive bird species by shooting, poisoning, or scaring them (e.g. with balloons). The use of short-term control measures (e.g. bioacoustics, chemical repellents, shooting) may well be successful in certain sites or over the short-term. Problems such as bird habituation with these measures make them less useful for long-term bird control. In every invasive bird species control program, public education is essential because people can release birds or feed them (Figure 6.15), thus facilitating the establishment of these species (www.earlham.edu/~biol/hawaii/birdsframe.htm). In order to achieve invasive management success, efforts are also needed to bring together various stakeholders (e.g. agencies, scientists, and people). A good case in point is an initiative in Florida Everglades to mitigate the invasive species, including birds (www.evergladescisma.org/species/birds.cfm). Internationally, the Global Invasive Species Programme (GISP) is also established to address the threat posed by

Figure 6.15 A poster warning residents that there is a fine for feeding birds in Hong Kong (China). Feeding by public is prevented so that populations of invasive birds do not increase due to excessive food availability. (Photograph by Navjot Sodhi.)

invasive species (www.gisp.org/about/index.asp). GISP aims to spread awareness on invasive species.

Below we discuss some of the measures that can be used to mitigate invasive bird species. Populations of invasive bird species can be reduced directly through removal (e.g. killing). However, direct methods to reduce bird populations are usually not effective in the long term. A common reason is that equilibrium in number of birds is reached when control mortality is balanced by recruitment and juveniles, either by reproduction or by immigration of juveniles from other populations, taking the places of those removed by the control program (Murton *et al.* 1972; Martin and Martin 1982; Haag-Wackernagel 1993). Further, the impacts of some methods (e.g. poisons, baits, and explosives) on the ecosystem need to be an adequate consideration in the control programs (see above), as some of the baits can harm non-target species. Use of guns is common in rural environments, but not always feasible in cities, which is why urban bird control is generally conducted through trapping or netting the birds, which are expensive, hands-on activities. Chemical products that either kill or stupefy the birds are likewise fractionally effective in most populations, being dose-dependent and dependent on the birds contacting toxic perches or feeding on toxic or stupefying baits. Detailed knowledge of the birds' biology is critical when using chemicals. Some repellents are toxic in excess and may kill birds that were intended to be merely stupefied. Toxic chemicals may also not kill immediately, but incapacitate the birds for some time prior to death. Dead or dying birds and the toxic substances themselves can be threats to the public health because children and household pets can contact them (Johnston and Janiga 1995). The dying birds can pose a public relations problem for the wildlife managers (Weber 1979). Further, effects of reduction or elimination of the pest on the community of animals, such as the replacement of the target by another (and possibly worse) pest species should be given thought (Dyer and Ward 1977). Use of sounds (e.g. playback of distress calls, periodic explosions from acetylene cannons) and scarecrow dummies (e.g. plastic owls, snakes) are known to be effective in scaring away the Feral Pigeon (*Columba livia*), White-vented Myna (*A. javanicus*), and Common Mynas for short periods of time, but birds habituate to such stimuli within a few days, and their use is ultimately unsatisfactory (Johnston and Janiga 1995).

Other methods such as chemical sterilization have been tried on invasive birds. Drugs such as nicarbazin and diazacon (Ornithol) have been used to reduce the reproductive success of pigeons, sparrows, and starlings (Sodhi and Sharp 2006). Similarly, chemical repellents are also used to protect crops from birds (see, for example, Cummings *et al.* 1994). The problems with field use of chemicals lie in the sometimes unknown efficacy of the chemical, difficulty in the delivery of the chemicals to the birds, and in the lack of species-specificity (Dyer and Ward 1977). Some chemicals may have side effects for birds, such as tumors (see Chapter 1).

Habitat manipulation has long been espoused as the proper way to manage wildlife species (Leopold 1939). Comprehensive knowledge about the biology of birds (e.g. the feeding, nesting, breeding, roosting requirements, and behavior) is needed for an effective habitat control program (Johnston and Janiga 1995). The type of information required would depend on the nature of the disturbance

Figure 6.16 The House Crow. (Photograph by Muhammad Mahdi Karim.) (Please see Plate 6.16 for the color version of this figure.)

caused by the invasive species. Habitat modification can take many forms, such as exclusion (e.g. blocking entrances to holes to prevent roosting or nesting, and installing netting to protect high-value crops), agricultural and horticultural methods (e.g. removing roost sites and planting bird resistant varieties of crops), and food removal (e.g. restricted feeding) (Fitwater 1994; Johnston and Janiga 1995; Yap *et al.* 2002). Currently, many wildlife management programs have shifted their emphasis towards ecological approaches to bird control, based on understanding the ecology of invasive bird species, especially their habitat and food requirements. Once the ecological knowledge is sound, the hope is that long-term control will be possible. Integrating sound science and habitat management into the long-term control of invasive bird species is a good wildlife management strategy and should be practiced.

However, avian invasive species remain poorly studied (Blackburn *et al.* 2009; Peh 2010). The biological knowledge may not be adequate to mount successful control of various invasive species. Below, we describe how integrating science and habitat management has been used to achieve successful long-term control of an invasive bird species, the House Crow. This species is widespread and considered as a noxious pest in Asia and Africa (Sodhi and Sharp 2006). This aggressive and gregarious black bird, weighing between 245 and 371 g (Figure 6.16), lives in close association with people, with an original range spanning southern Iran, Pakistan, India, Maynmar, and Sri Lanka. Its diet is varied and includes invertebrates, fruit, carrion, eggs, young birds, and human refuse. The latter is the principal food source in urban areas (Lim *et al.* 2003). The House Crow is capable of breeding throughout the year, though it favors the hotter and drier months. It can nest on a variety of trees and human-made structures (e.g. electricity poles) (Soh *et al.* 2002). The behavioral and life history flexibility predispose the House Crow to successful invasions of new areas (see above).

The House Crow has invaded throughout East and Southeast Asia (China, Malaysia, Singapore), Africa (South Africa, Kenya, Tanzania, Egypt), the Middle East (Oman, Yemen), and various oceanic islands over the past two centuries. This has been a result both of inadvertent introductions associated with increased global sea traffic and deliberate introductions (Sodhi and Sharp 2006). In the late 19th century it was introduced from Sri Lanka to Peninsular Malaysia as a possible biological control agent for caterpillars in coffee plantations. This was not a well-thought-out move as it did not, unfortunately, have the desired effects, and the crow soon spread to neighboring urban areas, and to Singapore in the late 1930s or early 1940s (Sodhi and Sharp 2006). There were only about 200–400 crows in Singapore in the late 1960s. However, the crow population expanded to over 4500 by the mid-1980s despite moderate, but consistent levels of culling since 1973 (Brook *et al.* 2003b). Representing a 30-fold increase, the crow population had exploded to over 130,000 by 2000 (Brook *et al.* 2003b).

This population density of the House Crow in Singapore was sufficiently high for it to be considered a major urban pest. For instance, large communal roosts of up to 3000 birds occurred close to residential areas, creating high levels of noise pollution and an accumulation of feces on vehicles and human property (Peh and Sodhi 2002). Nesting birds aggressively defended their nests, leading to a number of attacks reported on people. There was also human health risk because the House Crow is a possible vector of human pathogens such as *Cholera*, *Salmonella*, *Giardia*, *Enteroamoeba*, *Shigella*, and *Escherichia*. Additionally, the aggressive behavior of crows likely jeopardized native bird species through predation and/or competition (Brook *et al.* 2003b).

Previous attempts using poisoning, shooting, and nest destruction to control House Crow populations have largely resulted in failures, a copious outcome in invasive bird control. A germane example of the large-scale but ultimately unsuccessful House Crow eradication attempt comes from Aden (Yemen). In the late 1980s, over 240,000 birds were killed by poisoning during a 2-year government-funded program. However, a lack of comparable removal effort in neighboring Lahej and Abiyan meant that the population recovered through immigration (Jennings 1992). A similar management stalemate persisted in Singapore where despite efforts by the government to suppress the population through shooting, the House Crow population grew over the past 30 years (Figure 6.17; Brook *et al.* 2003b). Crow culling through shooting has also been unsuccessful in Malaysia (Sodhi and Sharp 2006). Common to all these failed management initiatives is a paucity of scientific knowledge (e.g. population size, reproductive success, immigration rate).

The bird management agency in Singapore recognized these problems and funded an initiative to study the autecology of the House Crows with a view to providing scientifically grounded guidelines to reduce the population. Regular population size and roost surveys, dissections of birds shot (to provide age structure and breeding status), detailed nest site observations, and monitoring of coastal dispersal were conducted over 2 years (2000–01) to determine the population ecology and behavior of the House Crow in Singapore. This information was used to develop a density-dependent population model to guide the culling rate (Brook *et al.* 2003b). The model suggested that at least 41,000 crows needed to be culled in the first year of a control program, and an equivalent effort maintained each year thereafter for 10 years so as to reduce the House Crow population from its 2001 density of 190 birds per km^2 to the reasonable management target of < 10 birds per km^2 (Brook *et al.* 2003b). The model also indicated that

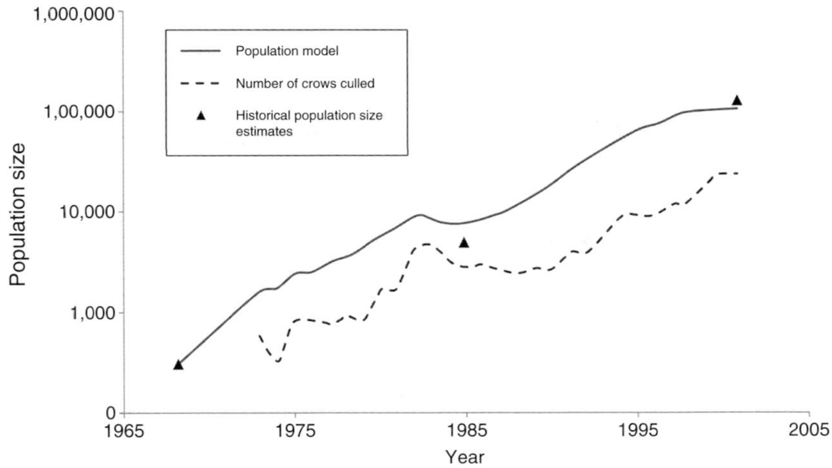

Figure 6.17 Logarithmic plot of the density dependent model (solid line) for the House Crow on Singapore Island in relation to historical population size estimates (triangles) and past culling rate (dashed line). (After Brook *et al.* 2003b.)

32,000 crows need to be removed if culling is combined with other management options such as resource limitation (e.g. roost tree removal or pruning) and the destruction of active nests. However, total eradication of the House Crow from Singapore may be unrealistic due to the difficulties of detection at low population and immigration from neighboring Malaysia. Shooting combined with the removal of active nests have suppressed the crow population to 10,000 in 2007 in Singapore (i.e. 7% of the population in 2001) (news.nationalgeographic.com/news/2007/01/070108-crows.html), and is possibly much lower today. Thus the success of House Crow management in Singapore illustrates the importance of integrating scientific knowledge in invasive species control.

7

Harvesting of Tropical Birds

Overview: An estimated 5–10 million wild birds are caught annually in developing countries for export to developed countries. Almost all parrot nests are poached from some tropical locations. Birds may not form a major component of bushmeat trade but many tropical countries have a tradition of keeping and trading birds. Harvesting jeopardizes the populations of numerous tropical birds such as parrots and hornbills. It is theoretically possible to estimate the sustainable harvesting rate for birds. To curtail bird extraction, bans on the import of wild-caught birds seem to alleviate bird captures. Assisting local communities in finding alternative livelihoods and investing in captive breeding programs may alleviate the hunting pressure on tropical bird populations.

In this chapter, we discuss the utilization of tropical birds by humans. First, we highlight the extent of extraction of wild birds. Second, we discuss some of the reasons why tropical birds are harvested. We also discuss the effects of harvesting on tropical birds and sustainable harvesting. Finally, we list some of the ways to mitigate bird trade and harvesting.

7.1 The volume of bird extraction

Globally, each year, 350 million live animals worth US$20 billion are traded, about a quarter illegally (Karesh *et al.* 2007). Included in this trade is between 5 and 10 million wild birds that are caught annually mainly in the developing

Conservation of Tropical Birds, First Edition. Navjot S. Sodhi, Çağan H. Şekercioğlu, Jos Barlow and Scott K. Robinson.

Published 2011 by Blackwell Publishing Ltd.

Figure 7.1 Traded birds are frequently kept in overcrowded and inhumane conditions. (Photograph by Navjot Sodhi.) (Please see Plate 7.1 for the color version of this figure.)

countries for export to developed countries (Gilardi 2006). Traded birds are frequently kept in overcrowded and inhumane conditions (Figure 7.1). Raising fears of increased harvesting pressure on wild bird populations, the demand for pets is also increasing in the developing world (Carrete and Tella 2008). For example, in the past decade, the import of parrots has increased by eightfold in Mexico (Cantaú *et al.* 2007; also see below).

Overall, about 42% of extant bird species (4173 of 9856 species) are utilized by humans (Butchart 2008). Most commonly, birds are used as pets or hunted for food (Figure 7.2; Butchart 2008). Other uses of birds include harvesting of feathers for bedding (e.g. Common Eider, *Somateria mollissima*), use as ornaments (e.g. feathers of Papuan Hornbill, *Aceros plicatus*, are used in headdresses, its bill being valued as a personal adornment, and the lower mandible is used in making spears), oil for fuel (e.g. Oilbird, *Steatornis caripensis*), skin for leather (Greater Rhea, *Rhea americana*), and use in traditional medicine (e.g. Hispaniolan Lizard-cuckoo, *Coccyzus longirostris*, is eaten to possibly cure indigestion) (Butchart 2008). Approximately 34% of extant bird species (3337) are traded internationally, mostly for the pet trade (Butchart 2008).

The high volume of trade of tropical birds is epitomized by Southeast Asia. One million birds were exported from Southeast Asia between 1998 and 2007; 269,000 of these were captured from the wild (Figure 7.3; Nijman 2010). The most commonly traded species was the Hill Myna (*Gracula religiosa*). The main exporters were China, Vietnam, and Malaysia, whereas the importers were the European Union (EU), Japan, and Malaysia (Figure 7.4; Nijman 2010). Malaysia apparently was re-exporting some of the imported birds. The Southeast Asian bird exports came almost to a complete halt in 2005 likely due to the EU

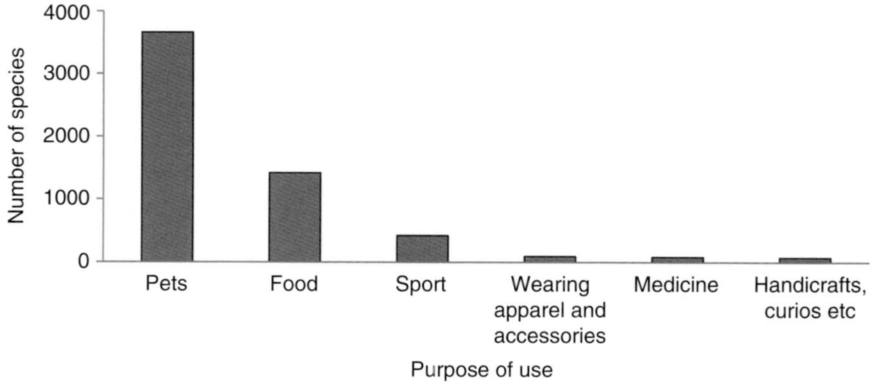

Figure 7.2 Number of bird species recorded for different purposes of use. (After Butchart 2008.)

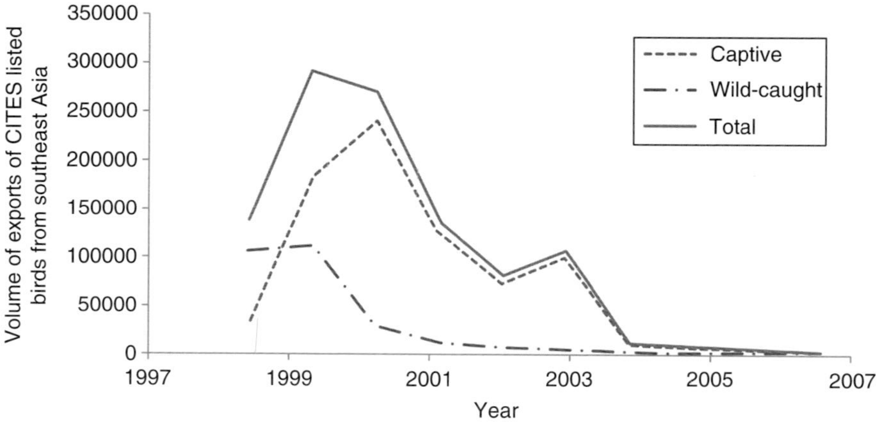

Figure 7.3 Volumes of exports of CITES (Convention on International Trade in Endangered Species of Wild Fauna and Flora) listed birds from Southeast Asia in the period 1998–2007. (After Nijman 2010.)

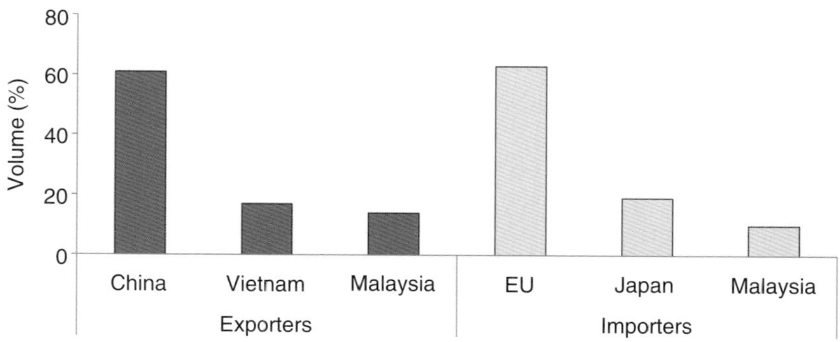

Figure 7.4 Major exporting and importing region/countries of CITES listed Southeast Asian birds. For definition of CITES see Figure 7.3. EU = European Union. (Data derived from Nijman 2010.)

restricting imports in response to avian influenza (H5N1) threat (Nijman 2010; see below). The traded bird numbers reported by Nijman (2010) are based on the Convention on International Trade in Endangered Species (CITES) of Wild Flora and Fauna data. This means, that these numbers are a gross underestimation of the birds captured as they do not include non-CITES species, domestic and illegal bird trade. Further, pre-export mortality in birds can be as high as 60% (Iñigo-Elias and Ramos 1991), and these number generally remain unrecorded in bird trade.

CITES was initiated in 1973 and has been signed by 175 countries to ensure that international trade does not harm wild species. CITES has three appendices listing 33,000 species (www.cites.org). CITES Appendix I lists species that are threatened by extinction and may be affected by trade. Trade of these species is not permitted other than in exceptional circumstances. CITES II lists species not currently threatened by extinction but may be with heavy exploitation. Trade of these species is permitted within predetermined annual quotas. CITES III lists species that are of national conservation concern; trade of these is permitted with the proof of origin certificate.

Parrots are frequently caught in tropical countries for bird trade. Between 1991 and 1996, 1.2 million parrots were traded, the majority originating from the Neotropics (Beissinger 2001). This number is also considered as a gross underestimation of the number taken from the wild because it does not include domestic and illegal trade. More realistically, between 400,000 and 800,000 parrots annually may be captured from the wild for the trade (Beissinger 2001). Thirty-six percent of the world's parrots are threatened, with trade impacting 55% of these (Pain *et al.* 2006). Trapping for trade was possibly the main cause of the recent likely extinction of the Spix's Macaw (*Cyanopsitta spixii*) from the wild (Pain *et al.* 2006; see Chapters 1, 3, and Figure 3.1). Parrot populations are especially vulnerable to exploitation because of low reproductive rates, late age of first reproduction, and breeding population limitation caused by lack of suitable nesting holes (Wright *et al.* 2001). Prospects for captive breeding parrots are also low because it is an expensive business proposition with limited success. High demand for parrots makes their nests vulnerable to poaching. On average, nest poaching occurred in 30% of the parrot nests in the Neotropics (Figure 7.5; Wright *et al.* 2001). Additionally, nest poaching was the most important cause of nesting failure. Similarly, 0–100% of parrot nests were poached in Africa, Asia, and Australia (Figure 7.6; Pain *et al.* 2006). These high numbers indicate that poaching definitely is an enormous threat to the populations of wild parrots.

Like parrots, hornbill populations are susceptible to unsustainable extraction because of the low reproductive output (Trail 2007). Hornbills are also critical for forest renewal because of their ability to disperse the seeds of primary forest trees (Hamann and Curio 1999; see Chapters 3 and 4). Hunting is a significant threat to four large-sized West African hornbills – the Brown-cheeked Hornbill (*Bycanistes cylindricus*), the Black-casqued Hornbill (*Ceratogymna atrata*), the Black-and-white-casqued Hornbill (*B. subcylindricus*), and the Yellow-casqued Hornbill (*C. elata*) (Kemp 2001). Despite this, 434 hornbill items (e.g. live birds, carcasses, body parts) were legally imported into the United States between 1999 and 2006 (Trail 2007). Of 219 live hornbills imported, only

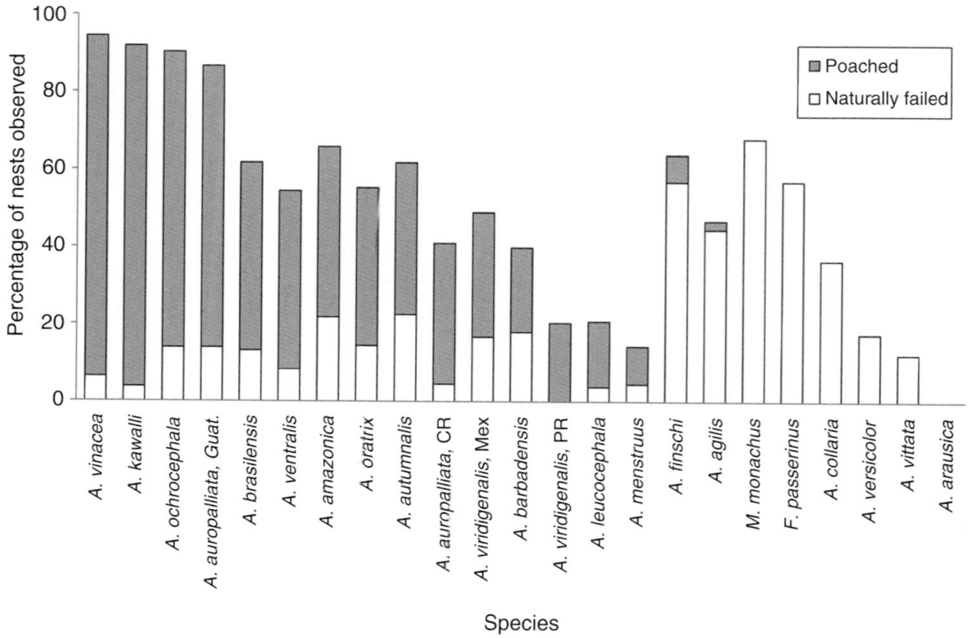

Figure 7.5 Nest mortality in the 23 studies used in the analysis of poaching rates of Neotropical parrots. Two species appear twice and they are identified by the country in which they were studied: CR = Costa Rica and Guat. = Guatemala; Mex. = Mexico and PR = Puerto Rico. (After Wright *et al.* 2001.)

< 5% were declared to be captive bred. The most important hornbill-exporting country was Zimbabwe (Figure 7.7; Trail 2007). It is likely that several undeclared, mostly illegal, importations of hornbills into the United States also occurred during this period (Trail 2007). Declining hornbill populations due to hunting and habitat loss dictate that their international as well as domestic trade needs to be curtailed (Trail 2007).

7.2 Utilization of birds

Below we briefly discuss some of the reasons why birds are captured or hunted by humans.

7.2.1 Birds as bushmeat

Bushmeat (meat of wild animals) may be an important protein source for poor rural people in the tropics (Milner-Gulland and Bennett 2003). In some tropical areas, as much as 70% of meat needs are met by bushmeat (Olupot *et al.* 2009). However, there is a growing realization that unsustainable bushmeat harvesting may be threatening wildlife populations and also jeopardizing the food security of numerous poverty-stricken tropical people, thus culminating in a "bushmeat crisis" (Milner-Gulland and Bennett 2003; Sodhi *et al.* 2007). Nonetheless, birds

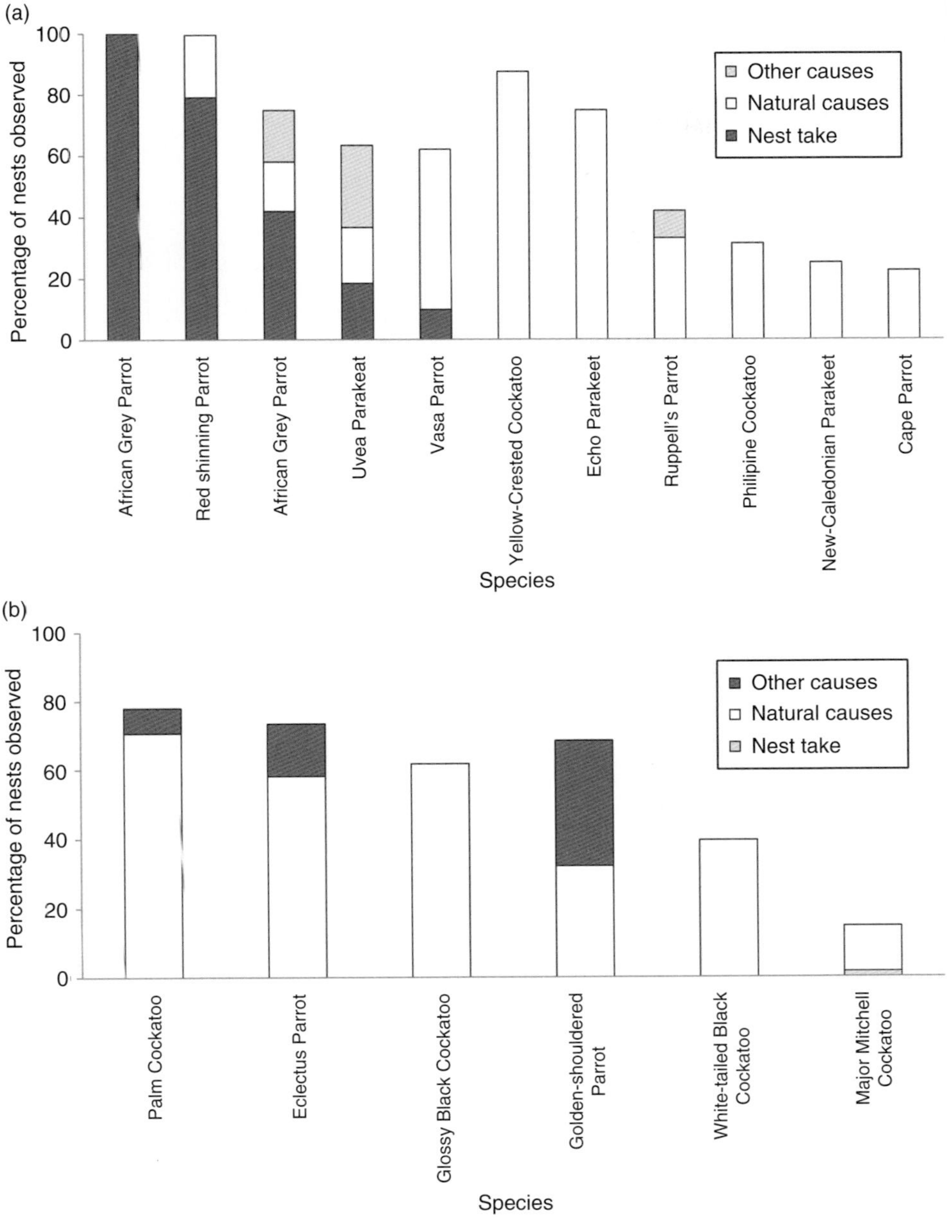

Figure 7.6 Causes of nest mortality in tropical parrots in (a) countries excluding Australia and (b) Australia. (After Pain *et al*. 2006.)

make up only a small proportion of bushmeat trade but it is likely that they are more often consumed by hunters and their families than being sold (Fa *et al*. 2000; Olupot *et al*. 2009). The low level of bushmeat trade of birds is exemplified by a market survey conducted in Bioko Island (Equatorial Guinea) – birds made up only < 1% of bushmeat sold there (Fa *et al*. 2000). Similarly, in a 35,000 km^2

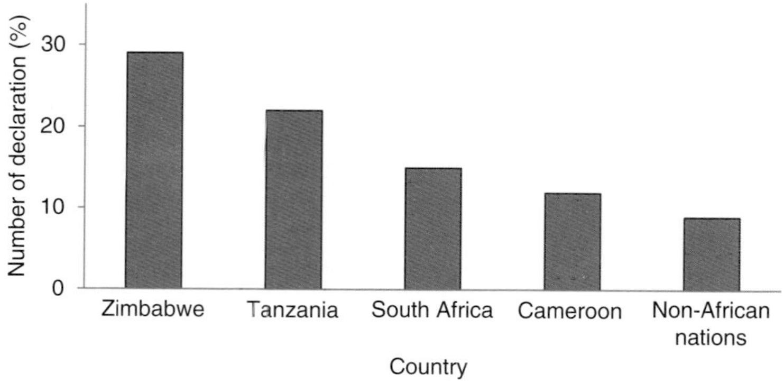

Figure 7.7 Major importers of African hornbill material to the USA. (Data derived from Trail 2007.)

area of the Cross-Sanaga Region of Nigeria and Cameroon, mammals represented >90% of the carcasses sold (Fa *et al.* 2006). However, an estimated 7603 bird individuals were annually extracted for meat in this region, with the Helmeted Guineafowl (*Numida meleagris*) being the most common prey (Figure 7.8; Fa *et al.* 2006). Even this seemingly low level of hunting can likely jeopardize the populations of slow breeding bird species such as hornbills (Trail 2007).

To illustrate the plight of birds hunted for food, we provide two examples below. Endemic to the islands of Sulawesi and Buton (Indonesia), the charismatic Maleo (*Macrocephalon maleo*) faces an imminent risk of extinction due to overharvesting. Maleos feed on insects, seeds and legumes on the ground in lowland forests (< 1200 m elevation). Weighing about 1.6 kg, the Maleo is a terrestrial megapode (Figure 7.9) that has a unique nesting behavior. Up to 100 pairs nest communally at traditional sites around geothermally heated soils in forests or in solar-heated sandy beaches or riverbeds. Using their powerful feet, adult birds excavate their individual burrows that are usually 60–70 cm deep. Each burrow contains one egg, with the depth at which it is laid dependent upon the surrounding soil temperature that ranges between 30 and 35°C. Eggs are covered with sand and are left to be incubated by solar or geothermal sources. Adults do not provide any parental care. Young are able to dig their way to the surface and can feed independently almost immediately after emergence.

Breeding between October and April, a pair of Maleo can lay up to 12 eggs in a year (Prawiradilaga 1997). At about 250 g each and with high yolk contents, Maleo eggs (16% of a female's body mass) are five times heavier than chicken eggs. This concentration of highly nutritious protein has long attracted humans to Maleo nesting grounds. Initially, local people used to collect Maleo eggs for traditional ceremonies. However, this situation has now changed, as Maleo eggs are considered both delicious and nutritious throughout Indonesia. In some local markets, Maleo eggs are sold for between Rupiah 2500 and 5000 (<US$0.60) (Sodhi and Brook 2006). Despite being listed on the CITES Appendix I and protected under Indonesian law, uncontrolled egg-collecting by the burgeoning human population of Sulawesi and Buton has led to declines in the Maleo population and abandonment of many traditional nesting sites (Gorog *et al.* 2005). Of 131 known nesting

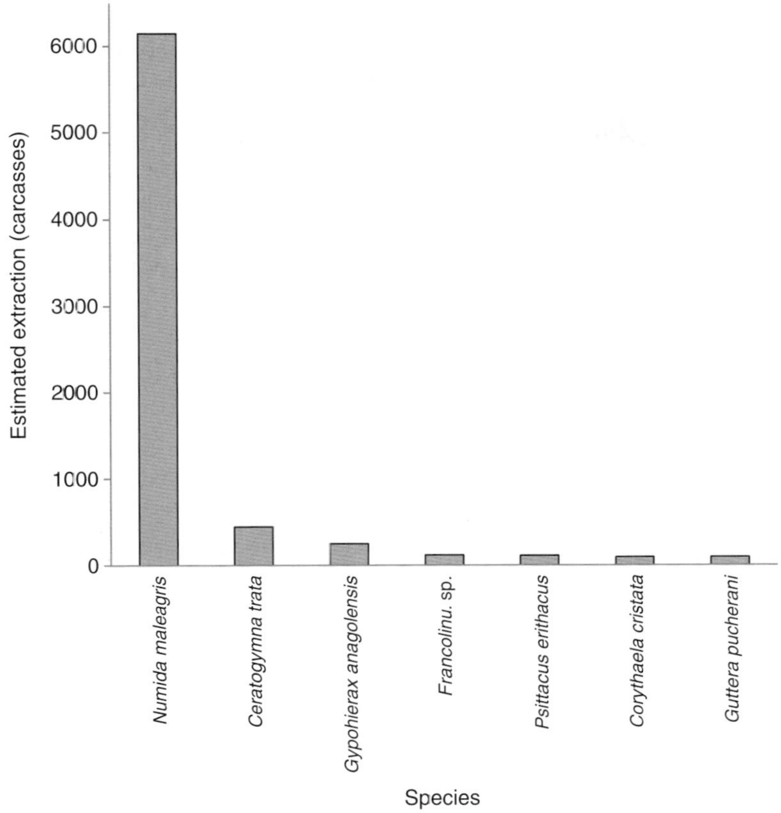

Figure 7.8 Estimated extraction of birds per year in the Cross-Sanaga rivers region, Nigeria, and Cameroon (Data derived from Fa *et al.* 2006.)

Figure 7.9 The Maleo. (Photograph by Scott Newell.) (Please see Plate 7.9 for the color version of this figure.)

sites, 42 have already been abandoned in recent times, with another 38 severely threatened, 12 of unknown status, and only five (4%) not yet considered threatened (Butchart and Baker 2000). The global population of the Maleo has been currently estimated to be between 4000 and 7000 breeding pairs (Sodhi and Brook 2006). However, this population is declining rapidly, with 90% declines in some areas since the 1950s (Argeloo and Dekker 1996).

In addition to heavy human exploitation, Maleos are also losing foraging habitats due to extensive deforestation (Whitten *et al.* 1987; see also Chapter 1). Deforestation around the nesting site increases the chances of its abandonment (Gorog *et al.* 2005). Clearly, the future of this species is in extreme jeopardy. There is a dire need to better understand the ecology (e.g. recruitment and site fidelity) of this unique bird, and the status of this species needs to be evaluated throughout its range with data collected over multiple years, in order to precisely pin-point those areas experiencing population declines. To halt the further population decline of this species, actions are needed immediately. Although over 50% of the known Maleo nesting sites are protected (Dekker 1990), such protection is inadequate and poorly enforced. Adequate protection can help in the survival of nesting sites (Gorog *et al.* 2005). There are numerous Maleo captive breeding programs in central Sulawesi (Christy 2002), but their success will be limited without the assurance that the released birds have a reasonable chance of surviving and reproducing in the wild. To achieve tangible conservation outcomes, social aspects need to be integrated fully into conservation plans (see Chapter 10). For instance, there is an urgent need to develop and promote sustainable egg-harvesting regimes with the full involvement of the local residents. Local communities should be assisted in exploiting alternative food sources, as has been done for a community in Pakuli (central Sulawesi). Here, locals are encouraged to rely on an invasive fish (*Osteochilus* sp.) so as to reduce their harvest of Maleo eggs (Sodhi and Brook 2006). This community is a custodian of their local Maleo nesting grounds. In our minds, similar approaches adopted across Sulawesi will work best to ensure the survival of maleos, and more generally, other exploited bird species.

Similarly, swiftlets in Southeast Asia provide another quintessential example of overexploitation. Three swiflet species are exploited commercially in Southeast Asia: the Edible-nest Swiftlet (*Collocalia fuciphaga*), Germain's Swiftlet (*C. germani*) and the Black-nest swiftlet (*C. maxima*). These birds nest both in coastal and inland caves. Male swiftlets weave nests from thin gelatinous strands secreted from their salivary glands. These strands are formed into a half-cup nest that is glued to the inside of a cave wall. For centuries, the Chinese have eaten the nests of swiftlets as a delicacy for soups or in jelly mixed with spices or sweets (Figure 7.10). It is believed traditionally that the nests contain aphrodisiac properties (Sodhi and Brook 2006). Chemically, however, swiftlet nests have been made mostly of proteins (50–60%; Kong *et al.* 1987). While there is glycoprotein in them that can promote cell division in the immune system, it may be lost during the cleaning process prior to eating, meaning that the nests probably have minimal medicinal properties. Nevertheless, the demand for swiftlet nests has soared over the past 30 years (Sodhi and Er 2000). Today, a top-quality swiftlet nest can sell for at least US$ 2500–4000 per kg. Indonesia is the major supplier

Figure 7.10 The bird nest soup. (Photograph by Stu Spivack.) (Please see Plate 7.10 for the color version of this figure.)

Figure 7.11 A house used to harbor a swiflet colony. (Photograph by Harry Kok.)

of swiftlet nests; other exporting countries include Thailand, Vietnam, Singapore, and Myanmar. To satisfy the increasing international appetite for the swiftlet nests, some people in Indonesia began to farm swiftlets in the early 1990s. Swiftlet farmers buy old houses with colonies of uniform swiftlets (*C. venikorensis*), and use them to cross-foster the eggs of Edible-nest Swiftlet (Figure 7.11). It is claimed that one-third of swiftlets exported from Indonesia now come from farmed swiftlets, but this claim is difficult to verify (Sodhi and Brook 2006). There is also the danger that inter-species hybridization in swiftlet farming will have spillover effects to wild populations.

Nest collection has been a traditional occupation in many Southeast Asian communities (Hobbs 2004). However, because of its high price, swiftlet nest collection is also a lucrative avenue of income for poachers. Primarily because of overexploitation, swiftlet populations are now under intense pressure. While concrete data are lacking, studies suggest that many swiftlet populations have been severely depleted, such as those in the Niah Caves of Sarawak (Indonesia) (Er *et al.* 1995). As local people in some communities rely upon swiftlet nest harvesting, declining yields can erode livelihoods (Valli and Summers 1990).

There have been some studies that attempt to evaluate quantitatively the question of how nest harvests might reduce reproductive success and subsequent recruitment (Kang *et al.* 1991; Tompkins 1999). Good harvest management can certainly result in stable or increasing populations of swiftlets, as illustrated by a study from south-central Vietnam (Casellini *et al.* 1999). Here, strict, sustainable nest-harvesting by a state-owned company has resulted in an increase in nest production of 3% per annum (Casellini *et al.* 1999). The nests are harvested in two phases. The first phase occurs when 10–15% of nests have eggs, as early harvesting means that most pairs can initiate a new nest. The second phase occurs after 160 days, when most nestlings have fledged. This is an excellent example of sustainable harvesting, as it minimizes the disruptive effects on the swiftlet population, whilst at the same time ensuring a high economic yield is delivered from the wildlife resource. Sodhi and Er (2000) recommended that governments, in partnership with industry, should work together towards the sustainable management of swiftlets in Southeast Asia. Better understanding of swiftlet taxonomy and ecology are needed to devise the most appropriate sustainable-harvest models (Hobbs 2004; see below). Social issues (e.g. alternative sources of income) may also need to be considered if large-scale sustainable harvesting of these imperiled species is to be achieved (see below). Nevertheless, both the Maleo and swiftlet examples illustrate that there are ways to mitigate the overexploitation of tropical birds.

7.2.2 Birds for cage

People in many tropical countries have a tradition of keeping and trading wild-caught birds (Jepson and Ladle 2005; Homberger and Beissinger 2006). These birds are kept as pets or are released to gain religious benefits (Homberger and Beissinger 2006; Brooks-Moizer *et al.* 2009; see Chapter 6). There are other benefits of keeping pet birds. Health-care professionals believe that pets may assist people in leading healthier and happier lives (Carrete and Tella 2006). Pets may also promote a positive attitude towards the protection of wild animals (Jepson and Ladle 2009).

Indonesia provides a good illustration of the magnitude and reasons for keeping pet birds. Pet trade in Indonesia is worth millions of dollars, most of which is illegal (Shepherd 2006; Jepson *et al.* 2008). Illegally obtained caged birds may act

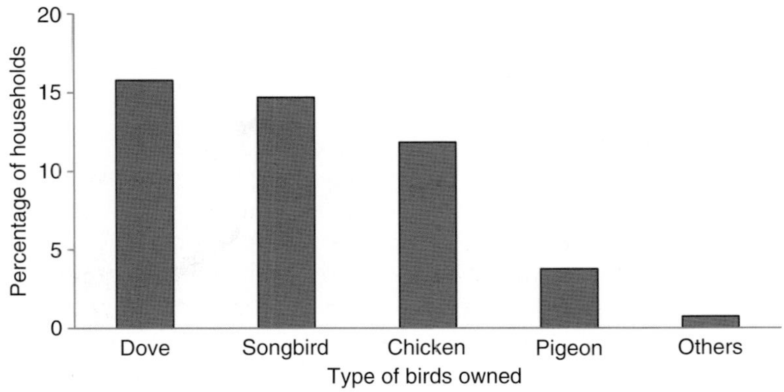

Figure 7.12 Percentage of households in six cities of Java and Bali keeping different types of bird pets. (After Jepson and Ladle 2009.)

as a symbol of prestige, affluence, and power in some countries such as Indonesia (www.indonesian-parrot-project.org/pdf_files/AAV_07_6–29–07.PDF). In Java and Bali, about 36% of households keep pet birds (Figure 7.12; Jepson and Ladle 2009). These include an estimated 2 million birds, only 58% of which were wild-caught. Indicating an increasing threat to bird populations, the wild-caught birds in Java increased from 738,518 individuals in 1999 to 1,086,692 individuals in 2006 (Jepson and Ladle 2009). The most popular native songbird pets were the Long-tailed Shrike (*Lanius schach*), Orange-headed Thrush (*Zoothera citrina*), and White-rumped Shama (*Copsychus malabaricus*). Household income was not significantly correlated with bird-keeping in Indonesia, suggesting that both poor and rich people keep pet birds (Jepson and Ladle 2009).

Overall, undoubtedly, a culture of keeping and trading wild birds is jeopardizing Indonesian birds, especially now with an expanding human population. The Bali Starling (*Leucopsar rothschildi*) epitomizes how overharvesting can endanger a bird species. With an estimated 24 individuals remaining in the wild, the Bali Starling is at the brink of extinction in Indonesia largely because of overharvesting (Figure 7.13; www.birdlife.org/datazone/species/index.html?action=SpcHTMDetails.asp&sid=6822&m=0). Because of high black-market price (US$ 2000 in 1990s), in 1999 an armed gang stole from a national park almost all the 39 captive starlings awaiting release into the wild (www.birdlife.org/datazone/species/index.html?action=SpcHTMDetails.asp&sid=6822&m=0).

To alleviate the overharvesting of wild birds, Jespon and Ladle (2009) recommended that the supply of captive birds should be increased and a campaign should be undertaken to encourage the populous to buy captive-bred birds.

7.2.3 Recreational bird hunting

People in many tropical countries have been traditionally hunting birds for recreation. For example, as a cultural practice, an estimated 500 million migratory birds are killed every year in the Mediterranean region and Northern Africa (Aaltola and

Figure 7.13 The Bali Starling. (Photograph by Tommy Tan.) (Please see Plate 7.13 for the color version of this figure.)

Oksanen 2002; www.birdlife.org). However, bird hunting tourism is also now being promoted by some tropical countries such as Mexico, Uruguay, and Argentina (birdhuntingmexico.com; www.losauchos.com). Overall, the effects of recreational hunting on tropical birds remain poorly understood, but as mentioned, some of the slow-breeding bird species may remain vulnerable. Loss of habitat and hunting may synergistically act to reduce tropical bird populations (see Chapter 1). Commendably, efforts are now being made in some tropical countries to prevent the hunting of migratory birds (us.oneworld.net/node/72981).

7.3 Effects of harvesting on birds

Indicating that hunting can culminate in bird losses, hunting by humans during prehistoric times has been the main cause of bird extinctions from some tropical islands (Duncan *et al.* 2002). Currently, bird species used by humans are less threatened than not being used (Figure 7.14; Butchart 2008). This may be because most of the bird species used by humans are common (Butchart 2008) or that harvesting of uncommon birds remains unreported. For instance, rare bird species remain vulnerable to human exploitation because they fetch high prices (Brook and Sodhi 2006; see above). Nonetheless, the utilized, as well as not-utilized species are increasingly being threatened with extinction (Figure 7.14).

There is evidence that some tropical bird species may suffer population decline due to harvesting. For example, unsustainable human hunting may be the main cause of decline in Pacific Pigeon (*Ducula pacifica*) between 1994 and 2004 on Niue, a South Pacific Island (Powlesland *et al.* 2008). Similarly, suggesting that hunting may depress populations, in French Guiana, the abundance of all game birds was lower in hunted than in non-hunted areas (Figure 7.15; Thiollay 2005). Hunted-

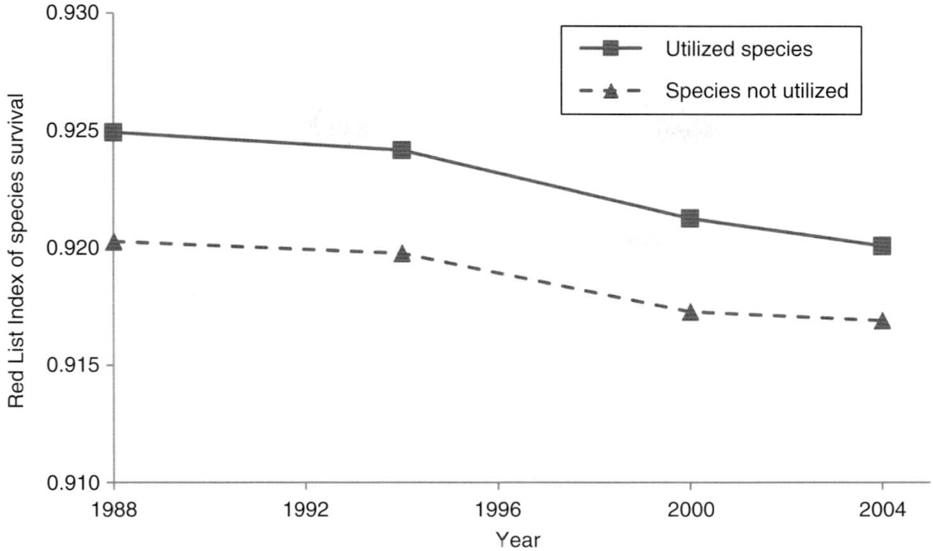

Figure 7.14 Red List Index for utilized and bird species not utilized by humans. The figure indicates that both these groups are becoming threatened with time. For definition of Red List Index see Figure 1.2. (After Butchart 2008.)

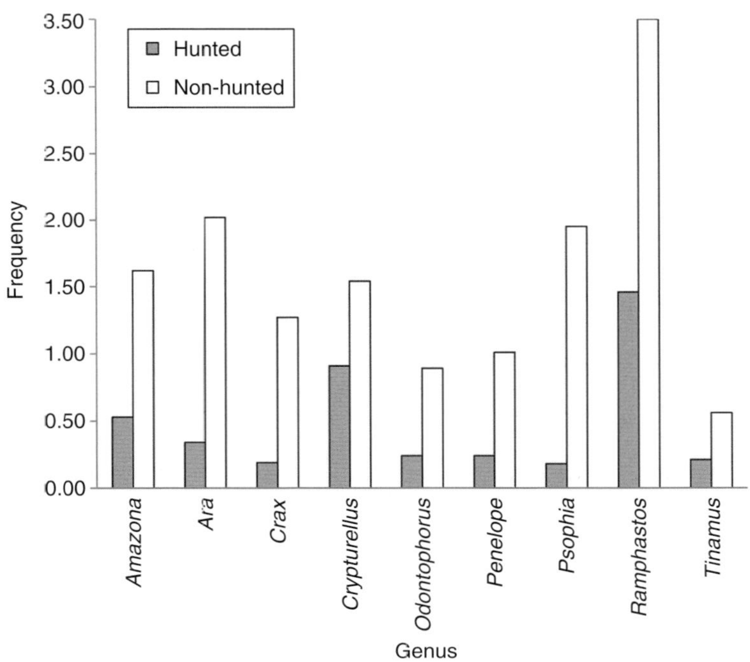

Figure 7.15 Mean frequency (number of birds recorded/10 hours) of different game bird genera in hunted and non-hunted areas of French Guinea. (Data derived from Thiollay 2005.)

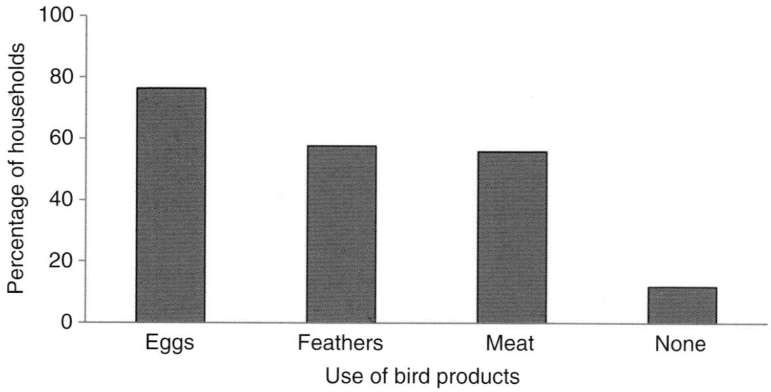

Figure 7.16 The use of ostrich products by people in Tanzania (Africa). (Data derived from Magige *et al.* 2009.)

mediated population declines may impact ecosystems. For example, hunting of frugivorous birds such as toucans in Eucador may reduce seed dispersal of a Neotropical tree (*Virola flexuosa*) (Holbrook and Loiselle 2009; see also Chapters 3 and 4).

Hunting can also alter bird behavior. In Tanzania, local people hunt birds, including Ostrichs (*Struthio camelus*), for various products (Figure 7.16; Magige *et al.* 2009). However, the densities of Ostriches were similar in the Serengeti National Park (SNP) compared to areas outside the park, where they were hunted. Nonetheless, this result may be an artifact of sampling design because transects were not placed near areas with high human density (Magige *et al.* 2009). However, the flight initiation distances to approaching humans were longer outside than inside the National Park, suggesting that ostriches were wary of humans probably because of hunting pressure (Magige *et al.* 2009).

Does hunting by traditional indigenous people substantially impinge on wildlife populations? Subsistence hunting of endangered the Blue Billed Curassow (*Crax alberti*) is contributing to their decline in Colombian Andes (Melo-Vasquez *et al.* 2008). Studies were also carried out on the impact of traditional subsistence harvest by indigenous people called the Wana on birds in Sulawesi (Indonesia) (Alvard and Winarni 1999). There were several thousand Wana living in and around the Morowali Nature Reserve in Sulawesi. These people have very little contact with the outside world, practice slash and burn horticulture, and obtain their protein by hunting and trapping animals. Their animal prey includes mice, birds, bats, reptiles, amphibians, ungulates, and primates. Blowguns are usually used for hunting birds (Figure 7.17). Alvard and Winarni (1999) found that the Wana's common bird prey included flowerpeckers, sunbirds, white-eyes, and parrots. The more abundant a bird species was the higher were its chances of being hunted by the Wana (Figure 7.18). Hunting does not seem to seriously affect the viability of bird populations in this area. Of the 46 prey species, 31 were more abundant at Wana-occupied sites than in regions not experiencing bird hunting. Although habitat heterogeneity and differential detectability can also influence such a result, this study does suggest that traditional hunting practices may not cause heavy declines in bird populations in some areas.

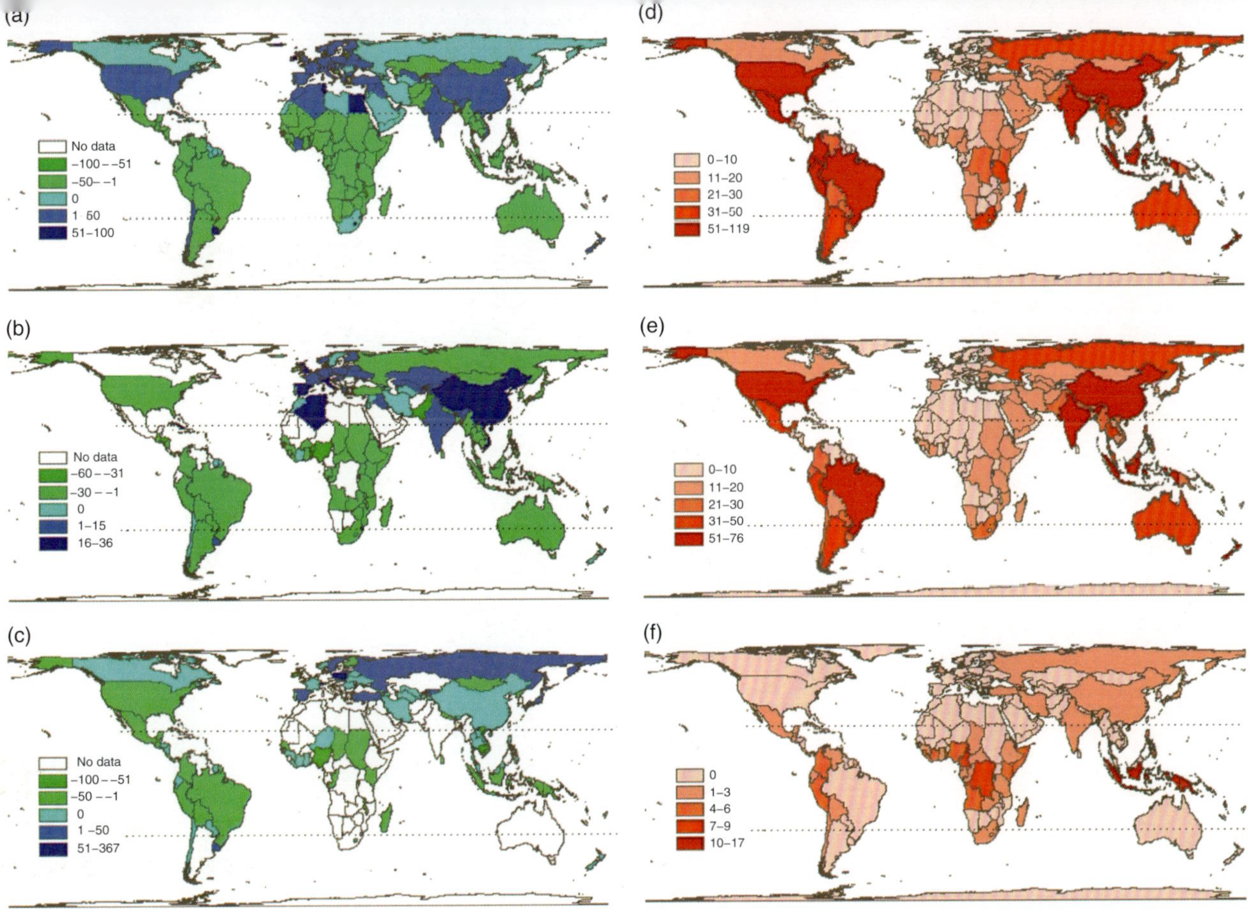

Plate 1.1 Global distribution of average annual forest cover percent change (1990–2005) and richness of threatened and data-deficient terrestrial birds among countries. Data sources for rate of forest cover change and terrestrial birds were obtained from Global Forest Resources Assessment (Food and Agriculture Organization [FAO] 2005) and IUCN Red List of Threatened Species (International Union for the Conservation of Nature and Natural Resources 2007), respectively. Definitions of (a) total, (b) natural, and (c) primary forest areas follow those of FAO (2005). (d) Globally threatened species (i.e. those classified on the IUCN Red List as "critically endangered," "endangered," or "vulnerable"). (e) Globally threatened species as defined for (d), but excluding those species listed under the Red List criteria A1c, B1b(iii), and B2b(iii) (i.e. criteria related to decline of the habitat area, in extent, quality, or both). (f) Data-deficient species (i.e. those classified on the Red List as "data-deficient"). Tropical countries are those in which all the land area is situated within ~23.5°N and ~23.5°S (indicated by the dotted lines). (After Sodhi *et al*. 2008.)

Plate 2.3 A narrow and degraded riparian forest strip in the São Felix region of the Brazilian Amazon. Evidence suggests these strips need to be at least 400 m wide to have a chance of conserving most tropical forest species (Lees and Peres 2008b; figure 2.4). (Photograph by Toby Gardner.)

Plate 2.5 Understory insectivores such as the White-plumed Antbird (*Pithys albifrons*) are among the most vulnerable species to forest fragmentation. (Photograph by Jos Barlow.)

Plate 3.1 (a) Spix's Macaw. (b) Hawaiian Crow. (c) Po'ouli. (Photographs by: (a) Mickey Muck; (b) Mark Sweep; (c) courtesy of San Diego Zoo.)

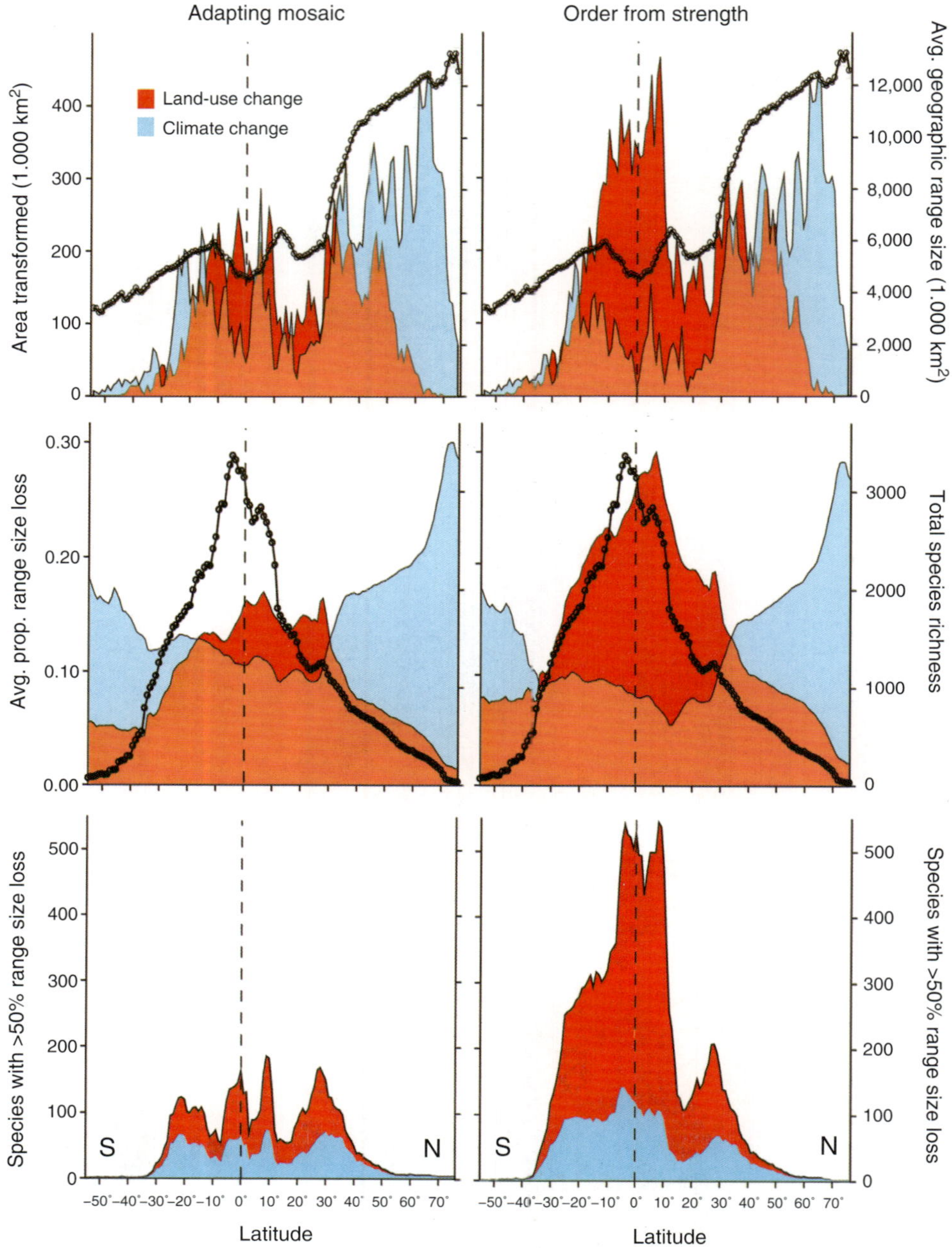

Plate 3.11 Climate and land-use changes and loss in avian species richness and range sizes. Changes are evaluated for two different scenarios – 'Adaptive Mosaic' and 'Order from Strength'. (After Jetz *et al*. 2007.)

Plate 4.1 Examples of the four main types of avian mobile links (Lundberg and Moberg 2003) and potential consequences of the lack of their services. Drawings are based on the cited references. (A) Genetic linkers. Loss of Philippine seed dispersers, such as Palawan Hornbills *Anthracoceros marcei*, can result in most seeds being deposited under the parent tree and being consumed by seed predators (Hamann and Curio 1999). (B) Resource linkers. Introduced foxes eliminating Aleutian seabirds, such as Tufted Puffins *Fratercula cirrhata*, can lead to reduced nutrient deposition triggering a shift from grassland to maritime tundra (Croll *et al*. 2005). (C) Trophic process linkers. Disappearance of scavenging Indian Long-billed Vultures *Gyps indicus*, can cause increases in the numbers of rotting carcasses and of attending mammalian scavengers (Prakash *et al*. 2003). (D) Trophic and non-trophic process linkers. Reduced numbers of Eurasian Three-toed Woodpeckers *Picoides tridactylus* in forest fragments can cause increases in spruce bark beetles (Fayt *et al*. 2005) and decreases in nesting holes used by other species (Daily *et al*. 1993). In addition to habitat loss that affects all avian functional groups, large frugivores are highly susceptible to exploitation, by-catch mortality and introduced species threaten seabirds, woodpeckers decline as a result of fragmentation, and vultures are particularly sensitive to chemicals. (Drawing by Darryl Wheye, © by Darryl Wheye/Science Art Birds.)

Plate 4.2 Examples of the four main types of mobile links and corresponding avian ecosystem services. From top left, clockwise: Chesnut-mandibled Toucan (*Ramphastos ambigius*), seed dispersing genetic linker; Black-browed Albatross (*Thalassarche melanophris*), nutrient depositing resource linker; Lappet-faced Vulture (*Torgos tracheliotus*), scavenging trophic process linker; Red-necked Woodpecker (*Campephilus rubricollis*), ecosystem engineering non-trophic process linker. (All photographs by Cagan H. Sekercioglu.)

Plate 5.1 Fire in two contrasting tropical reserves. (a) An understorey wildfire burns a private forest reserve in the northeast Brazilian Amazon. (b) A fast-moving grassland fire burns the Chapada das Mesas reserve in the Brazilian Cerrado. (Photographs by: (a) Jos Barlow; (b) Alex Lees.)

Plate 6.12 (a) Common Myna. (b) Red-vented Bulbul. (Photographs by J. M. Garg.)

Plate 6.16 The House Crow. (Photograph by Muhammad Mahdi Karim.)

Plate 7.1 Traded birds are frequently kept in overcrowded and inhumane conditions. (Photograph by Navjot Sodhi.)

Plate 7.9 The Maleo. (Photograph by Scott Newell.)

Plate 7.10 The bird nest soup. (Photograph by Stu Spivack.)

Plate 7.13 The Bali Starling. (Photograph by Tommy Tan.)

Plate 7.17 The use of a blowgun to hunt animals. (Photograph by Gao Jialiang.)

Plate 7.19 The Magpie Goose. (Photograph courtesy by Fir0002/Flagstaffotos.)

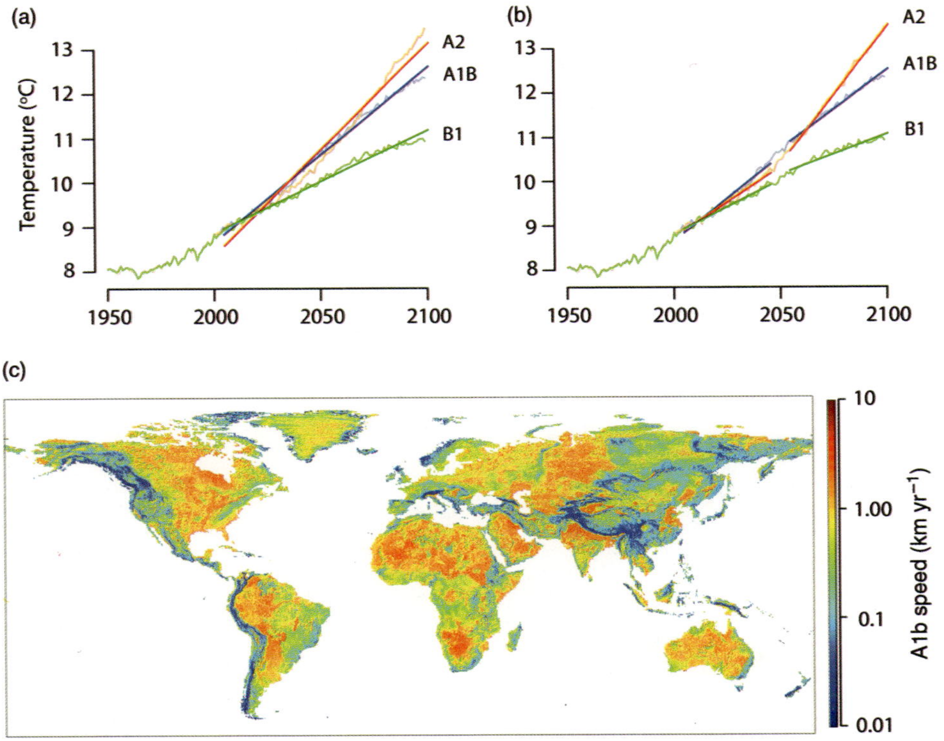

Plate 8.2 The velocity of temperature change globally. (a) Temporal gradients calculated from 2000–2100 across three emissions scenarios (A2, A1B and B1). (b) Temporal gradients calculated from 2000–2050 and 2050–2100 across three emissions scenarios. Trends plotted here are the average of the global land surface. (c) A global map of climate velocity calculated using the 2050–2100 Special Report on Emissions Scenarios (SRES) A1B emissions scenario temporal gradient. (After Loarie *et al.* 2009, copyright Scott Loarie.)

Plate 8.3 A White-ruffed Manakin (*Corapipo altera*). (Photograph by Cagan H. Sekercioglu.)

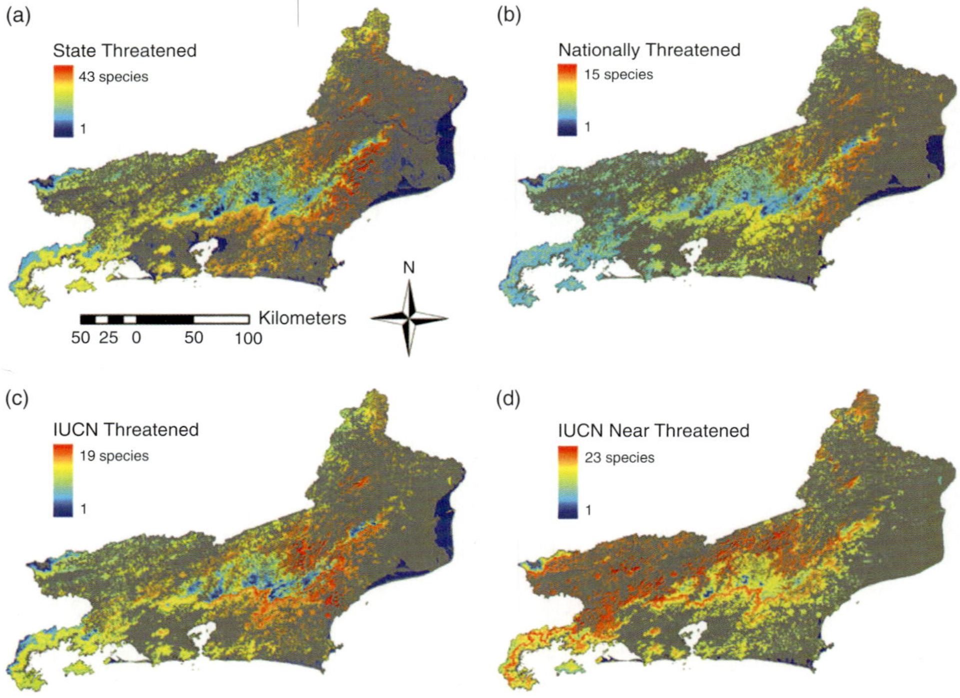

Plate 10.7 Threatened species richness using state (a), national (b), and global (c) red lists. (d) Richness of Near Threatened species according to the global IUCN red list. Gray areas are predicted to have no threatened species. (After Jenkins *et al*. 2010.)

Figure 7.17 The use of a blowgun to hunt animals. (Photograph by Gao Jialiang.) (Please see Plate 7.17 for the color version of this figure.)

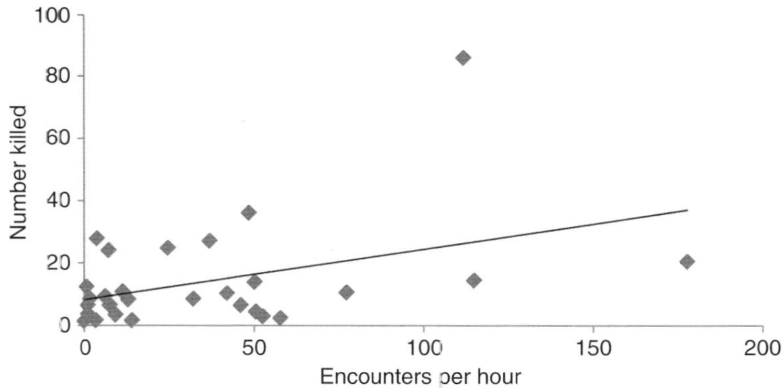

Figure 7.18 Relationship between species-abundance (as measured by the number of encounters per hour of transect) and the relative size of the harvest. Hunters tend to take more abundant species in greater number (After Alvard and Winarni 1999.)

7.4 Sustainable harvesting

Harvesting of nestlings usually causes less impact on bird populations because they have generally low survival probability (Beissinger and Bucher 1992). But how can nestlings be harvested sustainably? Put simply, the estimation of annual harvest rate depends upon how much the annual birth rate exceeds annual death rate (Beissinger and Bucher 1992). Theoretically, individuals hatched into a population more than is needed to compensate the deaths can be removed from a population. However in order to keep populations sustainable, the harvest rate should be well below the difference between the annual death minus birth rate to bet hedge against factors such as adverse environmental conditions (e.g. hurricanes; Beissinger 2001).

Figure 7.19 The Magpie Goose. (Photograph courtesy by Fir0002/Flagstaffotos.) (Please see Plate 7.19 for the color version of this figure.)

To develop a robust sustainable harvest rate for a bird species, the following information is required: (i) population size and range; (ii) habitat requirements; (iii) direct and indirect effects of human disturbance; (iv) mortality and reproductive rate; (v) effects of environmental variables on productivity and survivorship; and (vi) other factors that regulate the population (e.g. competition; Beissinger and Bucher 1992).

To illustrate the development of a sustainable harvesting model for tropical birds, we provide an example of the Magpie Goose (*Anseranas semipalmata*) that undergoes substantial harvest in tropical northern Australia (Figure 7.19). The geese are hunted by indigenous people for subsistence, and additionally, they generate income through tourism and recreational hunting (Brook and Whitehead 2005a). About 33,000 birds a year may be recreational off-take, while more than 288,000 are harvested by indigenous people annually (Brook and Whitehead 2005b). Although only recreational harvesting is regulated, Aboriginal people are willing to reduce off-take rates if it is jeopardizing the geese populations. There are an estimated 3.5 million geese in the wild. Using information on fecundity and survival, Brook and Whitehead (2005b) developed a maximum intrinsic rate of population growth. They found that previous maximum sustainable rates of 30–39% per annum to be excessive and recommended a lower rate of 5–14% or off-take of 94,000–245,000 birds per annum. This study shows that if good data on the demographics of a species is available, then it is possible to determine the level of sustainable harvest.

However, for most tropical birds, data needed to adequately determine the sustainable harvest rate are not available (Beissinger 2001). In the absence of detailed biological information, sustainable harvesting rates still can be calculated if it is clear that the population of a species is not declining (Beissinger 2001). For example, Beissinger (2001) developed a Conservative Sustainable Yield Model for species with poor biological information. One of the suggestions of this model is to increase productivity through management interventions (Figure 7.20;

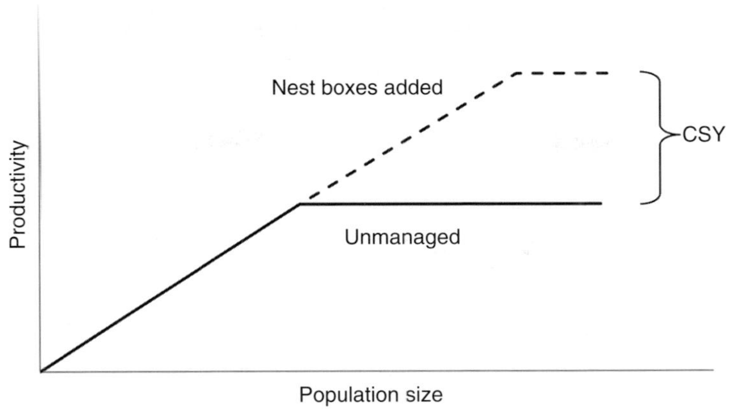

Figure 7.20 An example of the conservation sustainable yield model (CSY). The model depicts the effect of adding nest boxes on the productivity and yield of parrot nestlings under the CSY. (After Beissinger 2001.)

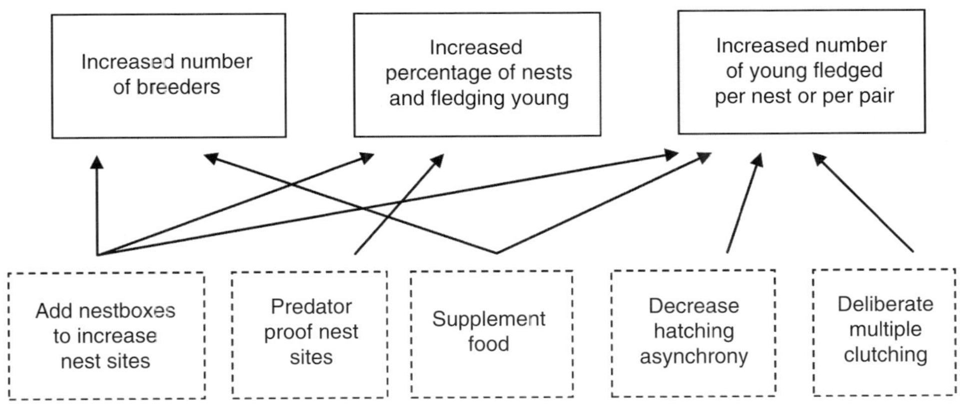

Figure 7.21 Ways to increase the productivity of parrot populations. (After Beissinger and Bucher 1992.)

Beissinger 2001). For instance, the productivity of harvested species can be increased by various means such as providing nest boxes (Figure 7.21; Beissinger and Bucher 1992). For example, the breeding population size and productivity of the Green-rumped Parrotlet (*Forpus passerinus*) increased with the provision of artificial nest boxes in Venezuela (Beisssinger and Bucher 1992). Harvesting of some of these "artificially produced" chicks will likely have less impact on wild populations (Beissinger 2001).

Field application of sustainable harvesting models will need support by local people. However, sustainable harvesting may disadvantage local communities because of factors such as boom-and-bust economics and changing preference in buyers (Homberger and Beissinger 2006). Therefore, it would be wise to assist local people to diversify their income, such as creating jobs through ecotourism or other sustainably harvested wildlife products (Beissinger and Bucher 1992). Beissinger and Bucher (1992) also cautioned that sustainable harvesting regimes should be carefully applied because sometimes even regimes developed with detailed biological knowledge can result in population decline. Further, it is usually difficult to verify if individuals claimed to be sustainably harvested really are (Beissinger 2001). More research is needed to develop genetic and isotope techniques to achieve a fool-proof system to verify whether birds are sustainably harvested or captive bred. However, in some cases, molecular techniques now are available to track the origin of the birds and can facilitate in determining the validity (e.g. origin or sex) of traded species (An *et al.* 2007).

7.5 Curtailing the bird harvesting and trade

Globally, only a few countries significantly benefit from the bird trade (Beissinger 2001). Ideally, bird trade should ensure that: (i) it is carried out in a sustainable manner and does not endanger wild populations; (ii) it does not pose threat of establishment of exotic populations or transmission of novel pathogens to humans and wildlife; and (iii) animal welfare is maintained for the traded species (Beissinger *et al.* 1991). However, these principals are rarely followed. To ameliorate bird harvesting and trade, Homberger and Beissinger (2006) proposed a number of recommendations: (i) create alternative livelihoods for local people involved in bird capture and trade; (ii) attempt captive breeding programs for selected species; (iii) create public awareness about the use of captive-bred birds; (iv) coordinate bird confiscations; and (v) institute import bans.

Traded birds remain a threat to people and ecosystems. Zoonotic disease outbreaks (e.g. avian influenza and West Nile Virus) resulting from the trade of birds can cause billions of dollars of damage (Karesh *et al.* 2007). Concentrating efforts to regulate, reduce, and even eliminate bird trade is a cost-effective way to decrease the risks of disease to humans and livestock (Karesh *et al.* 2007). Additionally, bird trade facilitates the spread of introduced birds, which can cause economic and environmental damage (Carrete and Tella 2008; see Chapter 6). Carrete and Tella (2008) recommended that banning wild-bird trade will be the most effective way to stop avian invasions. However, Cooney and Jepson (2006) argued that trade bans by developed bird exporting countries disadvantage local people because of loss of livelihoods and eliminate chances of sustainable management. They also stated that trade of wild birds is overstated and it may still persist illegally despite the ban. Further, Cooney and Jepson (2006) point out that the pet bird trade in Tanzania supports 40,000–80,000 families, and that the Argentinian government succeeded in regulating the trade of the Blue-fronted Amazons (*Amazon aestiva*). The latter provided livelihoods to famers and probably prevented agricultural expansion at the expense of forests.

Gilardi (2006) retorted Cooney and Jepson (2006). He pointed out that 5–10 million birds are caught from the wild every year for the trade and that it is destructive, unsustainable, and inhumane practice. Gilardi (2006) argued that none of the wild-caught traded species are sustainably harvested and that claims of sustainable harvesting of Blue-fronted Amazons cannot be scientifically

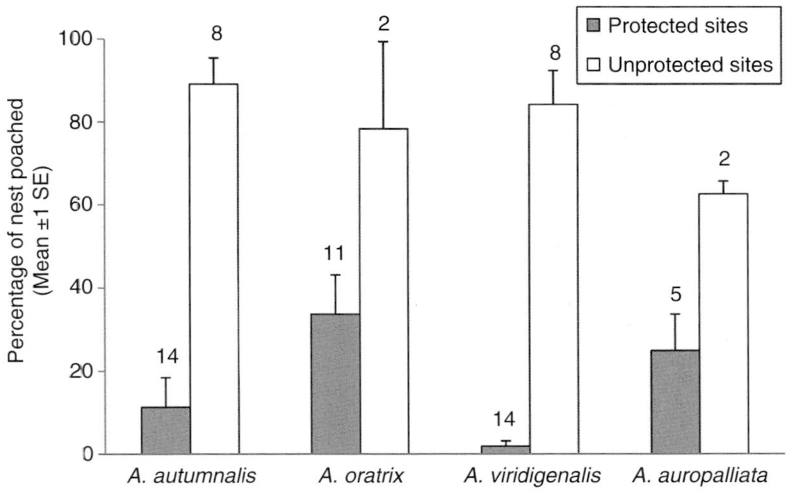

Figure 7.22 Percentage nests of *Amazona* species of parrots poached at protected versus unprotected sites. Sample sizes are above the bars. Means are shown with standard errors (SE). (After Wright *et al.* 2001.)

verified. He further suggested that pet bird trade represents exploitation of tropical people and biodiversity because local trappers receive < 2% of the proceeds. Supporting this claim, a local trapper in a tropical country receives only 2% of the price for an Amazon fledgling of that charged in a pet store in the United States (Beissinger and Bucher 1992). Further, only a small proportion of the population, for example only 0.02% in Tanzania, derives livelihoods from bird trade and programs such as ecotourism may provide many and more sustainable jobs (Gilardi 2006). For instance, a nest-sponsorship program for parrots in Argentina provides alternatives (e.g. ecotourism-related jobs) to harvesting chicks for the local people (Carrete and Tella 2008).

Additionally, there is evidence that trade bans can reduce bird harvesting. Australia was one of the countries first to embargo the import and export of wild birds in 1961 (Homberger and Beissinger 2006). In 1992, the United States enacted the Wild Bird Conservation Act (WBCA) that prohibits the import of birds listed in CITES I and II appendices, except those from licensed captive-breeding or sustainable-harvesting programs (Wright *et al.* 2001). Nest poaching of Neotropical parrots declined by 28% following the WBCA, suggesting that this act alleviated the off-take (Wright *et al.* 2001). Highlighting that protection is important, parrot nest poaching was higher in unprotected than in protected sites for four species of Neotropical parrots (Figure 7.22; Wright *et al.* 2001). Similarly, habitat protection increased nest success by three times for African, Asian, and Australasian parrots (Pain *et al.* 2006).

In addition to the United States, EU countries have been main importers of birds. For example, worldwide, 90% of traded wild-caught birds used to end up in EU countries (Cooney 2005). The trade of CITES-listed Asian birds came to a complete halt, possibly because of the EU ban on the import of wild-caught species in 2005 in response to an avian influenza outbreak (Figure 7.3; Nijman 2010). Since 2003, avian influenza has killed 262 people, millions of poultry, and unknown number of wild birds (www.who.int/csr/disease/avian_influenza/country/cases_table_2009_08_31/en/index.html). The Wildlife Conservation Society had unsuccessfully campaigned for a ban on all bird markets in Asia in response to the avian influenza outbreak (www.eurekalert.org/pub_releases/2004–02/wcs-afs020304.php). However, in addition to curtailing trade of wild-caught birds, the EU's ban could facilitate the enterprise in captive breeding of birds worth millions of dollars (Senior 2006; Carrete and Tella 2008). Economic forces can influence captive breeding operations; the high prices of some bird species can result in oversupply and eventually result in price decline (Robinson 2001), thus even benefitting the customers. To maintain high standards in the captive-breed bird trade, Jepson *et al.* (2008) suggested that a certification scheme should be introduced. The scheme should have the following objectives: (i) captive birds should be from certified facilities; (ii) these facilities should maintain animal welfare; and (iii) wild-caught birds should be replaced with captive-bred birds. Jepson *et al.* (2008) further suggested that this scheme can also be extended to individuals caught from degraded habitats such as agroforests.

Enforcement of wildlife laws and monitoring can even curtail domestic animal trade (Lee *et al.* 2005a). Between 2000 and 2007, there has been a 39% decline in bird trade in Hanoi (Vietnam) largely due to enforcement of a law in 2005 that prohibits the sale and movement of wild birds (Brooks-Moizer *et al.* 2009). However, 91% of the bird vendors surveyed did not perceive any risk of catching avian influenza from their birds (Brooks-Moizer *et al.* 2009), thus continuing their trade, albeit at reduced levels. Sometimes heavy fines fail to eliminate bird poaching. Despite, very steep fines (Singapore$50,000 (ca. US$35,000) or imprisonment) there have been cases of bird poaching in Singapore (www.wildsingapore.com/news/20070506/070526–4.htm), suggesting that reducing demand by public education may be needed. Therefore it is heartening to see programs such as the Campaign Against Illegal Wild Bird Trade in Bolivia (darwin.defra.gov.uk/documents/15030/18423/15–030%20AR3%20App%208%20Leaflet%20illegal%20bird%20trade%20in%20Bolivia.pdf). This campaign highlights the problems of illegal wild bird trade through exhibitions and media reports. Along the same lines, a facebook group has also been set up to stop the bird trade (apps.facebook.com/causes).

What should be done with confiscated birds? There have been a number of programs that rehabilitate and release confiscated bird species into the wild. For instance, there is a program that is partially sponsored by BirdLife in Indonesia that has wild released a number of species such as the Yellow-crested Cockatoo (*Cacatua sulphurea*) and the Hill Myna (www.indonesian-parrot-project.org/pdf_files/publicreport_SM.pdf). However, it is unclear if these releases have been successful or if the released birds are recaptured for the trade. Overall, much more needs to be done urgently to ensure that bird harvesting does not further endanger tropical birds.

8

Climate Change Effects on Tropical Birds

Overview: Climate change is a growing, but an understudied threat to tropical birds, with many uncertainties in extinction estimates due to the lack of data. Nevertheless, hundreds of species are expected to go extinct or become threatened as a result of climate change. In the worst case scenarios, projections of landbird extinctions approach 2500 species by the year 2100 and most of these are expected to be tropical birds. Tropical mountain birds, species restricted to flat lowland areas, coastal forest birds, and restricted-range species are especially vulnerable to climate change. Tropical forest bird species that experience limited temperature variation will be the most susceptible to the physiological effects of warming temperatures and heat waves. Tropical bird species have less demographic flexibility than their temperate counterparts, and will not be able to evolve as fast as temperate birds in response to rapid climatic change. Increased rainfall seasonality due to climate change will likely result in additional species losses in tropical bird communities. Some species are especially vulnerable to extreme weather events such as heat waves, cold spells, typhoons, and hurricanes. Sedentary birds, which make up most of tropical forest bird species, are five times more likely to go extinct than are long-distance migrants. However, most tropical bird species vulnerable to climate change are not currently considered threatened with extinction. Mostly unknown species' interactions, indirect effects, and synergies of climate change with other threats such as habitat loss, emerging diseases, invasive species, and hunting will influence the effects of climate change on tropical birds. Climate change effects will be especially exacerbated by ongoing habitat loss in the tropics, which poses an often insurmountable barrier to bird movements and range shifts. In some cases, worst

Conservation of Tropical Birds, First Edition. Navjot S. Sodhi, Çağan H. Şekercioğlu, Jos Barlow and Scott K. Robinson.

Published 2011 by Blackwell Publishing Ltd.

case scenarios of habitat loss can increase bird extinctions caused by climate change by nearly 50% compared to the best case scenarios. Depending on the amount of habitat loss, each degree of surface warming can lead to an additional 100–500 bird extinctions. Networks of protected areas need to be designed with climate change in mind, as 92% of currently protected areas are likely to become climatically unsuitable in a century. Protected area networks need to incorporate extensive topographical diversity, cover wide elevational ranges, promote connectivity to other protected areas, and integrate human-dominated landscapes into conservation schemes. Even though most birds species are tropical, the lopsided concentration of researchers in the temperate zone means that most of our understanding of climate change impacts on birds is based on temperate birds. Locally based, long-term tropical bird monitoring and conservation programs can help protect birds against climate change by combining biodiversity monitoring, habitat protection and restoration, environmental education, raising awareness, community-based conservation, and sustainable livelihoods.

Climate change is increasingly shaping up to be the most serious and widespread environmental phenomenon of the 21st century (IPCC 2007; Box 8.1). Even during the most rapid, natural climate change observed at the transition between the glacial and interglacial periods (Schneider 1989), the average temperature increased by about 5°C in 5000–7000 years (Huntley *et al.* 2006), but the rate of current climate change is extremely rapid, one or two orders of magnitude greater than that observed in the past.

As the best known order of organisms and with a multi-million-strong army of citizen scientist birdwatchers tracking them across the globe, birds have been among the first indicators of human-induced climate change (Kinzelbach 1995; Sparks and Crick 1999). Climate change has already led to large shifts in bird distributions (Gregory *et al.* 2009; Niven *et al.* 2009). Unchecked climate change, combined with habitat loss, may lead to the extinctions of hundreds and possibly thousands of bird species (Sekercioglu *et al.* 2008). Tropical bird species, which are mostly sedentary, are especially vulnerable to climate change, especially the endemic species isolated on tropical mountains (Janzen 1967; Sekercioglu *et al.* 2008). There are few studies of climate change effects on the bird communities of entire tropical forest regions and few tropical bird families have been assessed in their entirety (e.g. Box 8.1). Here, we review the known and potential effects of climate change on tropical birds, summarize what is known, and highlight areas where research is needed urgently.

8.1 Future projections and models

Sekercioglu *et al.* (2008) quantified the relative impact of habitat loss on bird extinctions caused by climate change by combining Intergovermental Panel on Climate Change (IPCC) 2007 climate change scenarios (IPCC 2007) with

Box 8.1 Climate change. (After Sekercioglu 2010a)

The global carbon cycle has been disturbed by about 13% compared to the pre-industrial era, as opposed to 100% or more for nitrogen, phosphorous, and sulfur cycles (Falkowski *et al.* 2000). Given the dominance of carbon in shaping life and in regulating climate, however, this perturbation has already been enough to lead to significant climate change, with worse likely to come in the future (IPCC 2007). Because gases like CO_2, methane (CH_4), and nitrous oxide (N_2O) trap the sun's heat, especially the long-wave infrared radiation that's emitted by the warmed planet, the atmosphere creates a natural "greenhouse" (Houghton 2004). Without this greenhouse effect, humans and most other organisms would be unable to survive, as the global mean surface temperature would drop from the current 14°C to −19°C (IPCC 2007). Ironically, the ever-rising consumption of fossil fuels during the industrial age and the resultant increasing emission of greenhouse gases have created the opposite problem, leading to an increase in the magnitude of the greenhouse effect and a consequent rise in global temperatures (IPCC 2007). Since 1750, atmospheric CO_2 concentrations have increased by 34% (Millennium Ecosystem Assessment 2005) and by the end of this century, average global temperature is projected to rise by 1.8–6.4°C (IPCC 2007). Increasing deforestation and warming both exacerbate the problem as forest ecosystems switch from being major carbon sinks to being carbon sources (Phat *et al.* 2004; IPCC 2007). If fossil fuel consumption and deforestation continue unabated, global CO_2 emissions are expected to be about two to four times higher than at present by the year 2100 (IPCC 2007). As climate and life have coevolved for billions of years and interact with each other through various feedback mechanisms (Schneider and Londer 1984), rapid climate change would have major consequences for the planet's life-support systems. There are now plans under way for developed nations to finance the conservation of tropical forests in the developing world so that these forests can continue to provide the ecosystem service of acting as carbon sinks (Butler 2008).

Millennium Ecosystem Assessment habitat loss scenarios (Millennium Ecosystem Assessment 2005). In the worst case surface warming estimate of 6.4°C by 2100, which may yet be an underestimate (Stainforth *et al.* 2005), the worst case habitat loss scenario resulted in an estimate of about 2500 bird extinctions, which was nearly a 50% increase (Figure 8.1) in bird extinctions compared to the best case. On average, 67% of the species predicted to go extinct in the projections of Sekercioglu *et al.* (2008) were not considered threatened or near threatened at the time of the study, indicating that the threat of climate change is not being incorporated effectively into conservation assessments (although this is now changing, e.g. BirdLife International 2010a). Sekercioglu *et al.* (2008) stated that tropical montane endemics will be especially vulnerable to extinctions induced by climate change. Similarly, projections of changes in birds' future ranges due to climate and land-use change indicate that most vulnerable species are restricted-range bird species endemic to the tropics (Jetz *et al.* 2007).

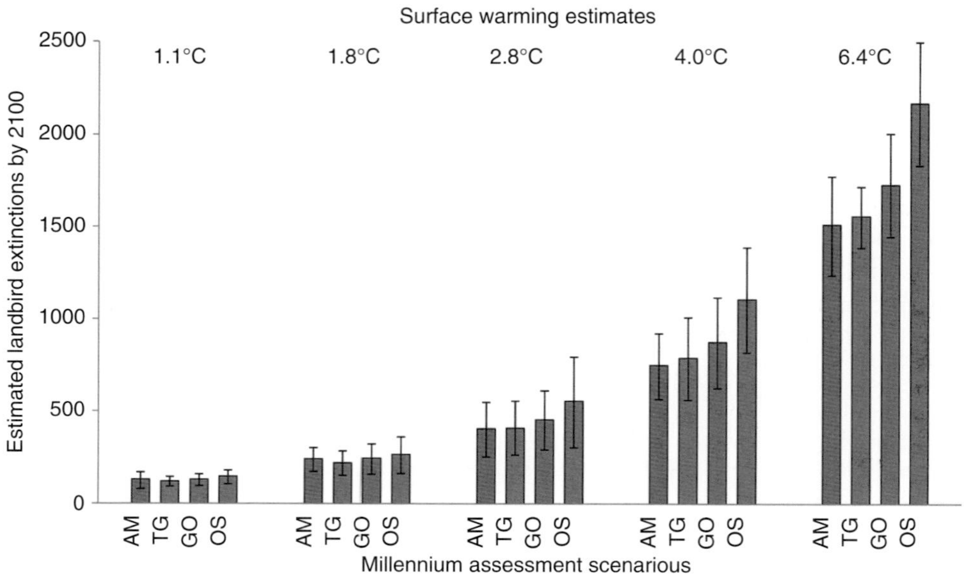

Figure 8.1 Number of world landbird species projected to go extinct (current baseline = 0) by 2100 based on the interactions of estimates of various surface-warming estimates (IPCC 2007), three possible shifts in lower elevational limits, and Millennium Assessment habitat-change scenarios (MA 2005; AM = adaptive mosaic; GO = global orchestration; OS = order from strength; TG = technogarden). Bars show the results of an intermediate amount of elevational shift, where lower limits of 50% of lowland (≤500 m) bird species and lower and upper limits of all other (>500 m) bird species are assumed to move up in response to surface warming. "Error bars" indicate best-case (0% none of lowland (≤ 500 m) and all of highland (>500 m) species move up) or worst-case (all species move up) climate-warming scenarios. (After Sekercioglu *et al.* 2008.)

Globally, species with restricted distributions are concentrated in the tropics and many of these species are already threatened with extinction (Stattersfield *et al.* 1998). Decreasing distribution size in birds increases the likelihood of bird extinction due to climate change (Schwartz *et al.* 2006). To make matters worse, model fit also declines for species with smaller ranges, resulting in high uncertainty in predicting climate change extinctions in these species (Schwartz *et al.* 2006). Excluding these species from conservation planning could result in their extinction, but mistakenly including species that are not actually threatened would mean the waste of scarce conservation resources. These uncertainties were observed in the well-known bird species of the eastern United States (Schwartz *et al.* 2006) and the problems predicting extinctions due to climate change in the little-known tropical bird species are much greater.

Despite the expectations for extensive range shifts and species turnover, different choices of both the climate and species distribution models can result in substantially different projections and large uncertainties remain in the forecasts of species turnover in response to climate change (Diniz *et al.* 2009; Soria-Auza *et al.* 2010), as well as in temperature and precipitation forecasts themselves (Vera *et al.* 2006). Climate-shaped models of geographical distribution are constantly improving, achieving finer resolution, improved representation of key processes, and more accurate depiction of oceanic circulations (Huntley *et al.* 2006). However, models are not useful if the underlying data themselves are not of good quality (i.e. "garbage in garbage out"). This is particularly a problem with tropical species, many of whom have little-known distributions, sometimes based on a handful of points from specimens collected a century ago. In fact, Jetz *et al.* (2008b) showed that the distributions of hundreds of bird species in relatively well-known North America, South Africa, and Australia are overestimates, and the overestimation is higher for threatened, range-restricted, and specialized birds, the very species that are most vulnerable to global change. Therefore, using detailed datasets that have good data on absences as well as presences is preferable (Huntley *et al.* 2006). In addition, many models confound occurrence with the probability of detection, but recently developed occupancy models can use basic presence/absence survey data from citizen science projects like bird atlases, while accounting for probability of detection (Altwegg *et al.* 2008). However, the kind of detailed data provided by bird atlases usually requires hundreds or thousands of dedicated, disciplined, knowledgeable, and well-trained amateur and professional ornithologists that are not available in most tropical countries where the data from visiting birdwatchers must be put to better use (Sekercioglu 2002b,c), with growing initiatives such as eBird (www.ebird.org) and Avian Knowledge Network (www.avianknowledge.net and www.worldbirds.org).

Another concern is that static climate models may be inadequate for mapping future ranges. In fact, even dynamic models may be inadequate in predicting species ranges, which are also influenced by ecological processes such as species interactions, adaptation, and flexibility in life history (Schwager *et al.* 2008). Models need to do a better job in considering slowly adapting processes and mechanisms (Schwager *et al.* 2008).

8.2 Where will climate change hit?

In Latin America, the tropical mountains of Mexico, the Andes, and the species rich biodiversity hotspot of Central America, are projected to be the most affected regions from climate change (Box 8.2), based on the changes and shifts in species ranges (Lawler *et al.* 2009). In 80% of the climate change scenarios, these regions are expected to experience at least 25–38% species turnover (Lawler *et al.* 2009). The Brazilian Atlantic forests and the southern and western boundaries of the Brazilian cerrado are also highlighted as areas likely to experience high turnover. Hundreds of restricted-range species are likely to be even more extinction-prone (Lawler *et al.* 2009). In sub-Saharan Africa, bird species whose ranges are expected to contract most live in the temperate areas of the Cape, the Horn of Africa, East Africa, montane and semi-arid habitats, whereas the distributions of species living in equatorial and moist tropical forest habitats should be less affected (Huntley *et al.* 2006). Using data from field guides, Peh (2007) showed that 92 common resident species of Southeast Asian forests potentially shifted their distributions upwards in the past three decades.

Tropical montane species are one of the most vulnerable groups to climate change (Sekercioglu *et al.* 2008). Warming temperatures force many montane species uphill and reduce their ranges, sometimes entirely (Shoo *et al.* 2005a). There are few data on the potential extent and magnitude of current and future shifts in the elevational limits of tropical bird species in response to climate change, with tropical montane forest bird communities in Australia and Costa Rica better known than others (Böhning-Gaese and Lemoine 2004; Shoo *et al.* 2006; Gasner *et al.* 2010). In the Tilaran Mountains of Costa Rica, populations of nearly half of the 77 forest species are expected to decline due to climate change in this century. About half of the declining species are endemic to Central America and seven of eight projected local extinctions are endemic to the highlands of Costa Rica and Panama (Gasner *et al.* 2010; see also Box 8.2).

Although elevational distribution is often used as a proxy for population size, in response to increasing temperatures highland bird populations can decline even faster than the areas they occupy (Shoo *et al.* 2005b). Most species' populations are not distributed uniformly across different elevations and climate-induced changes in abundance patterns have important conservation implications, but these changes have not been well studied (Shoo *et al.* 2005b).

Recent research suggests that high-elevation birds have life-history strategies similar to those of low-latitude birds (Tieleman 2009), further raising concern about tropical high-elevation species whose life histories may be particularly slow. Not only high-elevation birds are particularly susceptible to climate change, but most high-elevation endemic bird species are found in the tropics where hotter lowlands have isolated the populations of these sedentary birds (Janzen 1967). With limited capacity to shift their ranges across unsuitable habitat to other mountain ranges, tropical montane endemics are particularly vulnerable to climate change. However, because tropical mountains often have fewer people and more habitat than tropical lowlands, most tropical montane bird species are considered to be of least conservation concern despite the looming threat of climate change (Sekercioglu *et al.* 2008).

Effects of climate change on tropical coastal forests, such as mangrove forests, are known even less. Tropical coastal ecosystems are disappearing at a rapid rate and are prone to climate change (Waycott *et al.* 2009). In addition, they provide key ecosystem services (Sekercioglu 2010a), such as reducing the damage caused by tsunamis. Tropical forests on oceanic islands and their birds are especially vulnerable to climate change, as these communities have already suffered more from introduced species than any other habitat, many species have gone extinct due to invasive species and exploitation, ecosystems have been drastically modified, and still experience habitat loss (Kingsford *et al.* 2009). In addition, many oceanic islands are especially prone to rising sea levels due to climate change (Kingsford *et al.* 2009).

8.2.1 Lowlands in trouble: the Amazon and the Congo basins

Species in extensive low-lying areas face the added challenge of the lack of topographical diversity. As climate warms, many types of vegetation and dependent organisms have to shift their distributions to cooler conditions in order to track their preferred microclimates. This usually means moving north or moving up (Janzen 1967; Pounds *et al.* 1999; Shoo *et al.* 2006; Gregory et al. 2009; Niven

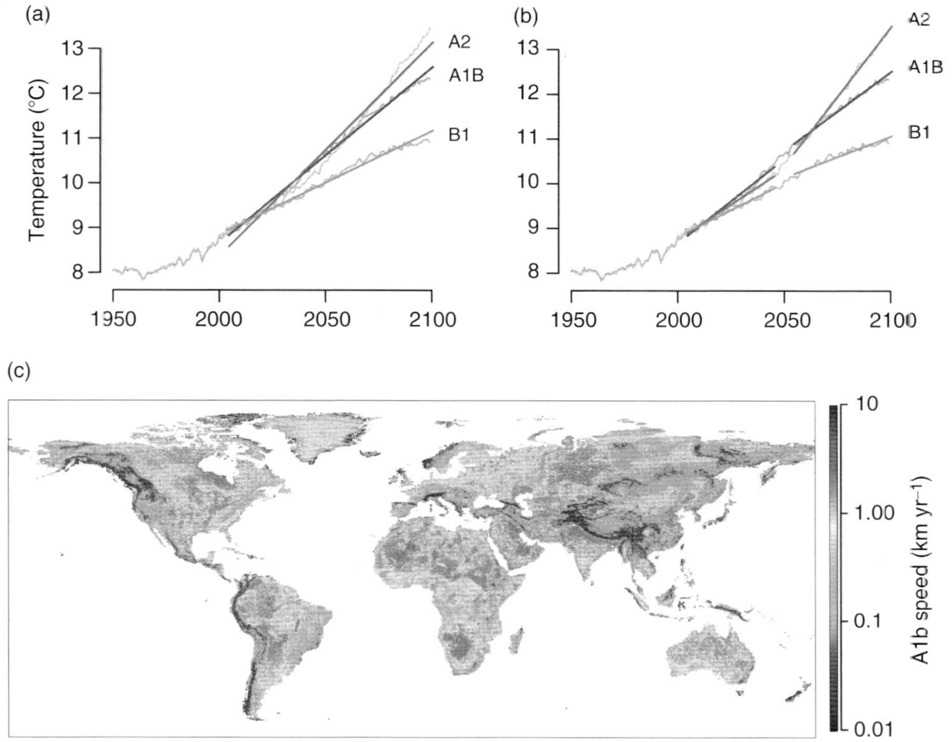

Figure 8.2 The velocity of temperature change globally. (a) Temporal gradients calculated from 2000–2100 across three emissions scenarios (A2, A1B and B1). (b) Temporal gradients calculated from 2000–2050 and 2050–2100 across three emissions scenarios. Trends plotted here are the average of the global land surface. (c) A global map of climate velocity calculated using the 2050–2100 Special Report on Emissions Scenarios (SRES) A1B emissions scenario temporal gradient. (After Loarie *et al.* 2009, copyright Scott Loarie.) (Please see Plate 8.2 for the color version of this figure.)

et al. 2009). If there are mountains nearby accessible for elevational expansion, the vegetation and dependent species need to move relatively short distances up and down the mountain. In extensive flat areas where mountains are few and far between, such as the Sahara Desert, the Amazon and the Congo basins, plant and animal communities may need to move much greater distances to track their preferred temperatures (Peterson *et al.* 2001, 2002; Sekercioglu *et al.* 2008; Anciaes and Peterson 2009; Loarie *et al.* 2009; Menon *et al.* 2009).

Loarie *et al.* (2009) recently calculated an index of the velocity required for ecological communities to keep up with climate change (Figure 8.2). This velocity is not uniform across habitats and regions, and is highest in flat regions and ecosystems. Among these are extensive tropical habitats such as mangroves, flooded grasslands, and deserts and most of the remaining lowland tropical forest wilderness (www.conservation.org), thought to be relatively pristine and less disturbed by human activities. Given the velocity of climate change, 92% of protected areas are likely to become climatically unsuitable in a century (Loarie

Figure 8.3 A White-ruffed Manakin (*Corapipo altera*). (Photograph by Cagan H. Sekercioglu.) (Please see Plate 8.3 for the color version of this figure.)

Box 8.2 Manakins – a case study

The family Pipridae (manakins) is endemic to the Neotropical region (Figure 8.3). The roughly 45 species in this family are limited to various forest habitats and nearby second growth. Ranging from lowlands to montane forest, manakins prefer relatively humid conditions, even in drier areas where they are limited to gallery forest (Snow 2004). Because they are limited to Neotropical forest habitats, consist of dozens of species with varied geographical distributions and ecological preferences, and range widely in elevational distribution, manakins comprise an ideal tropical family to model the effects of climate change on their distributions. Anciaes and Peterson (2009) used ecological niche modeling (Peterson *et al.* 2001) to model manakins' ecological niches based on their present distributions and climatic preferences and then projecting these models onto climatic projections for mid-21st century. The results indicate that the species limited to the lowland habitats of the Amazon basin and cerrado will be most affected and would need to move more to track their environmental niches. These species are expected to lose 80% of their habitable area and 20% of the cerrado manakin species are expected to go extinct. In other ecoregions, dominated by mountains, predicted loss of potential area (16–50%) and expected extinctions (0–5%) are much lower, once again showing the importance of topographic diversity in buffering against forced range shifts due to climate change. The authors also predicted more fragmented manakin distributions, highlighting the climate change conservation challenge that will face many tropical groups.

et al. 2009). The mean residence times of climatic suitability in protected areas, defined as the velocity of climate change divided by the diameter of the protected area, ranges from only 5.3 to 63.6 years in major tropical habitats (Loarie *et al.* 2009).

Many plant communities, especially in less seasonal tropical areas that experience limited temperature ranges, may be unable to shift their distributions as fast as required by climate change. The rate of recolonization in temperate Europe during the interglacial period is probably unrealistically high for tropical forest communities and we need a much better understanding of these rates in the tropics. In addition, in many tropical areas, large barriers will prevent habitats from shifting in response to climate change. Some of these are natural, such as rivers, oceans, and unsuitable soils. Many others are man-made, as larger and larger portions of the tropics are occupied by settlements, agriculture, pastures, and other expansions of civilization. These will form formidable and often insurmountable barriers to the shifts of habitats and their dependent species, creating "climate refugees."

8.2.2 Non-forest habitats

Even though earlier tropical studies focused mostly on tropical forest birds (e.g. Pounds *et al.* 1999; Shoo *et al.* 2005a), thousands of tropical bird species are found outside forests, in a diverse range of open habitats such as savanna, cerrado, grasslands, karoo, scrub, deserts, and human-dominated areas. Recently, Marini *et al.* (2009) have modeled the expected effects of climate change on the distributions of 26 widespread bird species of the cerrado, the largest extent of tropical savanna worldwide, with 44% endemism of its 10,000 plant species (see Chapter 1). Highly fragmented, threatened by agricultural conversion, and with only 2.2% of it protected, the cerrado and its inhabitants are particularly vulnerable to climate change because opportunities for their dispersal are greatly limited by human land use (Parmesan 2006). Projected range size contractions of up to 80% will especially affect forest and grassland-limited species and species with smaller ranges (Marini *et al.* 2009). Range shifts of 175–200 km are predicted towards southeastern Brazil, where high human population and development present barriers to dispersal.

Climatic models have largely focused on terrestrial species, but aquatic habitats face additional watershed problems. Patterns of aquatic bird species diversity, often high in coastal regions, are different than those of the terrestrial species (Ramirez-Bastida *et al.* 2008). Coasts will experience sea-level rises and coastal wetlands are often bounded by agriculture and settlements. Additional factors such as hunting, disease, and habitat loss further threaten tropical wetland species (Traill *et al.* 2009). Given their dependence on water, tropical aquatic birds will be particularly vulnerable to climate change.

Arid zone species are assumed to be resilient to high temperatures and low humidity. However, these species are already dealing with extreme conditions and are highly dependent on seasonal rains. Further increases in already high temperatures and/or changes in the rainfall regime may push them over the edge. We have limited knowledge of future changes in rainfall patterns. In addition, many "arid zone" species actually depend on aquatic habitats such as riparian forests, seasonal and permanent waterholes, natural and artificial lakes, and rivers (Schneider and Griesser 2009). The disappearance of such water bodies is likely to be devastating for these "oasis" communities.

Despite the fact that large expanses of the world have been heavily modified by humans, the tendency of ecologists to focus on relatively intact ecosystems means that our ability to predict climate change impacts on human-dominated ecosystems is also limited (Felton *et al.* 2009).

8.3 Protected areas

There is an urgent need to understand how climate change will affect the ability of protected areas to harbor species and communities (Willis *et al.* 2009; see also Chapter 10). By reducing food availability and, consequently, breeding success, droughts can result in widespread bird population declines even in large reserves (Mac Nally *et al.* 2009). Steep mountains with limited human use can be protected more easily, and higher and cooler climatic refugia on tropical mountains are disproportionately important for restricted range species. Such climatically diverse areas with rare climates have buffered cold-adapted species from extinction during past interglacial periods where global temperatures were unusually warm (Ohlemuller *et al.* 2008). However, climate change will disproportionately affect the rare climates characterizing these centers of species rarity, jeopardizing many restricted-range species (Ohlemuller *et al.* 2008).

Networks of protected areas designed with climate change in mind are critical for conservation. The Important Bird Area (IBA) network of Africa includes 1230 sites essential for maintaining populations of priority species (Hole *et al.* 2009). Climate-induced shifts in the distributions of the breeding birds of sub-Saharan Africa result in 42% of IBAs showing > 50% turnover (Hole *et al.* 2009). However, only seven or eight species' preferred climatic envelopes are entirely lost from the IBA network and 90% of priority species retain suitable climatic space.

Nevertheless, sophisticated conservation planning is of no use if it is not applied effectively in collaboration with local people and decision-makers (see Chapter 10). Projections of numerous range shifts in response to climate change highlight the importance and necessity of creating protected areas that cover a wide topographical/elevational range, increasing the connectivity to other protected areas via the use of habitat corridors, and integrating human-dominated landscapes into conservation schemes (Mac Nally *et al.* 2009).

8.4 Effects of life history

8.4.1 Slow lives and reduced speed of adaptation

Species' ecological niches are unlikely to evolve substantially over the decadal time-scales in which climate change is expected to take place (see references in Anciaes and Peterson 2009), especially for organisms with long generation times. This is particularly true for many tropical species that experience less seasonality and more stable temperature and humidity regimes than experienced in the temperate region, favoring birds to live "slower" lives, with "*r*-selected" life histories characterized by smaller clutch sizes (Jetz *et al.* 2008a), lower productivity, and longer life spans (Wiersma *et al.* 2007). Consequently, tropical bird species have less demographic flexibility than their temperate counterparts, and will not be able to evolve as fast as temperate birds in response to rapid climatic change.

However, longer life spans can buffer against climate variability, with long-lived birds potentially suffering less from climate variability (Morris *et al.* 2008). As long-lived homeotherms, the population growth rates of birds should be less affected by the variability in their demographic rates (e.g. survival, reproduction, growth) due to climate change than the growth rates of short-lived species like insects (Morris *et al.* 2008). Tropical birds, on average, are longer-lived than their temperate relatives (Wiersma *et al.* 2007). However, the advantage of longer living is likely to be negated by factors that make tropical forest birds more vulnerable, such as smaller clutch size (Jetz *et al.* 2008a), more sedentary habits (Sekercioglu 2007), and lower tolerance of climatic variability (Weathers 1997; Seavy 2006).

Because climate affects most aspects of an organism's life history, it can have far reaching consequences and even change sexual displays (Botero *et al.* 2009). Variable and unpredictable climates create strong selection pressure for improved cognition, so that organisms can compensate for a changing climate and resource landscape through learning and innovation. Songbird displays, given to attract females, also depend on learning and innovation, and increased climatic variability has been shown to result in more elaborate displays in mockingbirds (Botero *et al.* 2009). The implications of this for songbird evolution are yet to be determined.

In one of the few studies in the southern hemisphere of the changes in egg-laying dates in response to climate change, warming temperatures and reduced rainfall has led to earlier breeding in and possibly reduced clutch size of the critically endangered Helmeted Honeyeater (*Lichenostomus melanops cassidix*) in Australia, increasing the risk of further population decline (Chambers *et al.* 2008). Not all effects of climate change are negative and changes in temperature and precipitation regimes will benefit some species. The effects on bird life history may be unexpected and hard to predict. An example is the possible increase in egg volume of pied flycatchers with increasing temperature (Jarvinen 1994). Because larger eggs have better hatching success, warmer temperatures may change birds' reproductive strategies, make up for the negative demographic effects of climate change, and enable some species to expand to new areas (Jarvinen 1994). Nevertheless, many species will not benefit and there will be a range of responses, with climate change likely resulting in novel communities and even community disassembly, especially in diverse tropical communities with a wide range of life histories.

8.4.2 Physiology and microclimate

We need to improve our understanding of the interspecific variation in physiological traits and how these will influence tropical species' adaptability to climate change (Bernardo *et al.* 2007). In comparison to ectotherms, climate change effects on endothermic birds are more likely to act through indirect effects on vegetation than direct effects through physiology (Aragon *et al.* 2010). However, an increasing number of studies hint at the vital role physiology will play in birds' responses to climate change (McKechnie 2008; McNab 2009).

Birds living in hotter and drier habitats often show physiological adaptations, such as reduced thermal conductance (Weathers 1997), lower evaporative water loss (Weathers and Greene 1998), and higher heat tolerance (Weathers *et al.* 2001b). Open habitat sunbirds in Uganda have reduced thermal conductance, better insulation and better ability to tolerate fluctuating temperatures than forest sunbirds (Seavy 2006). Lowland and open country bird species, adapted to higher temperatures, are likely to tolerate temperature increases better than highland and forest interior species (Weathers 1997). Furthermore, climate change can favor the penetration of open-country species into climatically comprised forest remnants (Laurance 2004).

Temperature and humidity will also have indirect effects on avian activity and behavior. Increased avoidance of sites with unfavorable climates can reduce important activities such as feeding and breeding displays (Walsberg 1993). Many tropical forest species, such as Silver-throated Tanagers (*Tangara icterocephala*) in Costa Rica, spend a lot of time in the cool and humid forest interior, but display and forage in exposed areas, such as song perches and fruiting canopy trees. Intensive radio-tracking combined with constant monitoring of habitat temperature and humidity revealed that these birds track temperature and humidity differences as small as 0.8°C and 4.2%, respectively (Sekercioglu *et al.* 2007).

Physiological limitations also shape habitat selection. In tropical forests, tree holes are often critical resources that provide safe nesting cavities and microclimatic refugia from higher outside temperatures (Isaac *et al.* 2008). The importance of tree cavities is such that cavity-nesting has recently been shown to be a critical factor in determining avian clutch size (Jetz *et al.* 2008a). Increasing global temperatures mean tree hollows are more important than ever (Isaac *et al.* 2008), but large, older trees most likely to bear such holes are often the first ones to be logged in tropical forests, and many dead trees with holes are removed as firewood even in protected forests. The increasing scarcity of such trees, combined with the increasing need of temperature-sensitive species for tree holes means that birds will have more competition not only with other birds, but also with cavity-dwelling mammals (Isaac *et al.* 2008).

The basal metabolic rate (BMR) is the standard measure of the energy cost of maintenance and "the obligatory cost of living for endotherms" (Barcelo *et al.* 2009). BMR has critical importance in determining birds' physiological tolerance to climate change, but the knowledge of BMR and its variability in tropical birds is limited. BMR is not a fixed, taxon-specific trait, but has high phenological flexibility (McKechnie 2008). Compared to ectotherms, birds have more physiological flexibility in responding to temperature changes, but there is large variation between species (McKechnie 2008).

The fact that BMR is especially flexible in migratory birds (McNab 2009) indicates that sedentary tropical birds that live in habitats with low climatic variability are likely to have less flexible BMRs. This is supported by an analysis of the responses of 71 bird species to the 2003–04 heat wave in France (Jiguet *et al.* 2006). Bird species with the widest thermal range suffered the least whereas birds that had experienced the narrowest thermal variation suffered the largest population declines (Jiguet *et al.* 2006). The study's conclusion that a bird species' thermal range is a reliable predictor of its resilience to extreme temperatures suggests that thousands of tropical forest bird species that experience limited temperature variation will be the most susceptible birds to the physiological effects of warming temperatures and heat waves.

Birds in lower trophic levels such as frugivores and granivores have higher BMR than those at higher trophic levels like insectivores (Sabat *et al.* 2009). Migratory bird species, which experience a wider range of temperatures, also have higher BMR (McNab 2009). On the other hand, non-passerines, flightless birds, island birds, and tropical birds have lower BMRs (Wiersma *et al.* 2007; McNab 2009). Because higher BMR enables the use of a wider range of thermal environments, increases dispersal ability, and improves adaptability to climate change (Bernardo *et al.* 2007), birds with lower BMR are at a disadvantage. Because most studies of birds' responses to climate change focus on migratory, mainland songbirds of temperate habitats, our understanding of birds' adaptive capacity to climate change is likely to be more optimistic than is the case for most of the world's bird species.

The speed and magnitude of the BMR response may depend on the acclimation temperature (McKechnie 2008), suggesting that tropical montane birds that experience lower temperatures will show slower responses to climate change than lowland species. Even risk-taking and predator avoidance behavior has been shown to be influenced by the BMR (Møller 2009). Given the importance of BMR variability in shaping birds' responses to climate change, studies of the physiological responses of tropical birds to climate change constitute an urgent research priority.

8.5 Mobility and migration

Given their higher mobility, birds are better able to move in response to climate change in comparison to most other organisms. However, even seemingly rapid rates of tracking climate change may not be enough (Devictor *et al.* 2008). Most tropical bird species and their habitats will not be able to shift fast enough or far enough to keep up with rising temperatures, particularly in flat, lowland areas (Loarie *et al.* 2009).

The majority of bird species are sedentary and they are 2.6 times more threatened or near threatened with extinction than long-distance migratory bird species that comprise only a fifth of all bird species (Sekercioglu 2007). Due to higher mobility, migratory birds have a better capacity to deal with global change than sedentary birds. Based on 60 different scenarios to estimate bird extinctions that will result from a combination of climate change and habitat loss by the year 2100, Sekercioglu *et al.* (2008) showed sedentary birds are five times more likely to go extinct than are long-distance migrants.

Many migratory birds spend most of the year on their wintering grounds in the tropics. Migratory birds that winter in the tropics have their own unique challenges (Faaborg *et al.* 2010a,b; see also Chapter 9). They are exposed to the multiple effects of climate change on their breeding and wintering grounds and during their migrations (Both *et al.* 2006; Huntley *et al.* 2006; Ahola *et al.* 2007). Climate change can result in range reductions at both ends, wipe out key stopover sites, and make the migration longer (Huntley *et al.* 2006). Breeding bird surveys showed that while sedentary bird species and short-distance migrants in Denmark increased since 1974, long-distance trans-Saharan migrants declined 1.3% per year (Heldbjerg and Fox 2008).

Between Europe and Africa alone, 2.1 billion birds are estimated to migrate every year (Hahn *et al.* 2009; see also Chapter 9). Shifts and reductions in the wintering ranges of migratory birds are likely to have impacts on tropical bird communities. Barbet-Massin *et al.* (2009) project that due to climate change by 2100 the ranges of 37 of 64 trans-Saharan migrants will shrink, shifting on average by 500 km and resulting in major decreases in the richness of migrants in Africa (Barbet-Massin *et al.* 2009).

Migratory birds can experience increasingly severe food shortages in their wintering grounds due to reduced rainfall, which can affect non-breeding performance and influence the departure time of migratory birds for their breeding grounds (Studds and Marra 2007). Changes in tropical wintering habitats also disadvantage migrants on their breeding grounds. Female American redstarts (*Setophaga ruticilla*) wintering in high-quality habitat produce more young and fledge offspring weeks earlier than females from poor-quality wintering habitat (Norris *et al.* 2004). Changes in migratory timing are already having negative consequences for some species (Both *et al.* 2006; Ahola *et al.* 2007).

8.6 Evolution

Climate change also has evolutionary consequences (Bonaccorso *et al.* 2006). For species to successfully adapt fast enough to keep up with climate change, phenotypic plasticity, learning, maternal effects, and immigration will not be sufficient by themselves, and the rate of microevolution will be the primary determinant of adaptation to climate change (Visser 2008).

Tropical mountain areas have been centers of speciation and endemism due to their isolation (Voelker *et al.* 2010) and long-term habitat stability (Fjeldså and Lovett 1997; Graham *et al.* 2006). Pliocene climate change in Africa, by causing lowland forest retraction and rapid isolation of montane forests, has driven the speciation of African forest birds (Njabo and Sorenson 2009; Voelker *et al.* 2010). Climate change, which will further isolate tropical montane forests, is progressing faster than the average rate of evolutionary change in vertebrates. Features like long-term environmental stability and geographical isolation that promote endemism and speciation on actual and ecological islands (such as tropical montane forests), also make island species especially vulnerable to environmental change (Fordham and Brook 2010). Vertebrate evolution will mostly be unable to keep up with rapid, man-made climate change, exacerbated by extensive habitat loss (Njabo and Sorenson 2009), and the rate of bird species loss will be higher than the rate of evolution of new species. In the past 10,000–100,000 years, higher temperature change in an area led to fewer endemic species of terrestrial vertebrates and vascular plants (Jansson 2003). Range-restricted endemic species are concentrated in areas with smaller past climatic shifts, and past climatic change is a better predictor of endemism in birds than current annual range in temperature (Jansson 2003). This suggests that increasing temperatures in the tropics is likely to lead to the disappearance of many habitats and species (see also Chapter 3).

Mountainous areas covering a wide elevational range and harboring high topographical complexity have been important centers of species diversification and have played an essential role in the persistence of lineages (Fjeldså and Rahbek 2006). Areas with long-term climate stability not only help the long-term persistence of relict taxa, but also promote the diversification of radiating lineages, resulting in clusters of species with restricted ranges (Fjeldså and Lovett 1997). Such areas are essential for the conservation of endemic and range-restricted tropical species and must be prioritized in tropical conservation plans.

Equatorward populations, through their long-term persistence during periods of climate change, have accumulated more genetic variability, whereas populations in more temperate areas (and likely at higher elevations in the tropics) have been frequently eliminated during colder periods, have gone through more bottlenecks, and thus have reduced genetic variability (Hewitt 1996). It is expected that bird species' equatorward limits (and likely their lower elevation limits) (Colwell *et al.* 2008) will retreat roughly in equilibrium with climate change whereas their poleward (and higher elevation) limits are likely to lag behind climate change. This is especially because range expansion usually requires the concomitant expansion of bird habitat, which is linked to the range expansion of immobile, long-lived plants. Furthermore, tropical tree species require a long time to expand their ranges, even when there is suitable terrain that is not occupied by people. Together, these factors indicate that habitat expansion to newly suitable areas will not be fast enough to make up for habitat losses to climate change, rapid range reductions will result, and equatorward populations with higher genetic diversity will be among the first ones to go (Huntley *et al.* 2006; Sekercioglu *et al.* 2008).

8.7 Interspecific interactions and indirect effects

Most models of climate change effects on future species distributions do not take into consideration species' interactions, usually because related data are fewer than the basic data on species distributions. However, species do not exist in isolation and changes in biotic associations can be as important as changes in temperature and precipitation, if not more so (Dunn *et al.* 2009). Preston *et al.* (2008) showed that including biotic associations in climate change models reduced habitat availability for endangered California plant and butterfly species by 68–100% as opposed to a climate-only model. The authors indicated that incorporating biotic interactions in climate models is especially important for specialists and species that strongly depend on other species. Such species are typical of tropical bird communities, where mostly unknown interactions between species will influence the effects of climate change on tropical birds.

Indirect effects of climate change on bird communities through interspecific interactions have been little studied, but can be surprising and substantial (Traill *et al.* 2010). Climate-induced changes in the temporal partitioning of the breeding period can lead to increased interspecific competition with fatal consequences

(Ahola *et al.* 2007). Pied Flycatchers (*Ficedula hypoleuca*) increase their attempts to take over European Great Tit (*Parus major*) nests if there is greater overlap between the laying dates of these species (Ahola *et al.* 2007). These attempts can often be fatal for the Pied Flycatcher, whose arrival date from its breeding grounds, and hence its laying date, is affected by climate change (Both *et al.* 2006). In the tropical cloud forests of Monteverde, Costa Rica, increased temperatures led to a reduction in dry season mist frequency and the lifting of the cloud base (Pounds *et al.* 1999). This, in turn, led to some lowland bird species expanding their distributions upwards. Among these species, the Rainbow-billed Toucan (*Ramphastos sulfuratus*) is a cavity-nesting nest predator, which means that the montane forest specialist Resplendent Quetzal (*Pharomachrus mocinno*) now faces both competition from and nest predation by the toucans. In the same region, Jankowski *et al.* (2010) recently showed that interspecific aggression was a critical factor in constraining the elevational ranges of forest songbirds. The high elevation specialist Slaty-backed Nightingale-thrush (*Catharus fuscater*) was the most submissive of the species tested, indicating that *C. fuscater* is likely to lose ground to the more aggressive, lower elevation Black-headed Nightingale-thrush (*C. mexicanus*) if they expand to higher elevations in response to the warming temperatures in the region (Jankowski *et al.* 2010).

New interactions due to climate change are not limited to birds. One of the most fascinating and unexpected examples of negative climate change impacts on songbirds has been caused by Edible Dormice (*Glis glis*) in the Czech Republic (Adamik and Kral 2008). With increasing spring temperatures, the cavity-nesting dormice are waking up earlier from their winter hibernation. However, only one of four cavity-nesting bird species in the region advanced its breeding dates, leading to high brood losses caused by more nest predation by dormice, whose populations have increased due to favorable weather combined with good seed mast years. As Adamik and Kral (2008) point out, "changes in climate might affect organisms at various trophic levels with often unexpected outcomes" and "species most at risk are those at different trophic levels that do not shift at the same rate or in the same direction as their food resources, predators or competitors."

An interesting example of climate change-induced interactions affecting a tropical bird community comes from seabird nesting islands near the Great Barrier Reef of Australia. Effects of climate change on tropical seabirds have received little attention, but changes in sea surface temperatures can reduce marine prey availability for seabirds (Becker *et al.* 2007; Le Bohec *et al.* 2008; Watanuki *et al.* 2009). Reduced prey not only directly affects seabirds, but it can also influence the impoverished island ecosystems where seabird droppings (guano) provide a critical nutrient input to these ecosystems (Croll *et al.* 2005; Sekercioglu 2006a,b; see Chapter 4). A decrease in seabird prey due to an increase in sea-surface temperatures has also meant a reduction in nutrient input for plants on seabird nesting islands (Greenslade 2008). In turn, this reduction has put *Pisonia* trees under stress, resulting in a mobilization of nutrients that has led to the outbreaks of a sap-feeding herbivore and its attendant ant (Greenslade 2008). This outbreak further damages the *Pisonia* trees, an important nesting tree for seabirds, illustrating the complicated webs of interactions through which climate change effects reverberate.

8.7.1 Ecosystem services

Not only are tropical bird species involved in countless interactions with each other, they also provide key ecosystem services by interacting with other organisms as seed dispersers, pollinators, predators, nutrient dispersers, scavengers, and ecosystem engineers (Sekercioglu 2006a,b; see Chapter 4). Avian ecosystem services are important in many tropical communities, but the influence of climate change on these services is little understood. Bird pollination is more important in the tropics than in the temperate zone, with the exception of Australia where bird pollination reaches a peak (Ford 1985). Bird pollination has often been considered relatively unimportant in comparison to insect pollination. However, climate change may increase the importance of bird pollination, as exemplified by Caribbean islands where bird pollination increased with higher rainfall while insect pollination decreased (Gonzalez *et al.* 2009).

Ecosystem services become even more important with the increasing variability climate change brings (Traill *et al.* 2010). Mazia *et al.* (2009) showed that during the wet year of an ENSO cycle, excluding birds from tree saplings in an Argentinian *Nothofagus* forest resulted in leaf damage twice as much as was observed during a drought year. Bird exclusion also resulted in significant sapling leaf damage in the wet year, but not in the drought year (Mazia *et al.* 2009). These results indicate that large-scale climatic events can influence the strength of trophic cascades, but our understanding of such variable effects on ecosystem function, especially in the tropics, is limited (e.g. Van Bael *et al.* 2004).

Climate change is likely to influence thousands of such interactions through population changes, distributional shifts, physiological constraints, the creation of no-analog communities, and other mechanisms we can only guess at. Far more detailed, long-term research on tropical bird community ecology is needed to unravel the direct and indirect effects of climate change on the world's most diverse bird communities.

8.8 Deadly synergies

In addition to the major effects of climate change in physiologically stressing organisms, and forcing species and entire communities to shift their distributions, climate change will also affect biodiversity by synergistically interacting with and often exacerbating the effects of other environmental stressors, such as habitat loss, emerging diseases, invasive species, hunting, or pollution (Brook *et al.* 2008; Laurance and Useche 2009; Reino *et al.* 2009; see Chapters 1, 6, and 7). However, these synergies are often overlooked in climate change research, and a recent review found that about half of the papers in climate change literature considered climate change separately from other threatening processes (Felton *et al.* 2009). Tropical birds seem to be particularly vulnerable to synergisms between climate change and hunting and between climate change and habitat loss in the form of agriculture (Laurance and Useche 2009). Nevertheless, remaining uncertainties about tropical temperature and precipitation projections (Vera *et al.* 2006) and the responses of tropical forests to changes in CO_2, temperature, and rainfall

(Phillips *et al.* 1998; Feeley *et al.* 2007b; Laurance and Useche 2009) limit our ability to predict future synergies.

8.8.1 Habitat loss

Extensive habitat loss in the tropics will continue to interact with and exacerbate the effects of climate change on tropical birds, especially endemic and range-restricted species (Jetz *et al.* 2007; Sekercioglu *et al.* 2008; see Chapter 1). Based on a plausible "worst case" projection of 6.4°C surface warming by the end of this century (IPCC 2007), Sekercioglu *et al.* (2008) estimated that the worst case habitat loss estimates in the "Order from Strength" scenario of the Millennium Assessment (Millennium Ecosystem Assessment 2005) can increase bird extinctions from climate change by about 50% when compared with the best case estimates from the "Adaptive Mosaic" scenario (Figure 8.1). Sekercioglu *et al.* (2008) showed the sensitivity of bird extinctions to the combined effects of climate change and habitat loss to be quadratic, which means that the extinctions increase faster than would be expected from a one-to-one relationship with surface warming. The number of extinctions rapidly increase past 1.8°C of surface warming and so do the additional extinctions from climate change–habitat loss interaction effect, showing the critical importance of keeping global warming to under 2°C by 2100.

8.8.2 Hunting

Other factors, such as hunting, also interact with climate change and exacerbate its effects on tropical bird populations (see Chapter 7). Traill *et al.* (2009) developed one of the few spatially explicit population viability models for a tropical waterfowl species, the Magpie Goose (*Anseranas semipalmata*), to simulate population responses to the synergy of hunting, climate change, and increased disease prevalence due to climate change. Without hunting, the simulated disease outbreaks rarely threatened meta-population viability, and only with high mortality and regular outbreaks. With current hunting pressure, however, the population response switched from a threshold response to a linear one, and the threat to meta-population viability from a disease outbreak increased significantly.

8.8.3 Invasive species

While climate change is a crisis for many specialized tropical bird species, it presents opportunities to other, more invasive species that will expand their range into more temperate regions (Reino *et al.* 2009; see Chapter 6). The expansion of invasive species will result in new communities and ecological interactions with consequences that are often hard to predict, but likely to be negative for some species (Pounds *et al.* 1999).

8.8.4 Emerging diseases

Increases in the prevalence of infectious diseases provide some of the most important examples of a deadly synergy with climate change effects. Changes in temperature, humidity and precipitation will affect many pathogens, climate warming is likely to "increase pathogen development and survival rates, disease transmission, and host susceptibility, and most host–parasite systems are predicted to experience more frequent or severe disease impacts with warming" (Harvell *et al.* 2002). Some recent examples from the tropics include the altitudinal expansion of malaria with increasing temperature (Williams 2010), a rise in coral diseases (Harvell *et al.* 2002), and mass extinctions in Neotropical amphibians as a result of a fungal pathogen outbreak caused by warming temperatures in the highlands (Pounds *et al.* 2006). A similar expansion of malaria-carrying mosquitoes into Hawaiian highland forests threatens many Hawaiian endemic bird species (Freed *et al.* 2005; Atkinson and LaPointe 2009; see Chapter 6), which had no experience with malaria or mosquitoes before they were introduced to the islands in the 19th century (Freed *et al.* 2005). With increasing warming, the elevational distribution of the 13°C isotherm, above which malaria does not occur, and the 17°C isotherm, below which malaria is year-round and high-risk, are expected to move upslope (Atkinson and LaPointe 2009; see Figure 6.5). This is expected to result in large reductions in high-elevation forest habitat (up to 96%), where malaria is low-risk or is found only seasonally (Atkinson and LaPointe 2009). As avian malaria is found throughout the world, similar climate-related increases in disease prevalence among tropical birds will be widespread, but there are few long-term studies on the subject.

8.9 Seasonality in tropics

Even though the stereotypical view of tropical bird life history has been the lack of seasonality compared to that of temperate birds, detailed and long-term studies are revealing that many species show seasonality that is often tied to the wet–dry season cycles (Tye 1992; Wikelski *et al.* 2000, 2003a). There are few examples of the responses of tropical bird phenology to climate change, but Cameroon bird species that reverse from breeding in the wet season in the lowlands to breeding in the dry season in the highlands suggest that such responses are possible (Tye 1992). Heavy rainfall, high humidity, and low temperatures during the wet season in the mountains prevent birds from breeding, including lowland species which normally prefer to breed in the wet season in the drier and hotter lowlands (Tye 1992). It is likely that many tropical bird species will shift their breeding periods in response to changes in temperature and rainfall regimes. Changes in the frequency and severity of tropical storms also have important implications for the ecology and conservation of tropical birds (Boyle *et al.* 2010). If climate change results in a mismatch between a critical reproductive cue such as the photoperiod (Wikelski *et al.* 2000) and the optimal temperature and rainfall regime for reproductive success, population declines may be the result.

Research on the effects of climate change has almost entirely focused on changes in average temperature, and to a lesser extent, on changes in rainfall. However, changes in seasonality can be equally important, particularly in the tropics where seasonality is less pronounced than in the temperate zone and where organisms are adapted to fewer climatic fluctuations and extremes. In the tropics, temperatures show relatively little variation. However, even in the wettest tropical areas, rainfall shows some seasonality, and in many tropical regions, there is a distinct dry season that puts pressure on local populations. Increases in rainfall seasonality and consequent increases in resource bottlenecks are likely to exacerbate the expected impacts of changes in temperature and precipitation regimes (Williams and Middleton 2008). This is particularly the case for extended dry periods when birds experience a scarcity of food resources such as nectar, fruit, and insects. As many tropical birds time their breeding with increased resource abundance typical of the wet season, longer and less predictable dry seasons and droughts as a result of climate change can affect the migrations and reproductive performance of tropical birds and lead to population declines (Williams and Middleton 2008; Sekercioglu 2010b). Increased climatic seasonality, especially of rainfall, has led to lower bird numbers in Australian tropical rainforest (Williams and Middleton 2008), with the results supporting the hypothesis that more climatically stable areas are more diverse (Pianka 1966). Increased rainfall seasonality due to climate change will likely result in population reductions and in species richness declines in tropical bird communities.

Even though the effects of changes in seasonality is a major concern for migratory birds breeding in the temperate zone, many tropical bird species also exhibit breeding seasonality that is linked to seasonality, especially of rainfall (Altwegg and Anderson 2009). Changes in temperature and precipitation regimes can lead to mismatches between food and breeding in tropical birds whose resilience will partly be determined by their genetic heritage shaped by past climate change (Carey 2009).

8.9.1 Variability and extreme weather

Global warming, by increasing the amount of energy in the system, also increases climatic variability and is likely to increase the magnitude and frequency of extreme weather events, such as heat waves, "once-a-century" storms, cyclones, typhoons, and hurricanes (IPCC 2007). Such extreme weather events can be as destructive as higher average temperatures and changes in rainfall patterns, if not more so. It may be possible for a bird species to cope with a 2°C change in average temperature, but if that results in an increase in hurricane frequency and intensity that destroys critical habitat or makes foraging impossible (Boyle *et al.* 2010), the species can decline towards extinction (Martinez-Morales *et al.* 2009). Increases in cyclones, typhoons, and hurricanes will predominantly affect tropical bird communities, especially those in coastal and island habitats (Safford and Jones 1998; Lee *et al.* 2008b; Martinez-Morales *et al.* 2009). Furthermore, the recovery of damaged habitats may be very slow if regeneration is dependent on bird seed dispersal (Hjerpe *et al.* 2001).

The effects of El Niño and La Niña cycles on tropical forest birds can be dramatic, depending on the extent and severity of unseasonal rainfall and droughts that result from these weather cycles (Jaksic 2004). Tropical birds, especially granivores and insectivores, can respond rapidly to increases in plant and insect productivity as a result of El Niño-driven rainfall, and long-term studies are needed to understand how these responses affect bird population cycles (Jaksic 2004).

In estimating the effects of climate change on biodiversity, few models take into account the adverse effects of extreme weather events that are expected to become more frequent as a result of the predicted increase in climate variability. However, some species are especially vulnerable to extreme weather events such as heat waves, cold spells, typhoons, and hurricanes, and populations of these species are likely to decline more than expected based on changes in mean climate (Frederiksen *et al.* 2008). Most tropical bird species are adapted to narrower temperature ranges than temperate birds and tropical birds can track small differences in temperature (Sekercioglu *et al.* 2007). An increase in extreme temperatures will affect tropical birds more than temperate species.

8.10 Lack of knowledge

As is generally the case in ecology and conservation, most long-term studies of climate change and most modeling exercises on the changes in species distributions focus on the temperate zone. There are relatively few examples of tropical studies of the effects of predicted climate change on the distribution and conservation of tropical birds.

Despite increasing evidence of climate change effects on birds species worldwide, most research has been limited to the developed countries of the temperate world, particularly in North America and Europe (Felton *et al.* 2009; see also Chapter 1). Even though most birds species are tropical (Tscharntke *et al.* 2008) and sedentary (Sekercioglu 2007), the lopsided concentration of researchers and long-term datasets in the developed countries of the temperate zone (IPCC 2007) has meant that most of our understanding of climate change impacts on birds is based on studies of temperate birds that are largely migratory. Not only is this a major impediment for the conservation of world's birds, but it is also a significant obstacle to the advancement of ornithology, ecology, and conservation biology. The life histories of tropical birds differ from those of temperate birds in many aspects, such as clutch size, breeding success, longevity, basal metabolic rate, and the "pace of life" (Wiersma *et al.* 2007; Jetz *et al.* 2008a). In order to have a complete understanding of the interaction of climate change with avian ecology and of the consequences of changing climates for bird conservation, we must have many more long-term studies of tropical bird communities. Otherwise the predominance of the studies of temperate birds will continue to bias our outlook. However, long-term studies of climate change effects on tropical birds are few and far between (e.g. Pounds *et al.* 1999; Shoo *et al.* 2005a). Limited funding for long-term tropical ecological research, increasing emphasis on meta-analyses and molecular studies often at the expense of field research, and the reluctance of many funding agencies to fund long-term field research mean that conducting long-term tropical research on the impacts of climate change, essential for guiding policy, is a difficult prospect.

The exploding scientific interest in climate change impacts on birds has led to an interest in the papers on the subject, despite the scarcity of long-term studies. Again, we have much to thank for the interest in birds, but growing interest in climate change effects is also true for tropical birds (Marini *et al.* 2009). Creative scientists have analyzed differences in bird distribution data in field guides (Peh 2007), or the notebooks of 19th century amateur naturalists

(Willis *et al.* 2008; Primack *et al.* 2009), and the millions of data points collected by experienced birdwatchers volunteering for citizen science projects such as the US Breeding Bird Survey (Niven *et al.* 2009) or the European Breeding Bird Atlas (Hagemeijer and Blair 1997). However, amateur interest in birds is less common in tropical countries, most of whom have developing economies and less educated populations that are more concerned with survival then with birdwatching. On the other hand, the appeal of birds is universal, birdwatching can be a means towards a sustainable livelihood (Sekercioglu 2002b), and people in tropical countries will appreciate birds and learn to identify them rapidly if given the chance and motivation (Paaby *et al.* 1991). The low income in most tropical countries means that limited funding can go a long way and field guides make bird identification accessible to most of the world's population. The global birdwatching tourism market means ornithological training of local people can lead to combined careers in bird guiding and in collecting scientific data. Consequently, locally based, long-term bird monitoring programs in the tropics can combine biodiversity monitoring, environmental education, raising local awareness, community-based conservation, and ecotourism development. Such programs can be successful and cost-effective tools in poverty reduction in many developing countries.

8.11 Conclusion

Climate change has been a part of our planet's entire history. Past climate change that has often been of lesser magnitude and speed than we are currently experiencing has resulted in major upheavals in the planet's ecosystems and dependent organisms (Huntley *et al.* 2006). In the past 10 million years, the planet's average temperature has not increased by more than 1.5–2°C (Bartlein 1997), but it is expected to increase by 1.1–6.4°C in this century according to the International Panel on Climate Change (IPCC 2007). Due to the possibility of little-known or even unknown feedback loops, the upper limit is hard to predict, and even an unimaginable 11°C warming may be possible (Stainforth *et al.* 2005). As summarized by Huntley *et al.* (2006), "global climate is thus projected to be at least as warm, by the end of the present century, as it has been at any time during the evolution as species of most of the world's present diversity of organisms."

Even if one takes the unrealistically optimistic view that climate change has taken place in the past and species will simply adapt, it must be considered that in the past it has taken thousands of years for species to adapt to climate change, and also that organisms did not have to deal with 7 billion human beings that occupied the planet, consuming most of the planet's resources, and making it difficult or impossible for most species to shift their ranges in response to climate change. We need immediate societal change to reduce consumption, to control the emissions of greenhouse gases, and to halt climate change, in combination with having effective conservation areas that enable organisms to shift in response to the climate change that is already happening. Otherwise, we are looking at an out-of-control climate that will not only lead to enormous human suffering, but will also trigger the extinction of countless organisms, among which tropical birds will be but a drop.

9

Conservation of Migratory Birds in the Tropics

Overview: There are more than 500 species of birds numbering in the billions of individuals that spend the non-breeding season in the tropics. These birds occur in virtually all habitats, even severely human-modified habitats such as agroecosystems, and play crucial roles in tropical ecosystems. Conserving migratory birds has become a major focus of global avian conservation efforts aimed largely at preserving their abundance rather than preventing extinction, but is very difficult because loss and degradation of habitat on breeding, wintering and migratory stopover habitats can all cause population declines. Conserving migratory birds therefore requires large-scale conservation efforts that consider migratory connectivity between breeding and winter grounds, the possibility of effective conservation in at least some agroecosystems (e.g. shade coffee and cacao), and on transboundary acquisition of pesticides and pathogens.

9.1 Migration – an imperiled phenomenon

Migratory birds have been a major focus in avian conservation biology over the last three decades, as suggested by the large number of books, symposia, and multi-authored review articles published on their ecology and conservation (e.g. Keast and Morton 1980; Terborgh 1989; Crick and Jones 1992; Hagan and Johnson 1992; Mönkkönen *et al.* 1992; Rappole 1995; Martin and Finch 1995; Able 1999; Faaborg 2002; Greenberg and Marra 2005; Ralph and Rich 2005; Stutchbury 2007; Faaborg *et al.* 2010a,b). Terborgh's (1989) book, evocatively titled "Where Have All the Birds Gone?" exemplifies the high level of concern over

Conservation of Tropical Birds, First Edition. Navjot S. Sodhi, Çağan H. Şekercioğlu, Jos Barlow and Scott K. Robinson.

Published 2011 by Blackwell Publishing Ltd.

the conservation of migratory songbirds, which has led to huge intergovernmental and interagency task forces and conservation programs to promote their conservation (Rich *et al.* 2004; Faaborg *et al.* 2010a). Wilcove (2008) considers the problems faced by migratory songbirds to be emblematic of the more general phenomenon of the decline of the world's great animal migrations.

From a conservation perspective, migratory birds may face the worst of all possible worlds. Because they breed and winter in different geographical regions and must rely on stopover habitat during migration, human impacts in any of these regions can cause population declines (Greenberg 1980; Sherry and Holmes 1995; Sillett and Holmes 2005; Holmes 2007; Yamaura et al. 2009; see Figure 9.1). Habitat loss, degradation, and fragmentation as well as acid rain negatively

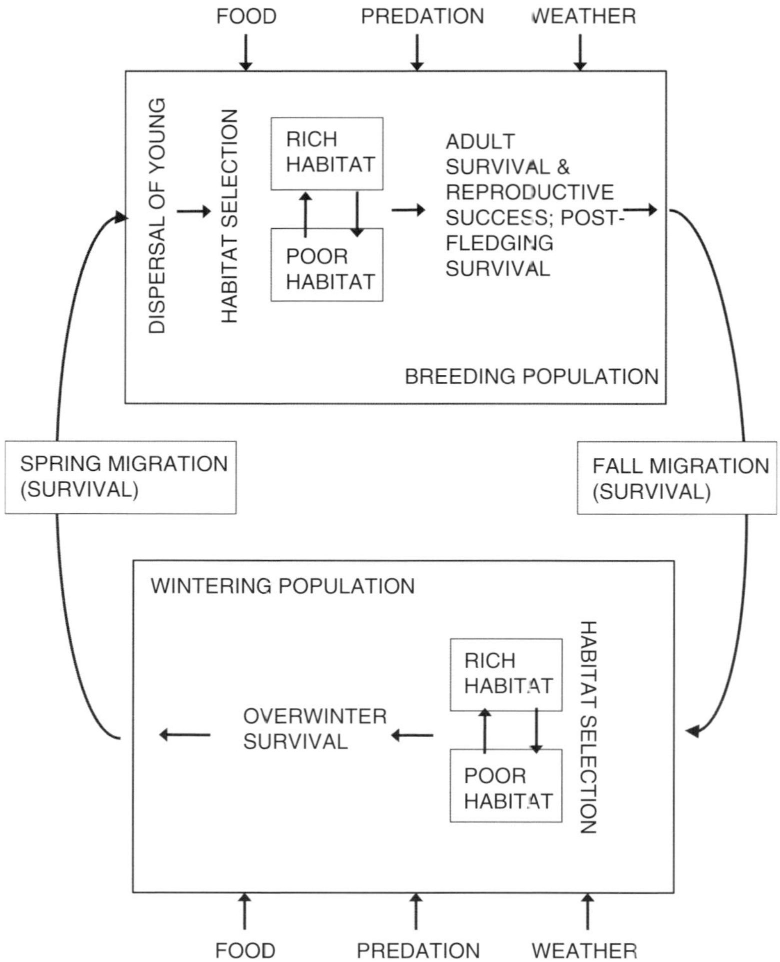

Figure 9.1 The often-reprinted Sherry–Holmes model of population regulation in migratory songbirds. This model demonstrates the many ways in which an event on often-distant breeding, overwintering and migratory habitats can regulate populations. (After Sherry and Holmes 1995.)

affect migratory birds on the breeding grounds (reviewed in Faaborg 2002; Stutchbury 2007; Faaborg *et al.* 2010a,b). Loss of critical stopover habitats is a severe and increasing cause of mortality (Moore 1999, 2000; Gauthreaux *et al.* 2005; Moore *et al.* 2005; Faaborg *et al.* 2010a,b), as is extensive trapping and mistnetting during migration, especially in Eurasia (Wilcove 2008). Mortality from tower kills (Jones and Francis 2003), windows, and wind turbines during migratory flights are also growing problems (Stutchbury 2007; Faaborg *et al.* 2010a,b). Migrants also face pathogens and environmental contaminants from widely different regions (Gard and Hooper 1995; Mora 1997; Rappole and Hubalek 2003). Trying to determine which of these threats is most important is a daunting task and there are very few studies that have definitively established whether populations are more limited by events on the migratory, breeding, or wintering grounds for any species except perhaps the American Redstart (*Setophaga ruticilla*) (reviewed in Faaborg *et al.* 2010a,b).

Initially, concern about conservation of migratory birds was spurred by evidence for large-scale population declines (e.g. Robbins *et al.* 1989; Sanderson *et al.* 2006). In North America, for example, the continent-wide Breeding Bird Survey (www.mbr-pwrc.usgs.gov/bbs) has been analyzed perhaps more extensively than any other dataset in recent times (e.g. Robbins *et al.* 1989; Askins *et al.* 1990; Peterjohn *et al.* 1995; James and McCulloch 1995; James *et al.* 1996), and shows evidence for steep declines, at least in some Neotropical migrants (species that breed in the Nearctic and overwinter in the Neotropics). Other datasets from long-term censuses in the tropics (Faaborg and Arendt 1992; Faaborg *et al.* 2007), Old World (e.g. Fuller *et al.* 2005; Sanderson *et al.* 2006; Hahn *et al.* 2009), and Australasia (Ford *et al.* 2001) have also suggested widespread declines in migratory (as well as resident) species. Although there is much debate about the magnitudes of these declines (James and McCulloch 1995; James *et al.* 1996; Robinson 1997), there is no doubt that some species have declined enough to send warning signals to conservationists (e.g. the Cerulean Warbler *Dendroica cerulea*: Robbins *et al.* 1992b; Hamel *et al.* 2004; Colorado *et al.* 2008; Bakermans and Rodewald 2009). When combined with other studies showing potentially major declines based on radar data (Gauthreaux 1999; Gauthreaux and Belser 2005) and the increasing absence of migratory songbirds from small habitat patches (e.g. Robbins 1979; Whitcomb *et al.* 1981; Robbins *et al.* 1992a; Faaborg *et al.* 2010a,b), there was enough concern to launch the massive Partners in Flight North American Bird Conservation Plan (Rich *et al.* 2004) and to fund millions of dollars of on-the-ground conservation efforts (Faaborg *et al.* 2010a).

Because of the geographical scale of migratory behavior, conserving migratory birds requires comparably large-scale conservation efforts (Martin *et al.* 2007). Conservation efforts in one region are unlikely to be sufficient to save species that spend much of their lives in other areas (Martin *et al.* 2007; Jones *et al.* 2008c; Yamaura *et al.* 2009). Migratory bird conservation has become a powerful tool for drawing attention to conservation problems in seemingly remote parts of the world. Loss of habitat in Central America (see Chapter 1) or drought in the Sahel, which might seem like problems of little relevance to birdwatchers in North America or Europe, become issues of direct concern when they lead to population declines of familiar breeding species in the backyards of people who live in the temperate zone (Wilcove 2008). As a result of this concern, the US Fish and Wildlife Service now funds research on migratory songbirds in the Neotropics, not just on their breeding grounds. Similarly, the conservation of declining migratory birds in Japan would require preserving habitats in heavily deforested Southeast Asia where they overwinter (Yamaura *et al.* 2009).

For the most part, migratory songbird conservation differs from programs designed mainly to prevent extinctions. Even some of the most dramatically declining species such as the Cerulean Warbler in eastern North America have likely populations numbering in the hundreds of thousands (Robbins *et al.* 1992; Hamel *et al.* 2004; Colorado *et al.* 2008), which does not yet qualify them for the endangered species list. Rather, the overriding concern of migratory songbird conservation efforts is to preserve abundance, both for our own enjoyment and to maintain the roles these species play in their ecosystems (Sekergioglu *et al.* 2002, 2004; Van Bael *et al.* 2008; Mooney *et al.* 2010; see Chapter 4). If conservation efforts aimed at reversing habitat loss in the tropics are not implemented, Terborgh (1980) argued that few migratory species would become extinct, but that our woodlots "would just not sound the same," a phenomenon that would lead to "future shock" among birders (Morton and Greenberg 1989).

Since the onset of massive conservation efforts aimed at migratory birds, there has been a renaissance of research into the phenomenon of migration (see Greenberg and Marra 2005; Faaborg *et al.* 2010b). Although there is general agreement that migrant populations are limited by a combination of factors acting throughout the life cycle of migratory birds (Figure 9.1), there is growing agreement that events on the tropical winter grounds are extremely important (Sherry and Holmes 1995; Runge and Marra 2005; Sherry *et al.* 2005). Indeed, many of those who first promoted conservation of migratory birds were primarily interested in using migratory birds and their connection with temperate breeding grounds to promote tropical conservation. Most papers in Keast and Morton (1980) and Hagan and Johnson (1992), for example, focused on conservation issues in the tropics. The shift in focus to temperate breeding grounds (e.g. Finch and Stangel 1993; Martin and Finch 1995), especially declining nesting success in relation to habitat fragmentation (reviewed in Faaborg 2002), seemed a bit like an unnecessary distraction to those who felt that most migratory birds were obviously tropical species that only left to spend a few months breeding in North America where there were abundant breeding opportunities and few predators (Rappole and MacDonald 1994; Rappole 1995; but see Latta and Baltz 1997). Indeed, the Neotropical origins of many, if not most migrants has been supported by studies from the fossil record (Steadman 2005) and from phylogenetic approaches (e.g. Böhning-Gaese and Oberrath 2003; Mila *et al.* 2006). More recently, however, the focus has shifted back to the interconnectedness of every aspect of migration (Holmes 2007; Faaborg *et al.* 2010a,b). Technological advances in stable isotopes, genetics, and telemetry are increasingly making it possible to unravel the complexities in the life cycle of migratory birds (Greenberg and Marra 2005), which is providing valuable information for the design of conservation strategies (Martin *et al.* 2007; Faaborg *et al.* 2010a).

In the rest of this chapter, we focus on the main issues in conserving tropical birds in terrestrial habitats in the tropics. We argue that one of the most compelling reasons to focus on migratory birds is that they are also excellent conservation indicators of the health of human-dominated landscapes in the tropics. As such, they are ideal targets for "reconciliation" or "win-win" ecology (Rosenzweig 2003), which seeks to maintain wildlife in landscapes that have already been converted to heavy human use. Before we describe the conservation problems faced by migratory songbirds, we will first review some aspects of their natural history, ecology, and behavior that are important to consider when identifying conservation issues that need to be addressed.

9.2 The natural history of migratory birds

9.2.1 The advantages of the migratory strategy

In many respects, migratory birds should be well adapted to cope with a changing world. From a life history and ecological perspective, if not necessarily from a conservation perspective, the migratory lifestyle may offer the best of all worlds (Able 1999; Greenberg and Marra 2005). Migratory birds that winter in the tropics spend their winters where it is warm and food is usually available year round. What's more, they do not have to breed in the tropics where nest predators abound (e.g. Oniki 1989; Robinson *et al.* 2000) and there are many potential competitors for resources (reviewed in Rappole 1995). In winter, migrants can switch to diets that would be marginal for raising young but are sufficient for keeping adults alive (Poulin and Lefebvre 1996) and many can nomadically follow seasonally and spatially variable resources (Morton 1971, 1979, 1980; Lefebvre *et al.* 1994) that would be difficult for residents to track, at least when breeding (reviewed in Rappole 1995). Instead, migrants leave the tropics to breed in northern latitudes where there is a flush of new foliage and fewer potential competitors and nest predators. Resident species in most northern latitudes are limited by mortality during the severe winter season (Newton 1998), which reduces potential competition with migratory birds during the breeding season at temperate latitudes (Rabenold 1978). Migrants can further minimize the costs of migration by timing their long-distance movements to coincide with peaks of resource availability such as the spring flush of new foliage and its attendant abundant folivorous insects (Graber and Graber 1983; Jonzen *et al.* 2007). Migratory species can also time their migratory flights to exploit optimal weather to minimize the energetic costs of migration (Able 1999; Wikelski *et al.* 2003b; Bowlin *et al.* 2005; Cochran and Wikelski 2005; Gauthreaux *et al.* 2005). Although migration is considered to be an extremely costly period in terms of mortality (Sillett and Holmes 2002), these estimates are complicated by between-year adult dispersal (e.g. Hoover 2003). At least one well-studied long-distance migrant, the Prothonotary Warbler (*Protonotaria citrea*), has a minimum annual survival rate of more than 80%, a very high value, even for a Neotropical resident species (Hoover 2003).

The advantages of migration may explain why there are so many migratory birds that breed in the northern hemisphere and winter in the tropics. There are more than 500 species of migratory birds that winter in tropical terrestrial habitats

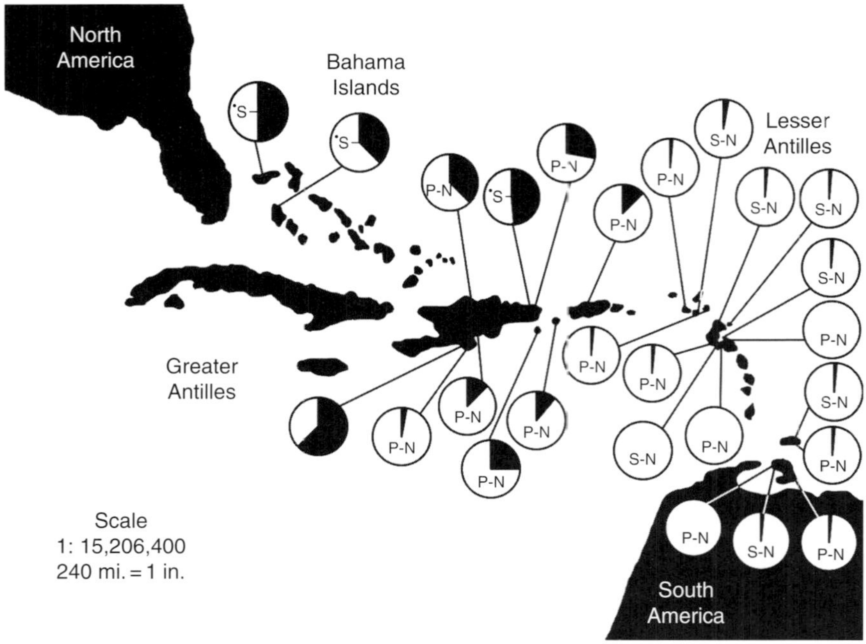

Figure 9.2 Proportion of forest mist-net samples consisting of Neotropical migratory birds in relation to distance from the breeding grounds. Note that the further from the breeding grounds, the lower the proportion of birds that consist of migrants. N = mistnet sample; P = primary vegetation; S = secondary vegetation. (After Terborgh 1989.)

and many billions of birds make these movements (Stutchbury 2007; Hahn *et al.* 2009). In many northern communities, more than 90% of the breeding species migrate to the tropics in winter (Böhning-Gaese 2005). In the northern tropics of the Americas, half or more of the birds present in some forest habitats during the northern winter are migratory (Figure 9.2; Terborgh 1989; Rappole 1995). Indeed, in some parts of Central America, many year-round resident species are often relegated to disturbed habitats and do not breed until the migratory birds have left to return to their breeding grounds, a period that also coincides with the onset of the wet season and an increase in fruit and insect resources (Rappole 1995).

Migratory birds occur in virtually all habitats in the tropics (Blake and Loiselle 1992; Robbins *et al.* 1992a; Petit *et al.* 1995; Rappole 1995; Estrada and Coates-Estrada 2005; Faaborg *et al.* 2010b). In general, the closer they are to their breeding grounds, the more likely it is that migrants will be found in the forest interior (Figure 9.2). In equatorial tropical South America (Robinson *et al.* 1988, 1995) and Africa (Brosset 1990), for example, migrants are largely restricted to the edges of forests and in early successional habitats such as those typically found along rivers and in areas of human disturbance. Tropical dry forests seem to be especially heavily used by migratory birds (e.g. Hutto 1992; Latta and Brown 1999; Rappole *et al.* 1999; Latta and Faaborg 2002; Strong and Sherry 2001b). Many migratory birds readily occupy severely modified habitats such as crop fields, pastures, heavily logged forests, urban areas, and plantations (reviewed in Rappole 1995). Severe habitat destruction such as that following hurricanes and fires seems to have few impacts on populations of most migratory birds (Lynch 1992; see Chapters 1 and 5) and may even be essential for some of the most endangered species such as the Kirtland's Warbler (*Dendroica kirtlandi*) (Wunderle *et al.* 2010). Migratory birds also exhibit great flexibility in their use of roost sites (Smith *et al.* 2008). Migratory birds therefore offer the potential for conservation even in landscapes that have been severely altered by human activity (see below).

9.2.2 Flexibility in social behavior

Social systems of migratory birds are highly variable, both within and among species (Greenberg and Salewski 2005; Brown and Sherry 2008; Faaborg *et al.* 2010a,b). Individuals of some species defend territories all winter and return to these territories between years (e.g. Greenberg 1979, 1984; Mabey and Morton 1992; Staicer 1992; Rappole 1995; Brown and Sherry 2008). In these same species, other individuals may wander nomadically or in flocks with large home ranges rather than defend a fixed territory (see also Rappole *et al.* 1989; Winker *et al.* 1990). Some individuals, especially females, may be forced into this nomadic lifestyle by behaviorally dominant conspecifics (Morton 1990; Holmes 2001). There is some evidence that these more nomadic individuals suffer increased mortality (Rappole *et al.* 1989; Winker *et al.* 1990; but see Conway *et al.* 1995). Territoriality seems especially well developed in forest species (Rappole 1995). Other species form large mono-specific flocks and move over large areas, especially in search of fruit and nectar, but also for some insects that occur in large concentrations such as aerial insects over water (Morton 1971, 1980; Rappole 1995).

9.2.3 Flexibility in diet and foraging behavior

Many migratory species change their diet and foraging ecology in subtler ways on their winter grounds, suggesting possible greater behavioral plasticity than in tropical resident species (Mettke-Hofmann and Greenberg 2005). Ovenbirds (*Seiurus auricapillus*), for example, switch from a generalized insectivore diet during the breeding season to a diet of mostly ants in winter (Poulin and Lefebvre 1996). Ants would be very difficult for nestlings to digest, which may explain why they are rarely eaten during the breeding season, but they apparently provide enough nutrition to enable adults to survive when not breeding. Ants are a resource exploited by very few year-round resident species, so this is a niche that would seem to minimize competition with resident species.

Some changes between breeding and winter grounds, however, may actually increase overlap in resource use with resident species. Many insectivores such as Golden-winged (*Vermivora chrysoptera*), Blue-winged (*V. pinus*), and Worm-eating warblers (*Helmitheros vermivorus*) switch from mostly searching live leaves during the breeding season to searching suspended dead leaves during the winter (Greenberg 1987). Dead leaves are a major foraging substrate in the tropics and this is a niche that has many specialist year-round residents (Remsen and Parker 1983). Suspended dead leaves are used as a hiding place by many insects such as spiders and katydids, and the competition between species that search these leaves can be intense (Rosenberg 1990). Other migrants join mixed-species flocks (e.g. Greenberg 1981, 1992; Staicer 1992) and forage with ant swarms (Willis 1967; Gram 1998), both niches in which they face severe competition from residents occupying the same niches. Most studies of these flocks suggest that the migrants are forced by behaviorally dominant residents to occupy marginal, subordinate positions (e.g. at ant swarms; Willis 1966). Therefore, the ecology of some migratory species in winter brings them into direct contact with year-round residents.

Migratory insectivores, however, may have morphological constraints (Winkler and Leisler 2005) that reduce dietary overlap with many tropical year-round residents. The main insect resources available to temperate birds during the breeding season are relatively small and soft-bodied species, especially the larvae of Lepidoptera. As a result, temperate birds tend to have smaller bills than tropical insectivores, which can also exploit the many larger insects, especially Orthoptera, that are available in tropical forests (Greenberg 1983; Rosenberg 1990). Large katydids (Orthoptera; Tettigoniidae), for example, are not available during much of the temperate breeding season (Remsen and Robinson 1990). In the tropics, on the other hand, katydids are available year-round and appear to offer excellent nutrition for an insectivore (Robinson *et al.* 1990). Especially in the forest understory, therefore, migratory birds and year-round residents appear to exploit different kinds of insects, which might reduce overlap in resource use.

Many migratory birds undergo huge changes in both their social behavior and diet between the winter and breeding grounds (Faaborg *et al.* 2010b). One of the best-documented cases is the Eastern Kingbird (*Tyrannus tyrannus*), which is an extremely aggressive, territorial insectivore during the breeding season and forms large mono-specific flocks during the non-breeding season that move among fruiting trees in South America (Morton 1971; Fitzpatrick 1980). Individually, kingbirds are subordinate to most resident frugivores that feed at these trees, but collectively they cannot be chased away by non-flocking resident birds (Fitzpatrick 1980; Robinson *et al.* 1988, 1995). Flocking behavior therefore makes it possible for the Eastern Kingbird to exploit concentrated resources such as fruits that are aggressively defended by many birds, including its larger and behaviorally dominant congener, the Tropical Kingbird (*Tyrannus melancholicus*). Another migratory congener, the Scissor-tailed Flycatcher (*Tyrannus forficatus*), also becomes more gregarious in winter when it gathers at areas where insects concentrate, such as along waterways at dusk, and catches flying insects. Territorial residents such as the Tropical Kingbird cannot chase all of them away. Instead, the year-round resident Tropical Kingbird is forced to tolerate these large flocks (S. Robinson, personal observation). The Orchard Oriole (*Icterus spurius*) is another bird that is mostly territorial on the breeding grounds, but switches to a nomadic lifestyle in which it moves over very large geographical areas in search of concentrations of flowering trees (Morton 1979). Orchard Orioles, in fact, are very successful in urban and agricultural areas where they exploit trees such as the jacaranda that are widely planted as ornamentals (Morton 1979). Yellow warblers (*Dendroica petechia*), which defend all-purpose territories during the breeding season, aggressively defend single sources of fruit against all species during the winter grounds (Greenberg and Salgado-Ortiz 1994). Whether this

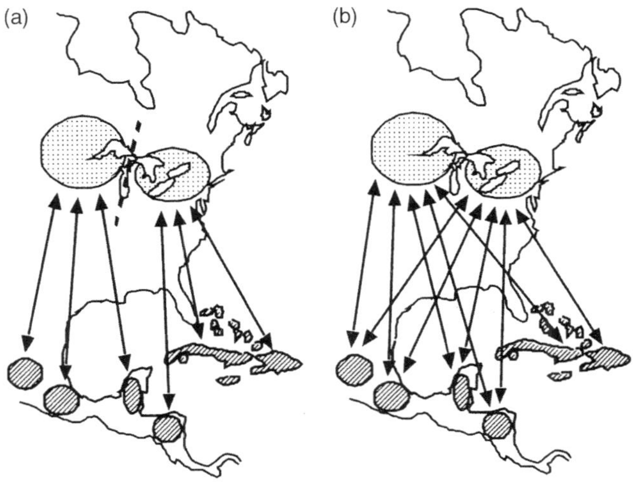

Figure 9.3 Hypothetical patterns of connectivity in migratory songbirds breeding in eastern North America and wintering in Central America and in the West Indies. Species with high levels of connectivity between specific wintering and breeding areas will require more geographically comprehensive conservation strategies. (After Webster and Marra 2005.)

flexibility will enable migratory songbirds to cope with increasing human alteration of habitats remains an open question.

The flexibility of many migrants suggests that they may also have increased cognitive capacity (Greenberg 1983; Mettke-Hofmann and Greenberg 2005; see also Chapter 6) relative to many tropical resident species, which tend to have relatively inflexible foraging behavior (Remsen and Robinson 1990). Such cognitive flexibility may be highly advantageous when adapting to landscapes that are undergoing changes as a result of human activities (Sol *et al.* 2008).

9.2.4 Migratory connectivity

Because of recent advances in technology (e.g. Hobson and Wassenaar 1997; Hobson 2005; Ricklefs *et al.* 2005; Smith *et al.* 2005; Wikelski *et al.* 2007; Stutchbury *et al.* 2009), we are beginning to understand where birds from different parts of their breeding ranges overwinter (Boulet and Norris 2006). Understanding migratory connectivity is vital to any long-term conservation planning (Figure 9.3; Martin *et al.* 2007; Jones *et al.* 2008c). If migratory populations show high fidelity to the same wintering and breeding areas, then conserving migratory songbirds will require that we identify these areas and ensure that breeding habitat is preserved in both winter and breeding grounds (Martin *et al.* 2007). In the Cerulean Warbler, for example, data from stable isotopes suggest that individuals from regions undergoing the steepest declines winter mainly in parts of South America such as in the mountains of Colombia where deforestation has been most severe (Figure 9.4). To date, however, most studies suggest only general segregation of breeding and winter populations into broad regions

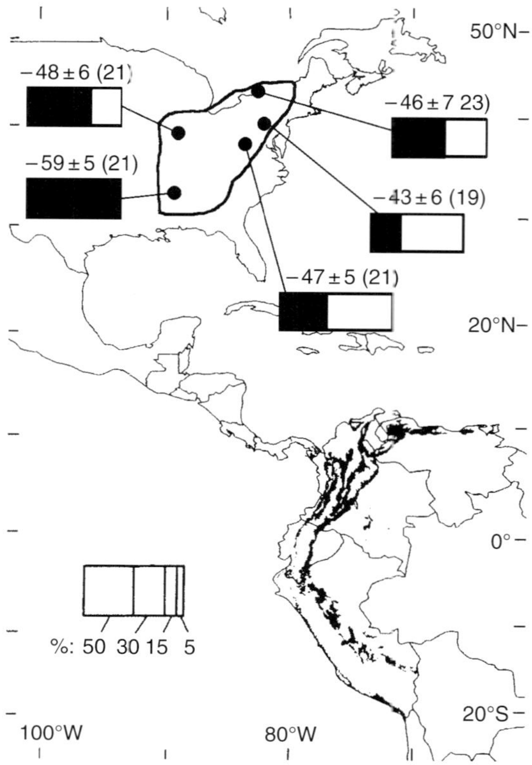

Figure 9.4 Migratory connectivity between wintering and breeding areas of the Cerulean Warbler estimated from stable isotope concentration in feathers. Note that eastern breeding populations tend to also winter further east and that western breeders winter further west. (After Jones *et al.* 2008c.)

(Faaborg *et al.* 2010a,b), and we still have a long way to go before we can firmly link the breeding and winter grounds of particular populations for most species. To date, we have only a general idea of where most breeding populations overwinter and vice versa (e.g. Rubenstein *et al.* 2002; Webster *et al.* 2002; Jones *et al.* 2008c), but with recent advances such as the use of isotopes (Hobson 2005), geolocators (Stutchbury *et al.* 2009), parasite communities (Ricklefs *et al.* 2005), genetics (see caveats in Smith *et al.* 2005) and satellite telemetry (Wikelski *et al.* 2007) to identify connections between breeding and wintering areas, we can expect rapid advances in our understanding of migratory connectedness.

9.2.5 Other migratory systems

In addition to the better-known migratory movements of birds between northern temperate and tropical latitudes, there are several other migratory systems in which birds move to or within the tropics for at least part of their life cycles. The second largest migratory system on earth is the austral migratory system

(Chesser 1994, 2005; Levey 1994; Joseph 1997; Jahn *et al.* 2004). Because land masses are much less extensive in the southern than in the northern hemisphere, the austral migratory system involves far fewer birds than the northern migratory systems. Nevertheless, there are many species of austral migrants and some habitats such as cerrado, chaco, and river edge vegetation may include a high proportion of migrants during the austral winter (Robinson *et al.* 1988, 1995; Jahn *et al.* 2004; Chesser 2005). There is an urgent need for studies of this system.

Altitudinal migrations have also been documented in several parts of the tropics (Hobson *et al.* 2003). Altitudinal migrations seem especially well developed in the Central American tropics (Loiselle and Blake 1991; Powell and Bjork 1994a,b; Chaves-Campo *et al.* 2008; Boyle 2010) and in the Eastern Arc mountains of Tanzania (Stevenson and Fanshawe 2002). Elevational migrants include some of the most charismatic of all tropical species such as the Resplendent Quetzal (*Pharomachrus mocinno*) and Three-wattled Bellbird (*Procnias nudicollis*) (Powell and Bjork 1994 a,b; Chaves-Campo *et al.* 2008) and some are highly endangered such as those breeding in the Eastern Arc Mountains of Tanzania (Stevenson and Fanshawe 2002). Elevational migration is especially prominent in subtropical mountains where there is far greater seasonality than in more tropical latitudes. There is some indication that elevational migrants move in response to changes in food abundance (Loiselle and Blake 1991; but see Boyle 2010), and there is also some evidence that elevational migrants may escape high levels of nest predation by moving upslope to breed (Boyle 2008).

Intratropical migrations also occur in some species that breed in the subtropics or in specific sections of the tropics and then migrate to other areas to breed (Morton 1981). In general, movements seem tied to variation in rainfall and attendant availability of fruit and nectar. Partial migration, with some individuals migrating and others remaining sedentary, is poorly understood in tropical birds (Jahn *et al.* 2010; Sekercioglu 2010b). However, a recent study in Tropical Kingbirds (*Tyrannus melancholicus*) showed that partial migration may be triggered by food shortage (Jahn *et al.* 2010).

9.2.6 Ecosystem services provided by migratory songbirds

Not surprisingly given the abundance of migratory species globally, migratory species play critical roles in ecosystem function (Van Bael *et al.* 2008; Mooney *et al.* 2010; see also Chapter 4). Within the tropics, migratory species are known to be key pollinators (Morton 1979), dispersers of fruit (Howe and DeSteven 1979; Wunderle 1997; Greenberg *et al.* 1995; Wenny and Levey 1998; Cordeiro and Howe 2003), and consumers of defoliating insects (Greenberg *et al.* 2000; Van Bael *et al.* 2003, 2008, but see Van Bael and Brawn 2005). In the canopy of drier forests in central Panama, which contain a high proportion of migratory songbirds during the northern winter, insectivorous birds consume a high proportion of defoliating insects and, as a result, significantly reduce losses of leaf area to herbivorous insects (Van Bael *et al.* 2003). In the forest understory and in the interior of wetter forests (which contain far fewer migrants: Karr 1971), insectivorous birds do not significantly reduce herbivory by insects (Van Bael and Brawn 2005). Although it is difficult to separate the effects of insect consumption by migrants and residents, these data suggest that migratory songbirds may play a significant role where they are abundant, which makes conserving them of potential economic importance (Sekercioglu *et al.* 2002, 2004). The ecosystem role of migrants as consumers of defoliating insects has also been demonstrated in both coffee and cacao agroecosystems (Van Bael *et al.* 2008).

9.2.7 Carryover effects of winter habitat on the rest of the life cycle

Another reason why tropical winter grounds are vital to the health of migratory bird populations is that tropical habitat quality has well documented carryover effects on migration, breeding, and dispersal (Runge and Marra 2005; Holmes 2007; see also Figure 9.1). Many migratory songbirds have strong, dominance-based behavioral interactions that determine how individuals are distributed along gradients of habitat quality (Sherry and Holmes 1996; Strong and Sherry 2001a; Marra 2000; Marra and Holmes 2001; Studds and Mora 2005). Many subordinate individuals, especially females, are forced to winter in habitats where their survival and body condition are reduced (Latta and Faaborg 2002; Latta *et al.* 2003; Brown and Sherry 2006; Johnson *et al.* 2006). Reductions in body condition are especially crucial in the late winter just before birds begin their migratory movements (Faaborg *et al.* 2010a,b).

Because habitat quality for many species such as the well-studied American Redstart and Black-throated Blue Warbler (*Dendroica caerulescens*) that winter in tropical wet forests is determined by moisture gradients, habitat quality can be inferred from stable isotope signatures related to soil moisture (Marra *et al.* 1998a); more generally, stress hormones may also provide an indicator of winter habitat quality as corticosterone stress hormones levels tend to be higher in poorer habitats (Marra *et al.* 1998b). American Redstarts that overwinter in lower-quality habitats as indicated by stable isotopes have been shown to have inferior body condition (Norris *et al.* 2004; Bearhop *et al.* 2004), arrive later (Both *et al.* 2006), breed further south (Studds *et al.* 2008), acquire less colorful plumage, and attract fewer mates (Reudink *et al.* 2009a,b; Germain *et al.* 2010). Thus, even though many tropical species may show extreme habitat breadth on their winter grounds (Lynch 1992; Petit *et al.* 1995; Petit and Petit 2003), the loss and degradation of some habitats, especially primary rainforest, may have severe negative effects of individual fitness and population growth.

9.3 Conservation problems faced by migratory birds in the tropics

9.3.1 Forest loss and fragmentation

Migratory songbirds are well integrated into tropical forests, especially at more northern latitudes closer to the breeding grounds (Figure 9.2). Many species defend territories and return annually to the same sites (Rappole 1995), which should increase their vulnerability to habitat loss (Warkentin and Hernandez 1996). Cleary, loss of primary rainforest will negatively affect populations of many species (see Chapter 1), especially those that pack into smaller winter ranges at higher population densities than they do on their breeding grounds (Terborgh 1989; Newton 1995; Stotz *et al.* 1996; Sillett and Holmes 2005). Loss of rainforest will be harmful to American Redstarts, for example, which leave the tropics in far better condition when they winter in wetter than in dry forests (Holmes 2007). Loss of dry forest, however, will also negatively affect those species that prefer drier conditions (Hutto 1989, 1992; Latta and Faaborg 2001; Strong and Sherry 2001b), just as loss of other, more restricted habitats such as mangroves (Warkentin and Hernandez 1996) will affect particular species such as the Prothonotary Warbler. Loss of forest habitat might also force more individuals to become wanderers, adopting essentially a fugitive lifestyle which may (Rappole *et al.* 1989; Winker *et al.* 1990) or may not (Conway *et al.* 1995) lead to higher mortality rates. Evaluating survival in nomadic birds is extremely difficult and remains a critical research need (Faaborg *et al.* 2010a).

The negative effects of habitat fragmentation, however, seem less extreme in overwintering migratory birds than during the breeding season or in resident species (Wunderle and Latta 2000; Faaborg 2002). Many migratory songbirds occur in even the smallest fragments (Robbins *et al.* 1992a; Petit *et al.* 1995; Petit and Petit 2003) and in gallery forest strips in mostly deforested landscapes (Turner and Corlett 1996). Perhaps this resilience to fragmentation is not surprising given the often-small territories defended by many migrants (Greenberg and Salgado-Ortiz 1994; reviewed in Rappole 1995) and the fact that overwintering migrants do not face the same constraints imposed by breeding species in the same habitats. Habitat fragmentation has extremely negative effects on year-round residents in the tropics (see Chapter 2). Indeed, one reason why migrants may be so successful in small fragments may be the lack of competition from residents (Dugger *et al.* 2004). Therefore, it may make more sense to focus on year-round resident species, which must breed and often occur at low population densities (Terborgh *et al.* 1990), when developing recommendations about minimum size of reserves (see Chapter 10). Nevertheless, there are at least some migrants that occur in lower densities in small fragments in the tropics (Rappole 1995). Migratory species may be excellent conservation foci for landscapes in which it may not be possible to restore large tracts, but in which some management to increase tract size is possible.

9.3.2 Corridors and altitudinal migration

In general, the extreme mobility of migratory birds should reduce the need for corridors connecting remaining habitat patches (see also Chapter 2). Many migrants have morphological adaptations that facilitate long-distance movements (Winkler and Leisler 2005). Many tropical species such as the Three-wattled Bellbird and the Resplendent Quetzal that engage in altitudinal migration, however, do not have these adaptations for long-distance flight and therefore might benefit from corridors connecting montane and lowland patches (Powell and Bjork 1994a,b). Such corridors are likely to be especially important in subtropical mountains such as the southern Himalayas where many terrestrial species, especially pheasants, engage in seasonal elevational migrations (Grimmett *et al.* 1999). For many altitudinal migrants, conserving intact altitudinal gradients might be especially critical.

9.3.3 Habitat degradation

For the same reasons that migrants might be less vulnerable to habitat fragmentation, degradation of tropical forests by silviculture may be less harmful to most migrant bird species than to resident species (see also Chapter 3). Even some of the most restricted species of migratory songbirds such as the Kirtland's Warbler appear to have at least some tolerance of anthropogenic habitat degradation (Wunderle *et al.* 2010). Migratory species persist even after the combined effects of hurricanes and fire have removed most of the vegetation from forests (Lynch 1992; see Chapter 1). As such, migratory birds may be excellent indicators of the value of even severely degraded forests, which are often converted to intensive agroecosystems such as oil palms in Southeast Asia (Koh and Wilcove 2008). Nevertheless, degraded forests adversely affect populations of many species (Rappole and Morton 1985; Rappole 1995) and declining body condition with habitat degradation might trigger the cascade of effects that result from declining habitat quality in the American Redstart (Figure 9.1; Holmes 2007). Migratory species therefore will also benefit from preservation of relatively intact tropical forest habitats.

9.3.4 Human-dominated landscapes

Migratory birds are often abundant in some of the most human-dominated landscapes in the tropics (e.g. Van Bael *et al.* 2007), which makes them an ideal target for conservation efforts in landscapes that have been severely altered by human activities ("reconciliation" or "win-win" ecology: Rosenzweig 2003; see also Chapters 1 and 2). Perhaps the best-studied agroecosystem is shade coffee plantations (Mas and Dietsch 2004). Shade coffee plantations contain far more birds, especially migrants, than sun coffee, which lacks a canopy of native species (Greenberg *et al.*, 1997a,b, 2000). In shade coffee plantations, birds provide an important ecosystem service by partially controlling folivorous insects and enhancing coffee crops (Perfecto *et al.* 2003; Van Bael *et al.* 2008). One of the fastest-declining songbirds, the Cerulean Warbler, occurs at higher population densities and has comparable estimated survival and body condition in shade coffee plantations than in primary forest in the Venezuelan Andes (Bakermans *et al.* 2009). Results of studies of birds have been critical in developing the specific set of management recommendations that determine if shade coffee is "certified" (Mas and Dietsch 2004). Studies of migratory birds in other agroecosystems (e.g. cacao: Van Bael *et al.* 2007) have the potential to help develop recommendations that greatly enhance the wildlife communities even in some of the most human-modified landscapes on the planet (Warkentin *et al.* 1995).

9.3.5 Disturbance-dependence

One of the most striking results of recent studies of North American birds is that the species that have undergone the steepest declines are not forest birds, but are species that depend upon disturbances to create habitat for them (Askins 2000; Brawn *et al.* 2001). Many tropical species also depend upon disturbances for the maintenance of their habitats. Kirtland's Warblers, one of the most critically endangered of all migratory species, may depend upon hurricane disturbance in winter just as they depend upon fire on the breeding grounds (Wunderle 1991). In South America, many Nearctic and austral migrants appear to depend upon disturbances created by rivers for the maintenance of their habitats (Remsen and Parker 1983; Robinson *et al.* 1988; Chesser 2005) and by fire (see Chapter 5), which historically maintained entire tropical ecosystems such as the cerrado where many austral migrants overwinter (Chesser 1994, 2005). Any changes in flooding and fire intensity will potentially adversely affect these species. Some of these species also use anthropogenic disturbances (Terborgh and Weske 1969), but we know little about the quality of these anthropogenic habitats for disturbance-dependent species. Based on what we know from the temperate zone, however, we should be wary of assuming that disturbance-dependent species will be able to maintain populations over the long term in human-altered habitats (Askins 2000; Brawn *et al.* 2001). Many formerly abundant species of human-disturbed habitats have declined so steeply that they are now on many regional lists of endangered species (Askins 2000).

9.3.6 Contaminants

Because pesticide and other pollutants vary greatly in their use among the nations of the world, migratory birds may be exposed to a much greater range of potentially harmful contaminants such as DDT (dichlorodiphenyltrichloroethane) that have been banned from many, but not all nations (Gard and Hooper 1995; Harper *et al.* 1996; Mora 1997; see Chapter 1; but see Capparella *et al.* 2003). We know little about the magnitude of this problem or about which pollutants are significant sources of mortality. This remains a fertile area for future study.

9.3.7 Pathogens

Just as migrants are exposed to "transboundary pollution," they may also potentially collect pathogens from many different regions (Latta 2003; Ricklefs *et al.* 2005; Kilpatrick *et al.* 2007). With the recent spread of avian pathogens such as bird flu and West Nile Virus (LaDeau *et al.* 2007; Kilpatrick *et al.* 2007), there is also a strong potential for migratory birds to spread these diseases over long distance in a very short time (Rappole and Hubalek 2003). The spread of pathogens by migratory birds is an area of intensive study, but there are few conclusive studies to date. Garvin *et al.* (2004), for example, found little evidence for local pathogen transmission to migratory birds in winter. Boone *et al.* (2010) also found no apparent effects of winter habitat and hematozoan infection in a migratory warbler. Nevertheless, parasite communities of birds do differ among breeding grounds, which offer at least some potential for the identification of where birds from different winter populations may breed (Ricklefs *et al.* 2005; Durrant *et al.* 2008).

9.3.8 Climate change

In some respects, migratory birds might be better able to cope with global climate change than many other groups of birds (see Chapter 8). Migratory songbirds disperse well, which should enable them to track new plant communities as they arise and they are likely to be able to tolerate a wider range of temperatures than many tropical resident species, which only experience a narrow range of temperatures throughout their lives (Janzen 1967). The foraging and social flexibility shown by many migratory songbirds may also predispose them to be able to adapt to new resources as they arise.

Nevertheless, there are also some ways in which migratory songbirds may be more vulnerable than resident species to global climate change (Jonzen *et al.* 2007; Sekercioglu *et al.* 2008). Perhaps the best-documented examples come from the increasingly great mismatch between the arrival of migratory songbirds from the tropics on their breeding grounds, and the timing of the spring resource flush that accompanies the first appearance of new foliage in the temperate zone (Graber and Graber 1983; Thomas *et al.* 2001; Strode 2003; Both *et al.* 2006; Jonzen *et al.* 2006, 2007). As global temperatures continue to warm, the spring resource flush is occurring earlier and earlier (see Chapter 8). Tropical birds, which must rely on cues such as increasing day length when deciding when to head to their breeding grounds, have no reliable cues on their winter grounds that enable them to track leafing phenology on the breeding grounds. As a result, tropical migrants may be unable to advance their migration enough to arrive during the resource peak (but see Jonzen *et al.* 2006, 2007). Presumably, year-round residents and short-distance migrants will benefit from many of these changes as they winter closer to their breeding grounds and can advance their breeding in response to local weather and food availability. Year-round residents may also experience reduced mortality as winters become less severe in the temperate zone, which might increase competition with migrants that overwinter in the tropics (Rabenold 1978). In the southern Appalachians where year-round residents are more abundant (presumably because of lower mortality rates during the milder winters), there are far fewer breeding migratory species (Rabenold 1978). Any increase in year-round residents and short-distance migrants and any tendency for migrants to winter further north as temperatures warm (La Sorte and Thompson 2007) could lead to reductions in the populations of migratory birds in the tropics (e.g. Faaborg *et al.* 2007), with possible attendant losses of the ecosystem services they provide (Van Bael *et al.* 2008; Mooney *et al.* 2010). Over the long term, migrants may evolve to leave the winter grounds earlier and earlier, but this process may not happen fast enough to match the rate of climate change (Jonzen *et al.* 2007).

Climate change may also have other, subtler effects on migratory birds (Niven *et al.* 2009). Changes in rainfall patterns may affect habitat quality and therefore the fitness of birds on their winter grounds, especially for species that depend upon more humid forest (Peach *et al.* 1991; Sillett *et al.* 2000; Studds and Mora 2001). An increase in drought frequency, especially in the Sahel zone of Africa, has already been associated with long-term declines of species of more open habitats in winter (Sanderson *et al.* 2006); any increase in the frequency and severity of droughts will likely cause continued declines of many species, regardless of how habitat is managed on the breeding grounds. Warming temperatures in montane habitats may push many species upslope (Sekercioglu *et al.* 2008; see Chapter 8) and lead to wholesale changes in predator communities as snakes, primates, and other less cold-tolerant predators are able to increase their populations and to move upslope. The possible increases in nest predation rates might create problems for altitudinal migrants if they do indeed move upslope to escape the very high nest predation rates at low elevations (Robinson *et al.* 2000).

9.3.9 Overabundant migrants

In the previous section, we discussed the possible implications of declines in migratory bird populations in the tropics that might accompany climate change. Inevitably, however, some species of migrants might benefit too much from human activities and cause problems on their tropical winter grounds. The best-studied example is the Dickcissel (*Spiza americana*), a species that has adapted well to many (but not all: Askins 2000) agricultural practices on their breeding grounds. Dickcissel populations are so large and concentrated into a few areas in the tropics that they have become significant crop pests and are even targets of pest control efforts in a few sites (Basili and Temple 1999; see also Chapter 6). The Rose-breasted Grosbeak (*Pheucticus ludovicianus*), a species that consumes mainly insects and breeds territorially in forests of eastern North America, also becomes a flocking granivore in winter where it can be at least an occasional crop pest (Rappole 1995).

The extent to which the huge changes that have occurred in temperate migratory bird populations since historical times (Askins 2000; Sanderson *et al.* 2006; Hahn *et al.* 2009) have affected tropical bird communities remains largely unknown. It is possible that some of the most abundant flocking species such as the Eastern Kingbird and Purple Martin (*Progne subis*), which simply numerically overwhelm local populations of year-round resident species, have also caused new problems for these resident species. Both the kingbird and the martin have benefitted from deforestation on the breeding grounds and, in the case of the martin, from active management to increase their populations. We lack the data to evaluate their impacts on local resident bird communities.

9.3.10 Hunting and trapping

One of the chief threats to migratory birds in some regions such as southern Europe and China is large-scale hunting and trapping for food (McCulloch *et al.* 1992; Wilcove 2008; see also Chapter 7). Much of this hunting occurs in the temperate zone, and is therefore not a problem that can be managed in the tropics. If hunting and trapping continue without regulation, however, they could severely reduce populations of migratory species, especially those that winter in the understory.

9.4 Towards a comprehensive strategy for migratory bird conservation

Even though there have been great advances in our knowledge about the ecology and behavior of migratory songbirds, there is still a great deal of debate about how best to conserve them (Faaborg *et al.* 2010a). Migratory songbirds can be preserved in virtually all habitats in the tropics and may play vital roles in both natural and human-modified landscapes (Van Bael *et al.* 2003, 2008; Mooney *et al.* 2010). The long-term future of many tropical species may depend upon developing management strategies in human-dominated landscapes that enable migrants to survive and to maintain sufficient body condition to breed successfully as well. Promoting shade coffee and cacao plantations, for example, will benefit large numbers of neotropical migrants, including the Cerulean Warbler (Bakermans *et al.* 2008).

Unlike many resident species, which are often geographically restricted to just a few sites, setting aside small tracts will likely be insufficient to meet the goal of the North American Bird Conservation Plan, which aims to preserve abundance, not just to prevent extinction. Migrants will likely use any reserves that are set aside; these reserves might act as "lifeboats" (Faaborg *et al.* 2010a; see also Chapter 10) while larger-scale land management plans are being developed, but only a few migrants are sufficiently endangered that they require the acquisition of particular parcels of land to ensure their survival. Rather, conserving migrants on their winter grounds requires large-scale efforts that maintain large areas of suitable habitat (Latta and Faaborg 2009; Faaborg *et al.* 2010a). As we gather more and more data on the extent to which migrants from different parts of their breeding ranges also winter in different areas (migratory connectivity), we will be better able to make sure that enough habitat is set aside or managed properly in each part of their winter range (Martin *et al.* 2007). Otherwise, many species will suffer steep local declines in parts of their breeding ranges associated with winter ranges that have been subjected to severe deforestation (Figure 9.4; Kelly *et al.* 2002; Martin *et al.* 2007). For now, however, there is a critical lack of data on connectivity for all but a few species (Faaborg *et al.* 2010a).

Although we know a great deal more about migrant birds than we did three decades ago (Greenberg and Marra 2005; Faaborg *et al.* 2010b), we are still a long way from knowing enough to develop comprehensive conservation plans (Faaborg *et al.* 2010a). Advances in methods of studying migrants will provide much better information, especially on migratory connectivity, but there are still enormous gaps in our knowledge and severe limitations in the technology for establishing connectivity (Faaborg *et al.* 2010b), especially for small birds that cannot carry transmitters. Conserving migrants will therefore require much additional research (Faaborg *et al.* 2010a).

10

Conservation Prospects for Tropical Birds

Overview: Maintaining, and even expanding, existing protected areas is critical for the survival of sensitive tropical bird species. However, funding is lacking even to maintain existing protected areas in the tropics. Nonetheless, areas that need avian conservation priority should be urgently identified in various tropical countries. If bird reintroductions are envisioned, they need to be carefully planned. Identification of Endemic Bird Areas by BirdLife International is a step in the right direction. In order to achieve tangible bird conservation, human livelihoods need to be integrated into conservation programs. There is an urgent need to build conservation capacity in the tropics. Additionally, knowledge on tropical birds needs to be urgently enhanced to improve conservation strategies.

In this chapter we highlight some of the issues involved in the conservation of tropical birds. We start by discussing the importance of protected areas. We then discuss some of the ways to manage birds in human-dominated landscapes in the tropics. Successes and failures of bird reintroductions are also briefly discussed. We stress that human livelihoods should be integrated into conservation programs to ensure success. We end by highlighting that there is an urgent need to build capacity in the tropics and that knowledge on tropical birds needs to be improved to devise better conservation recommendations. For some of the specific conservation recommendations, see previous chapters.

Conservation of Tropical Birds, First Edition. Navjot S. Sodhi, Çağan H. Şekercioğlu, Jos Barlow and Scott K. Robinson.

Published 2011 by Blackwell Publishing Ltd.

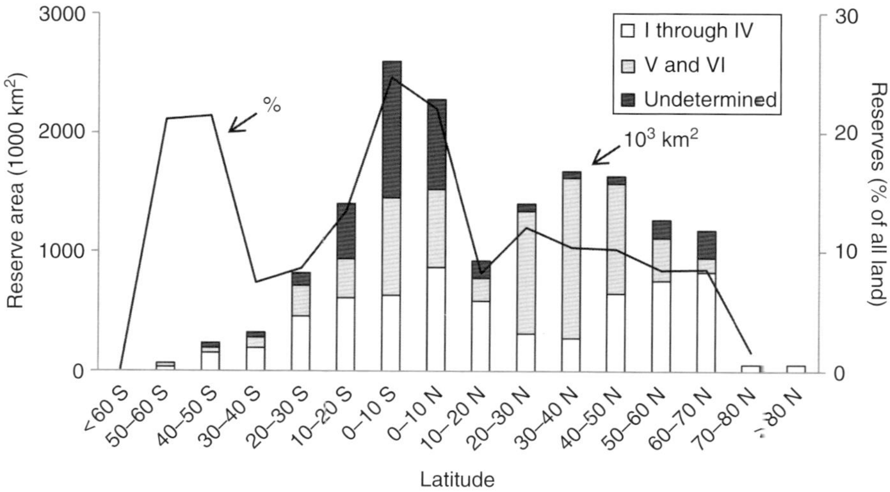

Figure 10.1 The latitude distribution of protected areas as a percentage of all land area (line) and as absolute area (histogram). Data are from the 2007 World Database on Protected Areas. The IUCN categories indicate management goals, wherein categories I–IV are strict nature reserves and categories V and VI are managed to preserve cultural features and ecosystem services. (After Brooks *et al.* 2009.)

10.1 Protection and triage

Protected areas (PAs; parks and reserves) cover about 12% of Earth's land surface (Chape *et al.* 2005; Jenkins and Joppa 2009) and PAs are growing on average at 0.1% annually, mostly in the Amazon (Jenkins and Joppa 2009; Peres *et al.* 2010). The greatest coverage of PAs is in the tropics (Figure 10.1; Brooks *et al.* 2009), but many of these do not receive strict protection (Curran *et al.* 2004, see below). Further, globally, only 8% of forests are protected (Schmitt *et al.* 2009), which is short of the 10% target proposed by the Convention on Biological Diversity (CBD), and many species have ranges that do not overlap with existing PAs (Rodrigues *et al.* 2004). Although 10% of the humid tropical forests are strictly protected (Gardner *et al.* 2009), it is unclear if these reserves receive adequate protection against activities such as illegal logging and poaching, and many are threatened by wildfires (see Chapter 5). Further it is unclear if existing PAs will be able to conserve species against the impacts of climate change (Lee and Jetz 2008; see Chapter 8).

Despite their limited spatial extent, PAs are critical for the survival of many tropical birds. For example, agricultural areas surrounding the protected savannas of Serengeti (Tanzania) contained only 28% of the species found in the latter (Sinclair *et al.* 2002). The density of insectivorous bird species was particularly low in agricultural areas compared to protected native savannas, presumably because of reduced insect abundance in agricultural areas (Figure 10.2; Sinclair *et al.* 2002). Because of a 5% expansion of human-dominated areas around the Serengeti every year, native bird species remain vulnerable. Planting of native

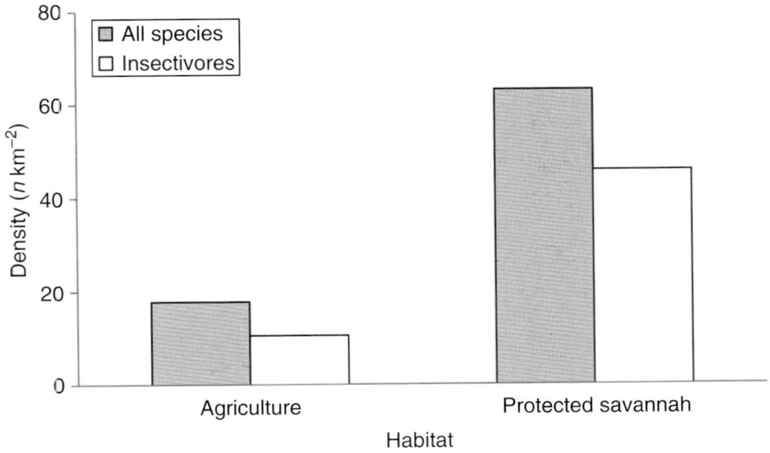

Figure 10.2 The density of all bird species and insectivorous bird species in agricultural areas and in protected savannah in the northern Serengeti (Africa). (Data derived from Sinclair *et al.* 2002.)

vegetation in agricultural areas, educating and providing monetary incentives for local people could alleviate habitat conversion and thus may ameliorate bird biodiversity loss in the Serengeti (Sinclair *et al.* 2002).

Similar results have been found on the Indonesian Island of Sulawesi. One-third of bird species are endemic on this island, but deforestation is rampant and it has lost 45% of its original forest cover (Sodhi *et al.* 2005a). Forested PAs in Sulawesi contained a higher number of forest and endemic forest bird species than did buffer areas (within 2 km) (Lee *et al.* 2007). Many of these species had significantly higher abundance in PAs than outside them (Figure 10.3; Lee *et al.* 2008a). These results suggest that PAs may be critical for the long-term survival of many endemic forest species on this island. Supporting this assertion, conservation programs perhaps saved 16 endangered bird species between 1994 and 2004 (Butchart *et al.* 2006), and the most common "extinction prevention" endeavor was habitat protection and management (Figure 10.4). For instance, the endangered Cebu Flowerpecker (*Dicaeum quadricolor*) has been a beneficiary of habitat protection (Butchart *et al.* 2006). Thus, conservation actions can alleviate the extinction of tropical forest birds (Figure 10.5; Brooks *et al.* 2009). Habitat protection was also found to increase the reproductive success of tropical parrots (see Figure 7.22).

Despite the likely conservation benefits, some "protected" forests in the tropics have become increasingly isolated, degraded, or deforested. Between the early 1980s and 2001, 25% of 198 protected areas containing tropical forests lost forest area within their boundaries, and 70% of these lost forest area in the surrounding buffer zones (Figure 10.6; DeFries *et al.* 2005). Buffer zones are important because they can safeguard PAs from catastrophic events such as fires (Gascon *et al.* 2000). Other studies illustrate similar predicaments faced by specific protected areas throughout the tropics (e.g. Sánchez-Azofeifa *et al.* 1999; Curran *et al.* 2004). For

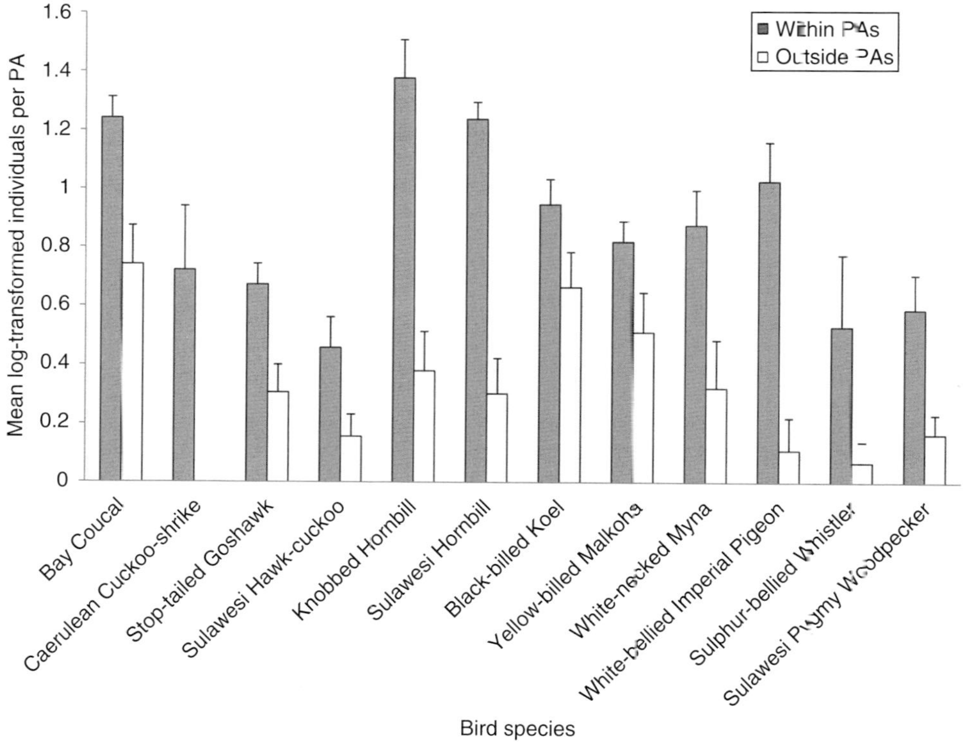

Figure 10.3 Mean log-transformed number of individuals of different bird species within and outside protected areas (PAs) in Sulawesi (Indonesia). Error bars represent standard errors. (After Lee *et al*. 2008a.)

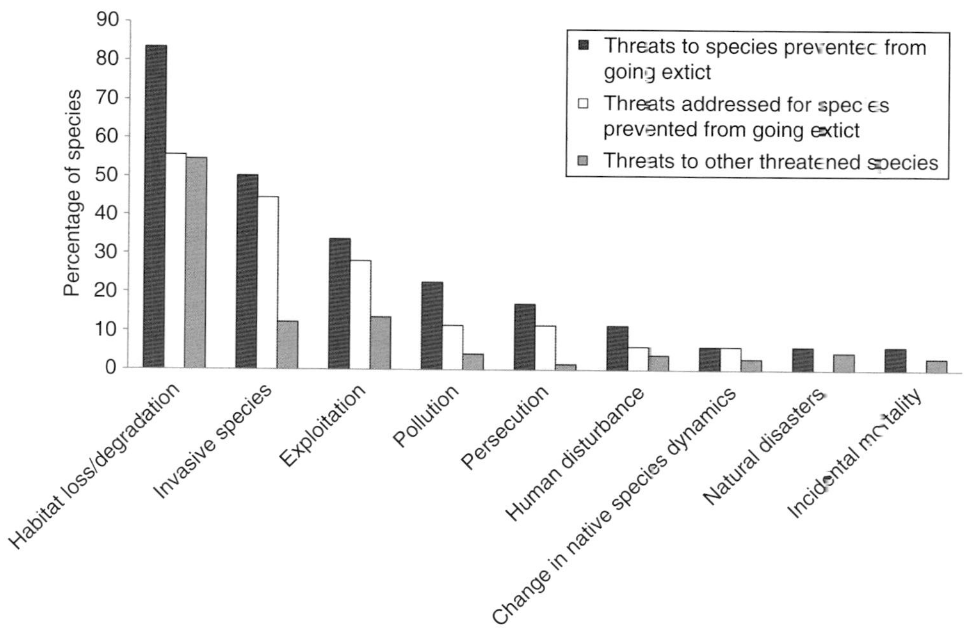

Figure 10.4 Threats to species whose extinction was prevented during 1994–2004 (n = 16), the threats addressed by conservation actions for these species, and threats to other threatened species (n = 1199). (After Butchart *et al*. 2006.)

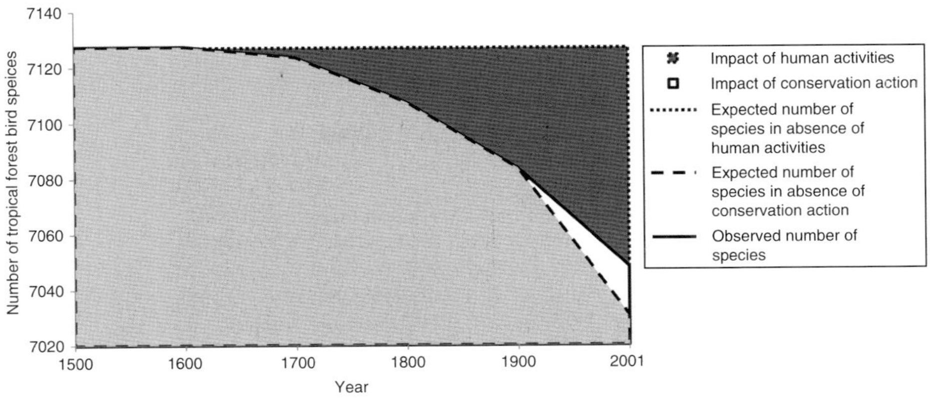

Figure 10.5 Estimated impact of conservation action to prevent extinctions of tropical forest birds compared with the impact of human activities on overall species richness of tropical forest birds. (After Brooks *et al.* 2009.)

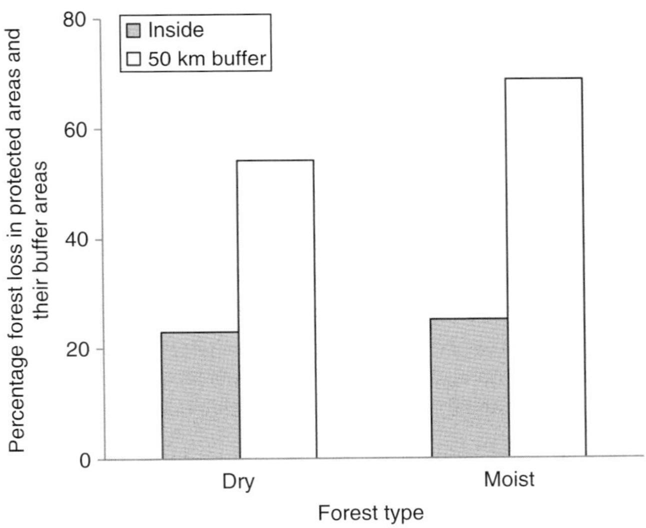

Figure 10.6 Percentage of tropical protected areas losing forest within their boundaries and surrounding buffer. (Data derived from DeFries *et al.* 2005.)

example, protected lowland forests of the hyper-biodiverse region of Kalimantan (Indonesian Borneo) have declined by 56% between 1985 and 2001, primarily due to intensive logging (Curran *et al.* 2004). This forest decline was not restricted to the parks and also occurred within the buffer zones. Similarly, due to the expansion of agriculture and cattle ranching in the Amazon, it is predicted that 40% of the forests in its protected areas could be logged by 2050 (Soares-Filho *et al.* 2006); PAs have become the new deforestation frontier in some regions of the Amazon where forests from outside reserves have already been lost (Pedlowski *et al.* 2005). However, there have also been some positive signs. Some PAs may remain protected due to inaccessibility and low human populations in surrounding areas (Brooks *et al.* 2009). Indeed, forest cover in and around PAs can approach 100% in areas of low human population densities (e.g. Western Amazonia and the Congo basins) (Joppa *et al.* 2008). Deforestation in PAs may also be slower than in surrounding buffer areas, suggesting at least temporary habitat protection (Bruner *et al.* 2001; Naughton-Treves *et al.* 2005; Nepstad *et al.* 2006; Nagendra 2008), although these studies often fail to take account of leakage (where reserve creation pushes deforestation elsewhere) (Ewers and Rodrigues 2008).

About US$13 billion may be needed annually to maintain PAs in the tropics (Bruner *et al.* 2004). However, currently only $1 billion may be spent on maintaining tropical areas per year – or $93/km²/year (James *et al.* 2001; Brooks *et al.* 2009). Clearly more funding is needed to maintain the existing PAs in the tropics. Programs such as REDD (Reducing Emissions from Deforestation and Forest Degradation in Developing Countries; www.undp.org/mdtf/un-redd/overview.shtml) can perhaps help fund the management of tropical PAs.

Conservation funds are limited, and land that could be set aside for conservation often has to compete against alternative land uses that would be more profitable in the short term. It is therefore critical to prioritize areas and species for conservation (Margules and Pressey 2000, reviewed in Brooks 2010). In Chapter 1, we discussed biodiversity hotspots, which identify critical areas for conservation. Similar approaches have been used to identify the areas that must be protected for tropical bird species. For example, Jenkins *et al.* (2010) mapped the distribution of 189 endemic and/or endangered bird species occurring in the Atlantic Forest in the state of Rio de Janeiro (Figure 10.7). Using these maps, habitats currently protected were assessed. More importantly, it was found that 10% of the unprotected areas in the state should be immediately protected for avian conservation (Jenkins *et al.* 2010). Previously Marsden *et al.* (2005) also pointed out that most forest reserves in the Atlantic lowland forest are too small to preserve the sensitive bird species. Similarly, over half of the 106 threatened bird species in sub-Saharan Africa were not covered by the existing PA network (De Klerk *et al.* 2004). They suggested filling this gap in bird protection by increasing the PAs to include sites such as Mt. Cameroon-Bamenda highlands (Cameroon) and the Angolan scarp (Angola), the latter of which has been severely neglected due to the three-decade Angolan civil war (Sekercioglu and Riley 2005).

Clearly, these studies suggest that the protected area network for tropical birds needs to be expanded. Such an expansion may also benefit other taxonomic groups. Birds can be good surrogates of overall biodiversity (see Chapter 1). Restricted-range bird species (ranges $< 50{,}000\,km^2$) can be good predictors of the national levels of endemism in plant and other animal groups (Balmford and Long 1995). The number of restricted-range species and the size of national protected areas are well correlated (Figure 10.8; Balmford and Long 1995). These results suggest that bird numbers, specifically of the restricted range species, can have high congruence with other endemic taxonomic groups and with the overall protection efforts by a nation. However, Balmford and Long (1995) also found that habitat protection for restricted-range species remains inadequate for countries such as the Philippines, Papua New Guinea, China, Jamaica, Fiji, Vietnam, and Myanmar.

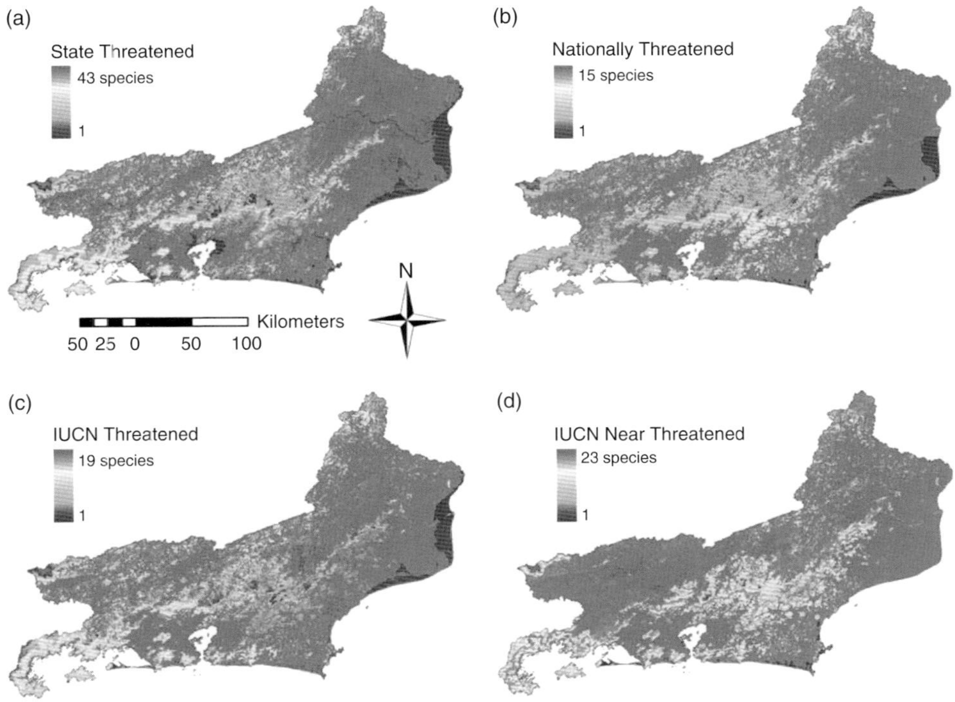

Figure 10.7 Threatened species richness using state (a), national (b), and global (c) red lists. (d) Richness of Near Threatened species according to the global IUCN red list. Gray areas are predicted to have no threatened species. (After Jenkins *et al.* 2010.) (Please see Plate 10.7 for the color version of this figure.)

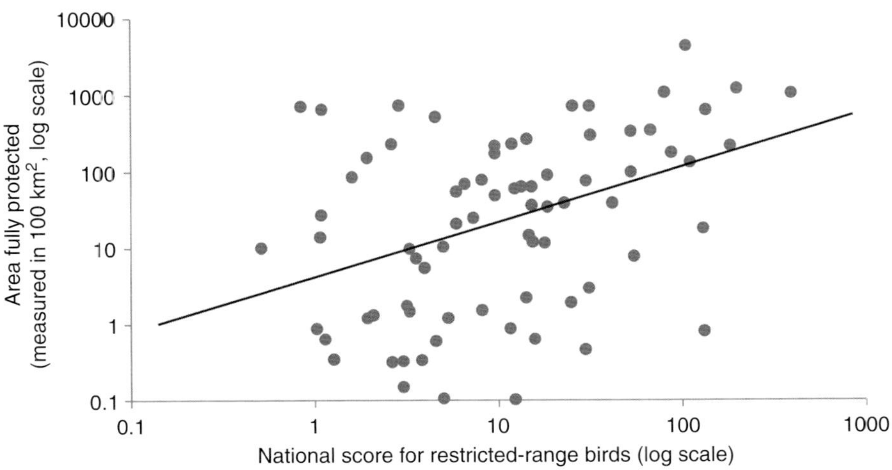

Figure 10.8 Overall extent of fully protected areas in each country versus national scores for restricted-range birds (After Balmford and Long 1995.)

To prioritize bird conservation efforts, BirdLife International has identified 218 Endemic Bird Areas (EBAs) that are critical for bird conservation (Stattersfield *et al.* 1998; BirdLife International 2010a). These EBAs contain the ranges of at least two restricted-range species and cover 4.5% of Earth's land area. About half of these species are threatened with extinction. Eighty-three percent of EBAs contain natural forests, and 77% are in tropical and subtropical regions. The countries containing at least 10 EBAs are Indonesia, Mexico, Brazil, Peru, Colombia, and Papua New Guinea. Because 70% of EBAs are also important for endemic plants, conserving native habitats in EBAs is likely to provide benefits for other taxonomic groups as well.

10.2 Bird conservation in human-dominated landscapes

Human footprint on the planet is expanding, with few areas in the tropics escaping human impact (Koh and Gardner 2010). Therefore, the future of tropical biodiversity may depend upon how we succeed in managing human-dominated landscapes (Gardner *et al.* 2009; see Chapter 1). Existing forest remnants in human-dominated tropical landscapes should be preserved to maintain the residual bird species, especially but not exclusively those species that overwinter in the tropics (see Chapter 9). For example, Brook *et al.* (2003a) projected that at least three of eight forest-dependent bird species could expire if forest reserves covering 2.5% of Singapore were to be obliterated. In addition to preserving remnant forests, native habitats in human-dominated landscapes should be augmented. Chan and Daily (2008) showed that planting native vegetation can enhance the richness of forest as well as non-forest bird species in various human-dominated habitat types in Costa Rica (Figure 10.9). Native vegetation such as windbreaks in coffee farms can protect crops thus benefiting farmers as well (Chan and Daily 2008). Enhanced insectivorous bird abundance is also likely to reduce insect pests in agricultural areas (e.g. Koh 2008; see Chapters 1 and 4). Therefore, habitat enhancements in human-dominated landscapes can benefit forest birds as well as the human enterprise.

In addition to habitat enhancement, the creation of habitat links (e.g. fence rows and windbreaks) between forest patches or reserves may facilitate dispersal, and thus buffer against local extinctions and population declines in fragmented landscapes (Sekercioglu *et al.* 2002; see Chapter 2). Data on the efficacy and adequacy of habitat corridors in the tropics are limited; however, it is known that high vegetation cover (both native and non-native) can enhance the attractiveness of corridors for forest species, thereby enhancing the likelihood of movement through these areas and facilitating recolonizations (Castellón and Sieving 2006). Castelletta *et al.* (2005) recommended that patches isolated about 18 years ago in Singapore should be re-connected to facilitate bird movements. On the negative side, corridors can be expensive to create and maintain, may not serve well in heavily fragmented landscapes, and could be counter-productive by facilitating the spread of predators and diseases (Simberloff and Cox 1987; Simberloff 1992b). Nonetheless, we conclude that the available evidence suggests more conservation advantages are derived in the creation of habitat linking corridors

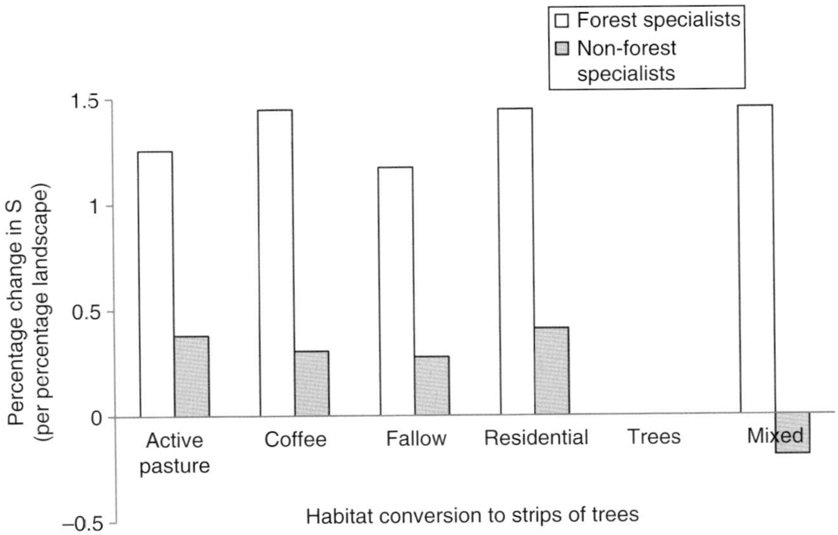

Figure 10.9 The impact of replacing habitat with strips of tress on species richness (S) of 18 forest specialist bird species relative to the impact on 96 non-forest specialist bird species. (After Chan and Daily 2008, copyright National Academy of Sciences, USA.)

than disadvantages; see Section 2.5.4), especially as wider corridors can be an important habitat for many tropical forest birds (see Section 2.4.3, Figure 2.4) and in montaine regions where corridors may facilitate elevational migrations (see Chapter 9).

Some management efforts in human-dominated landscapes can also be conducive for endangered bird species. For instance, appropriate fire management in grasslands can both benefit the birds and local people (Gray *et al.* 2007; Duca *et al.* 2009). The endangered Bengal Florican (*Houbaropsis bengalensis*) does not seem to be negatively affected by annual grassland burning by local communities in Cambodia (Gray *et al.* 2007). Villagers burn grasslands to provide fodder for livestock, and this action may increase arthropod abundance and thus benefit the floricans (Gray *et al.* 2007).

10.3 Population augmentation

Re-introductions are defined as the introduction of a species to previously occupied areas. The objective of re-introductions is to establish viable wild populations of a species that has become globally or locally extinct. Re-introduced individuals can come from captive breeding programs or can be translocated from wild populations. Re-introduction programs can be costly and should include a feasibility study, preparation time, release, and follow-up monitoring. A feasibility study should include estimation of carrying capacity and the potential for negative impacts on extant flora and fauna. It is also wise to restore original habitat if it has been heavily reduced and disturbed prior to re-introduction. In addition, it is essential that the underlying causes of the original extirpation be removed or suppressed prior to re-introduction for the probability of establishment success and persistence to be high. For example, if over-harvesting was the main cause of the original extirpation, efforts should be put in place to minimize hunting or restrict it to highly managed parcels of land away from the introduction site. In some cases, population augmentation may be needed after the first release. There are some high-profile cases where re-introductions have been successful in improving the conservation status of a threatened species. The Seychelles Warbler (*Acrocephalus sechellensis*) was once a highly threatened single-island species with only 26 individuals remaining in the wild. Conservation efforts resulted in population recovery to over 300 individuals, with successful translocations to two neighboring islands done to reduce the chance of total species extinction (Komdeur 1994).

There have been other conservation successes such as the population recovery by the endangered Mauritius Kestrel (*Falco punctatus*) through the release of captive-bred individuals (Jones *et al.* 2008a). Overall, captive breeding and release prevented the extinction of six bird species between 1994 and 2004 (Butchart *et al.* 2006). However, many captive-breeding programs such as the one to rescue the populations of threatened Maleo (*Macrocephalon maleo*) have failed likely because of lack of suitable nesting sites and high human harvesting (Sodhi and Brook 2006; see Chapter 7). Therefore, as mentioned, prior to any captive-release program, it is critical to determine if released individuals will have a legitimate chance of surviving and reproducing in the wild.

10.4 Livelihoods and bird conservation

Funds generated from birdwatching can support local economies and conservation efforts. Birdwatching-related activities contributed US$23 billion in 1996 to the American economy and created jobs for 800,000 people (Sekercioglu 2002b). Birdwatchers also probably contribute significantly to tropical economies. For example, in 1999, 41% of US$1 billion tourism-related income in Costa Rica was related to bird-watching (Sekercioglu 2002b). Thus birdwatching can sustain local livelihoods and can translate into conservation benefits by possibly reducing the extraction of natural resources (Table 10.1). There have been good examples showing how local people could benefit from bird watching tourists. For example, there have been efforts to involve local people to preserve bird watching sites in the Andes. Land has been bought and employment is given to local people to preserve the most visited lek of the Andean Cock-of-the-Rock (*Rupicola peruvianus*) (www.tropicalnature.org). Birdwatching fairs can also generate funds for conservation in tropical areas (BirdLife International 2010a). However, birdwatchers need to be mindful of local cultures and exert minimum negative impact on the environment (Table 10.1).

Conservation actions can sometimes disadvantage local communities. Protected areas in the tropics will inevitably cause some disruption to the livelihood of indigenous communities. For example, the creation of nature reserves may be detrimental to local inhabitants who are reliant on the forest for sustenance

Table 10.1 Impacts of birdwatching and some recommendations to minimize disturbance and maximize local involvement. (After Sekercioglu 2002b)

Positive impacts of birdwatching	*Negative impacts of birdwatching*	*Recommendations for optimal birdwatching*
A link between avian diversity and local income	Disturbing birds by playing tapes and by approaching	Adhere to and insist on ethical birding conduct
A financial incentive to conserve wildlife	Increased nest predation and nest abandonment	Avoid nests and young as much as possible
Less impact and more income than typical tourism	Increased disturbance of rare and/or threatened birds	Show particular care with threatened and rare species
Increased local control due to unique bird species	Visitor-related pollution and habitat destruction	Minimize tape use and try to minimize being seen
Visitation of areas outside traditional tourist itineraries	Cash leaks from local communities	Do not approach further once a bird notices you
Protection of unprotected areas with desired species	Resentment by excluded locals	Stick to established roads/trails/walkways
Valuation of local natural history knowledge	Cultural degradation associated with tourism	Use scopes for observation and photography
Education and employment of local guides		Educate locals about birds and their financial benefits
Generation of funds for bird conservation		Support local and low-impact establishments
Contribution to ornithological knowledge		Contribute to NGOs active in bird conservation

(e.g. bush meat and traditional agriculture), and have few available substitutes (Milner-Gulland and Bennett 2003). Continued conflicts around certain tropical protected areas indicate that social issues need better consideration for sustainable conservation (Bawa *et al.* 2004; Sodhi *et al.* 2006c). Nature conservation may be seen as a luxury and short-term needs induced by poverty usually override conservation needs, so impoverished rural communities generally need to be compensated when their livelihoods are compromised by conservation actions. Suggestions have been made to give direct payments to rural communities to safeguard biodiversity (du Toit *et al.* 2004). Costa Rica, for example, began a Payment for Environmental Services programme in 1997 in which private landowners are paid for preserving forests. Such forest preservation is considered to provide the following benefits: (i) biodiversity conservation; (ii) carbon sequestration; (iii) hydrological services; and (iv) preservation of scenic beauty that attracts ecotourists (Pagiola 2008). Direct payments may work in some instances but wherever possible, efforts should be made to find long-term and sustainable solutions that involve the development of compatible employment opportunities (e.g. resource management) or the introduction of alternative food sources for the impacted communities (see Chapter 7). Further, direct payments to conserve biodiversity can only work if they result in durable conservation practices. The due consideration of such socioeconomic forces is paramount to the success of any conservation exercise (Adams *et al.* 2004; Sachs *et al.* 2009). Outside assistance, however, may be critical to alleviate poverty around protected areas. For example, money from carbon-credit funds (see Chapters 8) can be channelled to developing alternatives for rural communities so that they reduce forest cutting (Laurance 2006). REDD (or further modifications of REDD) can be another mechanism through which local communities can be financially compensated for conserving habitats (www.undp.org/mdtf/un-redd/overview.shtml). However, REDD must work towards conserving habitats important for biodiversity (Grainger *et al.* 2009).

Despite rather insurmountable conservation hurdles, there is hope with local ornithologists taking initiatives, such as those taken by Poonswad *et al.* (2005) to protect hornbills in Thailand. With mostly locally generated funds (68%; hornbill family adoption for US$120 each), they attempted to integrate 28 known hornbill poachers into monitoring hornbill nests, and such endeavor increased, over 3 years, the number of nests with fledglings by 39%. In South America, chicks of the Hyacinth Macaw (*Anodorhynchus hyacinthinus*) had been trapped for pet trade but recently former poachers have been protecting them in hopes of earn ecotourism dollars (Sodhi and Smith 2007). Similarly, to protect the endangered Blue-billed Curassow (*Crax alberti*) in Colombia, land is being purchased to establish a protected area, and locals are being educated on sustainable harvesting of curassows, and ecotourism is being promoted to ensure that the protected area remains sustainable (www.abcbirds.org/abcprograms/international/action/curassow.html). This is a good example of using a multi-pronged approach to conserve a species.

10.5 Education and capacity building

Knowledge of birds can influence conservation support. A survey in Brisbane (Australia) found that people's knowledge of tropical Australian birds was generally poor (Wilson and Tisdell 2005). However, when people were "educated" about the birds, they were willing to donate more money to protect endemic and threatened Australian tropical bird species (Figure 10.10; Wilson and Tisdell 2005). This study shows that people are more likely to give greater economic and other support to species that are known to them. Clearly, there is a need to educate people about the plight of tropical birds to gain public support and funds. Charismatic endangered bird species (e.g. parrots and raptors) generate more public appeal and possibly support (Butchart *et al.* 2006). Therefore, special efforts may be needed to educate the public about the plight of less charismatic bird species.

One of the ways to educate people and build conservation capacity is to involve people through research projects. For instance, research projects on tropical birds in Puerto Rico and Dominican Republic trained students and wildlife professionals and promoted conservation awareness at the local as well as the national level (Latta and Faaborg 2009). Eighty-five percent of interns trained on Hispaniola were employed in conservation-related work or were university students (Figure 10.11; Latta and Faaborg 2009). Some of these

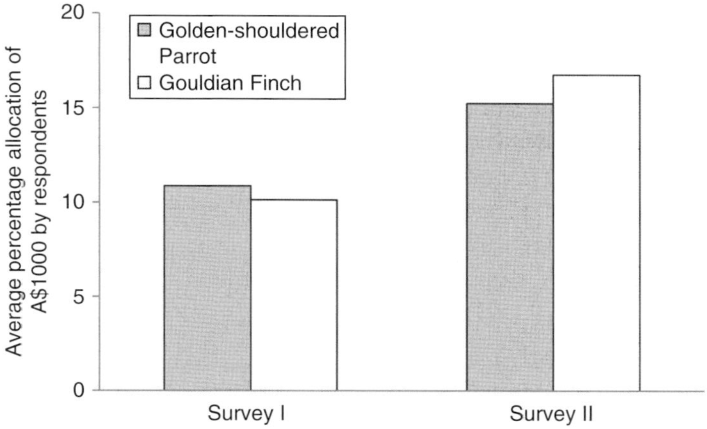

Figure 10.10 After people were "educated" about the birds (before Survey II), they were willing to donate more money to the conservation of endemic and threatened Australian tropical bird species. (Data derived from Wilson and Tisdell 2005.)

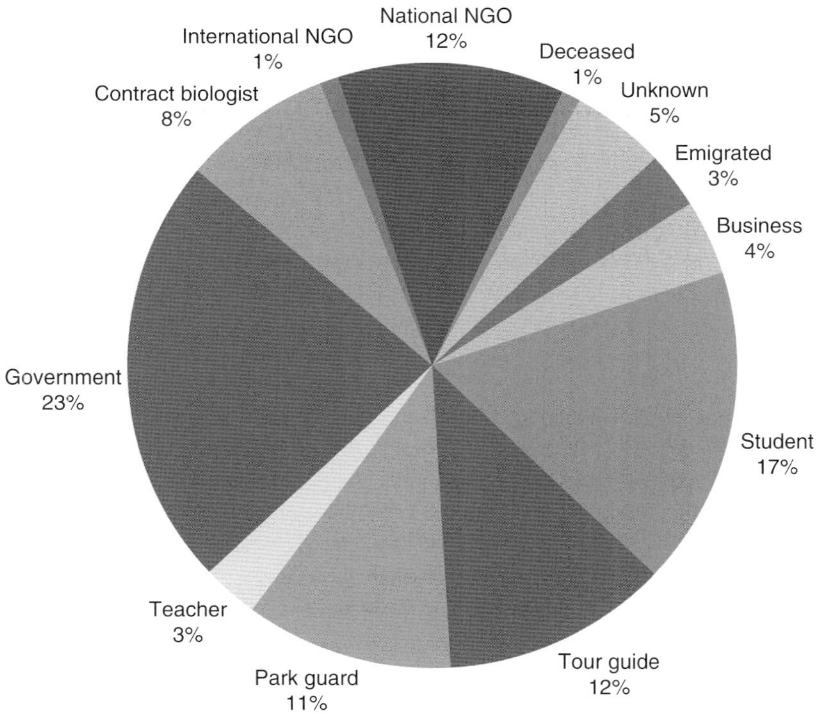

Figure 10.11 Current (late 2007) employment of 73 interns who received training in avian monitoring and field-research techniques, 1990–2007. NGO = Non-governmental Organization. (After Latta and Faaborg 2009.)

interns assisted in efforts to augment the Sierra de Bahoruco National Park, and building educational and training centers, and expansion of ecotourism industry. Conservationists also influenced children as they turned in their slingshots to ornithologists and community leaders and thus reduced the hunting on native birds (Latta and Faabord 2009). This project should inspire other ornithologists to make tangible differences in tropical bird conservation.

10.6 Enhancing knowledge of tropical birds

As mentioned in Chapter 1, much remains unknown about tropical birds (Brito and Oprea 2009). We identify the following urgent needs for tropical avian conservation research: (i) basic ecological data on numerous bird species (e.g. distribution, habitat selection, population trends, and reproductive biology); (ii) long-term studies to understand the effects of habitat loss and climate change on birds and their resilience; (iii) better information on the population sustainability of birds in regenerating and degraded habitats, including information on their dispersal; (iv) better understanding of the loss of ecosystem services to humanity through bird losses; (v) better understanding of socioeconomic and political drivers of habitat loss and bird harvesting; and (vi) understanding and integration of indigenous conservation regimes, local knowledge and livelihoods into bird conservation planning. Scientific knowledge on tropical birds needs to be urgently improved if we are to devise robust bird conservation measures.

References

Aaltola, E. and Oksanen, M. (2002) Species conservation and minority rights: the case of springtime bird hunting in Aland. *Environmental Values* **11**, 443–460.

Able, K. P. (1999) *Gatherings of Angels: migratory birds and their ecology*. Comstock, Ithaca, USA.

Abramsky, Z., Strauss, E., Subach, A., Kotler, B. P. and Riechman, A. (1996) The effect of Barn Owls (*Tyto alba*) on the activity and microhabitat selection of *Gerbillus allenbyi* and *G. pyramidum*. *Oecologia* **105**, 313–319.

Abramsky, Z., Rosenzweig, M. L. and Subach, A. (2002) The costs of apprehensive foraging. *Ecology* **83**, 1330–1340.

Achard, F., Eva, H. D., Stibig, H.-J., Mayaux, P., Gallego, J., Richards, T. and Malingreau, J.-P. (2002) Determination of deforestation rates of the world's humid tropical forests. *Science* **297**, 999–1002.

Adamik, P. and Kral, M. (2008) Climate- and resource-driven long-term changes in dormice populations negatively affect hole-nesting songbirds. *Journal of Zoology* **275**, 209–215.

Adams, W. M., Aveling, R., Brockington, D., Dickson, B., Elliott, J., Hutton, J., Roe, D., Vira, B. and Wolmer, W. (2004) Biodiversity conservation and the eradication of poverty. *Science* **306**, 1146–1149.

Adeney, J. M., Ginsberg, J. R., Russell, G. J. and Kinnaird, M. F. (2006) Effects of an ENSO-related fire on birds of a lowland tropical forest in Sumatra. *Animal Conservation* **9**, 292–301.

Ahola, M. P., Laaksonen, T., Eeva, T. and Lehikoinen, E. (2007) Climate change can alter competitive relationships between resident and migratory birds. *Journal of Animal Ecology* **76**, 1045–1052.

Alencar, A. A. C., Solorzano, L. A. and Nepstad, D. C. (2004) Modeling forest understory fires in an eastern Amazonian landscape. *Ecological Applications* **14**, S139–S149.

Aleixo, A. (1999) Effects of selective logging on a bird community in the Brazilian Atlantic Forest. *Condor* **101**, 537–548.

Allesina, S. and Bodini, A. (2004) Who dominates whom in the ecosystem? Energy flow bottlenecks and cascading extinctions. *Journal of Theoretical Biology* **230**, 351–358.

Almeida, S., Louzada, J., Sperber, C., Barlow, J. (in press) Subtle land-use change and tropical biodiversity: dung beetle communities in Cerrado grasslands and exotic pastures. *Biotropica*

Altwegg, R. and Anderson, M. D. (2009) Rainfall in arid zones: possible effects of climate change on the population ecology of blue cranes. *Functional Ecology* **23**, 1014–1021.

Altwegg, R., Wheeler, M. and Erni, B. (2008) Climate and the range dynamics of species with imperfect detection. *Biology Letters* **4**, 581–584.

Conservation of Tropical Birds, First Edition. Navjot S. Sodhi, Çağan H. Şekercioğlu, Jos Barlow and Scott K. Robinson.

Published 2011 by Blackwell Publishing Ltd.

Alvard, M. S. and Winarni, N. L. (1999) Avian biodiversity in Morowali Nature Reserve, Central Sulawesi, Indonesia and the impact of human subsistence activities. *Tropical Biodiversity* **6**, 59–74.

An, J., Lee, M. Y., Min, M. S., Lee, M. S. and H., L. (2007) A molecular genetic approach for species identification of mammals and sex determination of birds in a forensic case of poaching from South Korea. *Forensic Science International* **167**, 59–61.

Anciães, M. and Marini, M. A. (2000) The effects of fragmentation on fluctuating asymmetry in passerine birds of Brazilian tropical forests. *Journal of Applied Ecology* **37**, 1013–1028.

Anciaes, M. and Peterson, A. T. (2009) Ecological niches and their evolution among Neotropical manakins (Aves: Pipridae). *Journal of Avian Biology* **40**, 591–604.

Anderson, S. H., Kelly, D., Ladley, J. J., Molloy, S. and Terry, J. (2011) Cascading effects of bird functional extinction reduce pollination and plant density. *Science*, in press.

Anderson, W. B. and Polis, G. A. (1999) Nutrient fluxes from water to land: seabirds affect plant nutrient status on Gulf of California islands. *Oecologia* **118**, 324–332.

Andrade, G. I. and Rubio-Torgler, H. (1994) Sustainable use of the tropical rain forest: evidence from the avifauna in a shifting-cultivation habitat mosaic in the Colombian Amazon. *Conservation Biology* **8**, 545–554.

Andrén, H. (1994) Effects of habitat fragmentation on birds and mammals in landscapes with different proportions of suitable habitat – a review. *Oikos* **71**, 355–366.

Anonymous (2009) A Chirp-less Guam becomes a Living Laboratory. http://www.artsci.washington.edu/Newsletter/Dec09/Guam.asp

Antongiovanni, M. and Metzger, J. P. (2005) Influence of matrix habitats on the occurrence of insectivorous bird species in Amazonian forest framents. *Biological Conservation* **122**, 441–451.

Antonovics, J. and Levin, D. A. (1980) The ecological and genetic consequences of density-dependent regulation in plants. *Annual Review of Ecology and Systematics* **11**, 411–452.

Antunes, A. Z. (2007) Endemic Atlantic forest bird diversity and dynamics in a semideciduous forest fragments in southeastern Brazil. *Revista Brasileira de Ornithologia* **15**, 61–68.

Aragão, L. and Shimabukuro, Y. E. (2010) The incidence of fire in Amazonian forests with implications for REDD. *Science* **328**, 1275–1278.

Aragão, L., Malhi, Y., Roman-Cuesta, R. M., Saatchi, S., Anderson, L. O. and Shimabukuro, Y. E. (2007) Spatial patterns and fire response of recent Amazonian droughts. *Geophysical Research Letters* **34**, L07701.

Aragon, P., Lobo, J. M., Olalla-Tarraga, M. A. and Rodriguez, M. A. (2010) The contribution of contemporary climate to ectothermic and endothermic vertebrate distributions in a glacial refuge. *Global Ecology and Biogeography* **19**, 40–49.

Aratrakorn, S., Thunhikorn, S. and Donald, P. F. (2006) Changes in bird communities following conversion of lowland forest to oil palm and rubber plantations in southern Thailand. *Bird Conservation International* **16**, 71–82.

Archibald, S., Bond, W. J., Stock, W. D. and Fairbanks, D. H. K. (2005) Shaping the landscape: fire-grazer interactions in an African savanna. *Ecological Applications* **15**, 96–109.

Arcos, I. T., Jimenez, F., Harvey, C. A. and Casanoves, F. (2008) Richness and abundance of birds in riparian forest belts of varied breadths at the Sesesmiles river microwatershed, Copan, Honduras. *Revista De Biologia Tropical* **56**, 355–369.

Argeloo, M. and Dekker, R. W. R. J. (1996) Exploitation of megapode eggs in Indonesia: the role of traditional methods in the conservation of megapodes. *Oryx* **30**, 59–64.

Arizmendi, M. C., Dominguez, C. A. and Dirzo, R. (1996) The role of avian nectar robber and of hummingbird pollinators in the reproduction of two plant species. *Ecology* **10**, 119–127.

Armesto, J. J., Diaz, I., Papic, C. and Willson, M. F. (2001) Seed rain of fleshy and dry propagules in different habitats in the temperate rainforests of Chiloe Island, Chile. *Austral Ecology* **26**, 311–320.

Askins, R. A. (2000) *Restoring North America's Birds: lessons from landscape ecology*. Yale University Press, New Haven, Connecticut, USA.

Askins, R. A. and Ewert, D. N. (1991) Impact of Hurricane Hugo on bird populations on St. John, US Virgin Islands. *Biotropica* **23**, 481–487.

Askins, R. A., Lynch, J. F. and Greenberg, R. (1990) Population declines in migratory birds in eastern North America. *Current Ornithology* **7**, 1–57.

Atkinson, C. T. and LaPointe, D. A. (2009) Introduced avian diseases, climate change, and the future of Hawaiian honeycreepers. *Journal of Avian Medicine and Surgery* **23**, 53–63.

Atkinson, C. T., Lease, J. K., Dusek, R. J. and Samuel, M. D. (2005) Prevalence of pox-like lesions and malaria in forest bird communities on leeward Mauna Loa Volcano, Hawaii. *Condor* **107**, 537–546.

Avila, M. L., Hernandez, V. H. and Velarde, E. (1996) The diet of Resplendent Quetzal (*Pharomachrus mocinno mocinno*: Trogonidae) in a Mexican cloud forest. *Biotropica* **28**, 720–727.

Awkerman, J. A., Huyvaert, K. P., Mangel, J., Shigueto, J. A. and Anderson, D. J. (2006) Incidental and intentional catch threatens Galapagos Waved Albatross. *Biological Conservation* **133**, 483–489.

Bakermans, M. H. and Rodewald, A. R. (2009) Think globally, manage locally: the importance of steady-state forest features for a declining songbird. *Forest Ecology and Management* **258**, 224–232.

Bakermans, M. H., Vitz, A. C., Rodewald, A. H. and Renjifo, C. G. (2009) Migratory songbird use of shade coffee in the Venezuelan Andes with implications for conservation of the Cerulean Warbler. *Biological Conservation* **142**, 2476–2483.

Baldi, A. (1996) Edge effects in tropical versus temperate forest bird communities: three alternative hypotheses for the explanation of differences. *Acta Zoologica Academiae Scientiarum Hungaricae* **42**, 163–172.

Balmford, A. and Long, A. J. (1995) Across-country analyses of biodiversity congruence and current conservation effort in the tropics. *Conservation Biology* **9**, 1539–1547.

Balmford, A., Bruner, A., Cooper, P., Costanza, R., Farber, S., Green, R. E., Jenkins, M., Jefferiss, P., Jessamy, V., Madden, J., Munro, K., Myers, N., Naeem, S., Paavola, J., Rayment, M., Rosendo, S., Roughgarden, J., Trumper, K. and Turner, R. K. (2002) Economic reasons for conserving wild nature. *Science* **297**, 950–953.

Banks-Leite, C., Ewers, R. M. and Metzger, J. P. (2010) Edge effects as the principal cause of area effects on birds in fragmented secondary forest. *Oikos* **119**, 918–926.

Barbet-Massin, M., Walther, B. A., Thuiller, W., Rahbek, C. and Jiguet, F. (2009) Potential impacts of climate change on the winter distribution of Afro-Palaearctic migrant passerines. *Biology Letters* **5**, 248–251.

Barcelo, G., Salinas, J., Cavieres, G., Canals, M. and Sabat, P. (2009) Thermal history can affect the short-term thermal acclimation of basal metabolic rate in the passerine *Zonotrichia capensis*. *Journal of Thermal Biology* **34**, 415–419.

Barlow, J. and Peres, C. A. (2004a) Avifaunal responses to single and recurrent wildfires in Amazonian forests. *Ecological Applications* **14**, 1358–1373.

Barlow, J. and Peres, C. A. (2004b) Ecological responses to El Niño-induced surface fires in central Amazonia: management implications for flammable tropical forests. *Philosophical Transactions of the Royal Society of London B* **359**, 367–380.

Barlow, J. and Peres, C. A. (2008) Fire-mediated dieback and compositional cascade in an Amazonian forest. *Philosophical Transactions of the Royal Society of London B* **363**, 1787–1794.

Barlow, J., Haugaasen, T. and Peres, C. A. (2002) Effects of ground fires on understorey bird assemblages in Amazonian forests. *Biological Conservation* **105**, 157–169.

Barlow, J., Lagan, B. O. and Peres, C. A. (2003) Morphological correlates of fire-induced tree mortality in a central Amazonian forest. *Journal of Tropical Ecology* **19**, 291–299.

Barlow, J., Peres, C. A., Henriques, L. M. P., Stouffer, P. C. and Wunderle, J. M. (2006) The responses of understorey birds to forest fragmentation, logging and wildfires: an Amazonian synthesis. *Biological Conservation* **128**, 182–192.

Barlow, J., Mestre, L. A. M., Gardner, T. A. and Peres, C. A. (2007) The value of primary, secondary and plantation forests for Amazonian birds. *Biological Conservation* **136**, 212–231.

Barlow, J., Gardner, T. A., Louzada, J. N. C. and Peres, C. A. (2010) Measuring the conservation value of tropical primary forests: the effect of occasional species on estimates of biodiversity uniqueness. *PloS One* **5**, e9609.

Barnett, J. M., Carlos, C. J. and Roda, S. A. (2005) Renewed hope for the threatened avian endemics of northeastern Brazil. *Biodiversity and Conservation* **14**, 2265–2274.

Barquero, M. D. and Hilije, B. (2005) House Wren preys on introduced gecko in Costa Rica. *Wilson Bulletin* **117**, 204–2005.

Bartlein, P. J. (1997) Past environmental changes: characteristic features of quaternary climate variations. In: *Past and Future Rapid Environmental Changes: the spatial and evolutionary responses of terrestrial biota* (eds. Huntley, B., Cramer, W., Morgan, A. V., Prentice, H. C. and Allen, J. R. M.), 11–29. Springer, New York, NY.

Basili, G. and Temple, S. A. (1999) Dickcissels and crop damage in Venezuela: defining the problem with ecological models. *Ecology* **80**, 732–739.

Bawa, K. S., Kress, W. J., Nadkarni, N. M. and Lele, S. (2004) Beyond paradise – meeting the challenges in tropical biology in the 21st century. *Biotropica* **36**, 437–446.

Beard, K. H. and Pitt, W. C. (2005) Potential consequences of the coqui frog invasion in Hawaii. *Diversity and Distributions* **11**, 427–433.

Bearhop, S., Hilton, G. M., Votier, S. C. and Waldron, S. (2004) Stable isotope ratios indicate that body condition in migrating passerines is influenced by winter habitat. *Proceedings of the Royal Society of London Series B* **271**, S215–218.

Becker, B. H., Peery, M. Z. and Beissinger, S. R. (2007) Ocean climate and prey availability affect the trophic level and reproductive success of the marbled murrelet, an endangered seabird. *Marine Ecology – Progress Series* **329**, 267–279.

Beehler, B. M. and Dumbacher, J. P. (1996) More examples of fruiting trees visited predominantly by birds of paradise. *Emu* **96**, 81–88.

Beier, P. and Noss, R. F. (1998) Do habitat corridors provide connectivity? *Conservation Biology* **12**, 1241–1252.

Beier, P., Van Drielen, M. and Kankam, B. O. (2002) Avifaunal collapse in West African forest fragments. *Conservation Biology* **16**, 1097–1111.

Beissinger, S. R. (2001) Trade of live wild birds: potentials, principles and practices of sustainable use. In: *Conservation of Exploited Species* (eds. Reynolds, J. D., Mace, G. M., Redford, K. H. and Robinson, J. G.), 182–202. Cambridge University Press, Cambridge, UK.

Beissinger, S. R. and Bucher, E. H. (1992) Can parrots be conserved through sustainable harvesting. *BioScience* **42**, 164–173.

Beissinger, S. R., Snyder, N. F. R., Derrickson, S. R., James, F. C. and Lanyon, S. M. (1991) International trade in live exotic birds creates a vast movement that must be halted. *Auk* **108**, 982–984.

Bennett, P. M. and Owens, I. P. F. (1997) Variation in extinction risk among birds: chance or evolutionary predisposition? *Proceedings of the Royal Society of London Series B* **264**, 401–408.

Berger, A. J. (1975) Red-whiskered and Red-vented Bulbuls on Oahu. *Elepaio* **36**, 16–19.

Bernardo, J., Ossola, R. J., Spotila, J. and Crandall, K. A. (2007) Interspecies physiological variation as a tool for cross-species assessments of global warming-induced endangerment: validation of an intrinsic determinant of macroecological and phylogeographic structure. *Biology Letters* **3**, 695–698.

Berry, N. J., Phillips, O. L., Lewis, S. L., Hill, J. K., Edwards, D. P., Tawatao, N. B., Ahmad, N., Magintan, D., Khen, C. V., Maryati, M., Ong, R. C. and Hamer, K. C. (2010) The high value of logged tropical forests: lessons from northern Borneo. *Biodiversity and Conservation* **19**, 985–997.

Bhatt, D. and Kumar, A. (2001) Foraging ecology of Red-vented Bulbul *Pycnonotus cafer* in Haridwar, India. *Forktail* **17**, 109–110.

Bierregaard, R. O. Jr. and Lovejoy, T. E. (1989) Effects of forest fragmentation on Amazonian understory bird communities. *Acta Amazonica* **19**, 215–241.

Bierregaard, R. O. Jr., Lovejoy, T. E., Kapos, V., Dossantos, A. A. and Hutchings, R. W. (1992) The biological dynamics of tropical rain-forest fragments. *BioScience* **42**, 859–866.

Bildstein, K. L., Schelsky, W., Zalles, J. and Ellis, S. (1998) Conservation status of tropical raptors. *Journal of Raptor Research* **32**, 3–18.

Bird, J. A., Pettygrove, G. S. and Eadie, J. M. (2000) The impact of waterfowl foraging on the decomposition of rice straw: mutual benefits for rice growers and waterfowl. *Journal of Applied Ecology* **37**, 728–741.

BirdLife International (2000) *Threatened Birds of the World*. Barcelona and Cambridge.

BirdLife International (2004) *State of the World's Birds 2004: indicators for our changing world*. BirdLife International, Cambridge, UK.

BirdLife International (2008a) *Critically Endangered birds: a global audit*. BirdLife International, Cambridge, UK.

BirdLife International (2008b) *State of the World's Birds: indicators for our changing world*. BirdLife International, Cambridge, UK.

BirdLife International (2009) BirdLife's species work. BirdLife International, Cambridge, UK. http://www.birdlife.org/action/science/species/index.html

BirdLife International (2010a) BirdLife's Online World Bird Database: the site for bird conservation. BirdLife International, Cambridge. http://www.birdlife.org/datazone

BirdLife International (2010b) Egyptian Vulture. BirdLife International, Cambridge, UK. http://www.birdlife.org/datazone/species/index.html?action=SpcHTMDetails.asp&sid=3371&m=0

BirdLife International (2010c) Species factsheet: *Amytornis dorotheae*. Downloaded from http://www.birdlife.org

BirdLife International (2010d) Species factsheet: *Geophaps smithii*. Downloaded from http://www.birdlife.org

BirdLife International (2010e) Species factsheet: *Psephotus chrysopterygius*. Downloaded from http://www.birdlife.org

Blackburn, T. M. and Jeschke, J. M. (2009) Invasion success and threat status: two sides of a different coin? *Ecography* **32**, 83–88.

Blackburn, T. M., Cassey, P., Duncan, R. P., Evans, K. L. and Gaston, K. J. (2004) Avian extinction and mammalian introductions on oceanic islands. *Science* **305**, 1955–1958.

Blackburn, T. M., Cassey, P. and Lockwood, J. L. (2008) The island biogeography of exotic bird species. *Global Ecology and Biogeography* **17**, 246–251.

Blackburn, T. M., Lockwood, J. L. and Cassey, P. (2009) *Avian Invasions: the ecology and evolution of exotic birds*. Oxford University Press, Oxford, UK.

Blake, J. G. and Loiselle, B. A. (1992) Habitat use by Neotropical migrants at La Selva Biological Station and Braulio Carrillo National Park, Costa Rica. In: *Ecology and Conservation of Neotropical Migrant Landbirds* (eds. Hagan, J. M. III. and Johnston, D. W.), 257–272. Smithsonian Institution Press, Washington, DC.

Blanco, G. and Tella, J. L. (1997) Protective association and breeding advantages of choughs nesting in Lesser Kestrel colonies. *Animal Behaviour* **54**, 335–342.

Blankespoor, G. W. (1991) Slash-and-burn shifting agriculture and bird communities in Liberia, West Africa. *Biological Conservation* **57**, 41–71.

Blanvillain, C., Salducci, J. M., Tutururai, G. and Maeura, M. (2003) Impact of introduced birds on the recovery of the Tahiti Flycatcher (*Pomarea nigra*), a critically endangered forest bird of Tahiti. *Biological Conservation* **109**, 197–205.

Bleher, B. and Böhning-Gaese, K. (2001) Consequences of frugivore diversity for seed dispersal, seedling establishment and the spatial pattern of seedlings and trees. *Oecologia* **129**, 385–394.

Bock, C. E., Bock, J. H. and Grant, M. C. (1992) Effects of bird predation on grasshopper densities in an Arizona grassland. *Ecology* **73**, 1706–1717.

Bogliani, G., Sergio, F. and Tavecchia, G. (1999) Woodpigeons nesting in association with Hobby falcons: advantages and choice rules. *Animal Behaviour* **57**, 125–131.

Böhning-Gaese, K. (2005) Influence of migrants on temperate bird communities. In: *Birds of Two Worlds: the ecology and evolution of migration* (eds. Greenberg, R. and Marra, P. P.), 143–153. The Johns Hopkins University Press, Baltimore, MD.

Böhning-Gaese, K. and Lemoine, N. (2004) Importance of climate change for the ranges, communities and conservation of birds. In: *Birds and Climate Change* (eds. Møller, A., Fiedler, W. and Berthold, P.), 211–236. Elsevier Academic Press, Oxford.

Böhning-Gaese, K. and Oberrath, R. (2003) Macroecology of habitat choice in long-distance migratory birds. *Oecologia* **137**, 296–303.

Boinski, S., Kauffman, L., Westoll, A., Stickler, C. M., Cropp, S. and Ehmke, E. (2003) Are vigilance, risk from avian predators and group size consequences of habitat structure? A comparison of three species of squirrel monkey (*Saimiri oerstedii*, *S. boliviensis*, and *S. sciureus*). *Behavioural Ecology* **140**, 1421–1467.

Bonaccorso, E., Koch, I. and Peterson, A. T. (2006) Pleistocene fragmentation of Amazon species' ranges. *Diversity and Distributions* **12**, 157–164.

Bond, W. J. (1994) Do mutualisms matter – assessing the impact of pollinator and disperser disruption on plant extinction. *Philosophical Transactions of the Royal Society of London Series B* **344**, 83–90.

Bond, W. J. and Keeley, J. E. (2005) Fire as a global 'herbivore': the ecology and evolution of flammable ecosystems. *Trends in Ecology and Evolution* **20**, 387–394.

Bond, W. J. and Parr, C. L. (2010) Beyond the forest edge: ecology, diversity and conservation of the grassy biomes. *Biological Conservation* **143**, 2395–2404.

Boone, A. T., Rodewald, P. G. and DeGroote, L. W. (2010) Neotropical winter habitat of the Magnolia Warbler: effects on molt, energetic condition, migration timing, and haematozoan infection during spring migration. *Condor* **112**, 115–122.

Botero, C. A., Boogert, N. J., Vehrencamp, S. L. and Lovette, I. J. (2009) Climatic patterns predict the elaboration of song displays in mockingbirds. *Current Biology* **19**, 1151–1155.

Both, C., Bouwhuis, S., Lessells, C. M. and Visser, M. E. (2006) Climate change and population declines in a long-distance migrant bird. *Nature* **441**, 81–83.

Boulet, M. and Norris, D. R. (2006) The past and present of migratory connectivity. *Ornithological Monographs* **61**, 1–13.

Bourhy H., Dautry-Varsat, A., Hotez, P. J. and Salomon, J. (2010) Rabies, still neglected after 125 years of vaccination. *PLoS Neglected Tropical Diseases* 4, e839.

Bouton, S. N. and Frederick, P. C. (2003) Stakeholders' perceptions of a wading bird colony as a community resource in the Brazilian Pantanal. *Conservation Biology* **17**, 297–306.

Bowlin, M. S., Cochran, W. W. and Wikelski, M. C. (2005) Biotelemetry of New World Thrushes during migration: physiology, energetics and orientation in the wild. *Integrative and Comparative Biology* **45**, 295–304.

Bowman, D. M. J. S., Woinarski, J. C. Z., Sands, D. P. A., Wells, A. and McShane, V. J. (1990) Slash-and-burn agriculture in the wet coastal lowlands of Papua New Guinea: response of birds, butterflies and reptiles. *Journal of Biogeography* **17**, 227–239.

Bowman, D., Balch, J. K., Artaxo, P., Bond, W. J., Carlson, J. M., Cochrane, M. A., D'Antonio, C. M., DeFries, R. S., Doyle, J. C., Harrison, S. P., Johnston, F. H., Keeley, J. E., Krawchuk, M. A., Kull, C. A., Marston, J. B., Moritz, M. A., Prentice, I. C., Roos, C. I., Scott, A. C., Swetnam, T. W., van derf Werf, G. R. and Pyne, S. J. (2009) Fire in the Earth System. *Science* **324**, 481–484.

Boyer, A. G. (2008) Extinction patterns in the avifauna of the Hawaiian islands. *Diversity and Distributions* **14**, 509–517.

Boyer, A. G. (2010) Consistent ecological sensitivity through time in Pacific Island avian extinctions. *Conservation Biology* **24**, 511–519.

Boyle, W. A. (2008) Can variation in risk of nest predation explain altitudinal migration in tropical birds? *Oecologia* **155**, 397–403.

Boyle, W. A. (2010) Does food abundance explain altitudinal migration in a tropical frugivorous bird. *Canadian Journal of Zoology* **88**, 204–213.

Boyle, W., Norris, D. and Guglielmo, C. (2010) Storms drive altitudinal migration in a tropical bird. *Proceedings of the Royal Society of London Series B* **277**, 2511–2519.

Bradshaw, C. J. A., Giam, X., Tan, H. T. W., Brook, B. W. and Sodhi, N. S. (2008) Threat and invasive status in legumes is related to opposite extremes of the same ecological and life history attributes. *Journal of Ecology* **96**, 869–883.

Brady, N. C. (1996) Alternatives to slash-and-burn: a global imperative. *Agriculture Ecosystems & Environment* **58**, 3–11.

Brash, A. R. (1987) The history of avian extinction and forest conversion on Puerto Rico. *Biological Conservation* **39**, 97–111.

Brawn, J. D., Robinson, S. K. and Thompson, F. R. (2001) The role of disturbance in the ecology and conservation of birds. *Annual Review of Ecology and Systematics* **32**, 251–276.

Brito, D. and Oprea, M. (2009) Mismatch of research effort and threat in avian conservation biology. *Tropical Conservation Science* **2**, 353–362.

Broadbent, E. N., Asner, G. P., Keller, M., Knapp, D. E., Oliveira, P. J. C. and Silva, J. N. (2008) Forest fragmentation and edge effects from deforestation and selective logging in the Brazilian Amazon. *Biological Conservation* **141**, 1745–1757.

Brook, B. W. and Sodhi, N. S. (2006) Rarity bites. *Nature* **440**, 555–557.

Brook, B. W. and Whitehead, P. J. (2005a) Sustainable harvest regimes for magpie geese (*Anseranas semipalmata*) under spatial and temporal heterogeneity. *Wildlife Research* **32**, 459–464.

Brook, B. W. and Whitehead, P. J. (2005b) Plausible bounds for maximum rate of increase in magpie geese (*Anseranas semipalmata*): implications for harvest. *Wildlife Research* **32**, 465–471.

Brook, B. W., Sodhi, N. S. and Ng, P. K. L. (2003a) Catastrophic extinctions follow deforestation in Singapore. *Nature* **424**, 420–423.

Brook, B. W., Sodhi, N. S., Soh, M. C. K. and Lim, H. C. (2003b) Abundance and projected control of invasive house crows in Singapore. *Journal of Wildlife Management* **67**, 808–817.

Brook, B. W., Sodhi, N. S. and Bradshaw, C. J. A. (2008) Synergies among extinction drivers under global change. *Trends in Ecology and Evolution* **23**, 453–460.

Brooke, R. K., Lockwood, J. L. and Moulton, M. P. (1995) Patterns of success in Passeriform bird introductions on Saint-Helena. *Oecologia* **103**, 337–342.

Brooks, M. L., D'Antonio, C. M., Richardson, D. M., Grace, J. B., Keeley, J. E., DiTomaso, J. M., Hobbs, R. J., Pellant, M. and Pyke, D. (2004) Effects of invasive alien plants on fire regimes. *BioScience* **54**, 677–688.

Brooks, T. M. (2010) Conservation planning and priorities. In: *Conservation Biology for All* (eds. Sodhi, N. S. and Ehrlich, P. R.), 199–219. Oxford University Press, Oxford, UK.

Brooks, T., Tobias, J. and Balmford, A. (1999a) Deforestation and bird extinctions in the Atlantic forest. *Animal Conservation* **2**, 211–222.

Brooks, T. M., Pimm, S. L. and Oyugi, J. O. (1999b) Time lag between deforestation and bird extinction in tropical forest fragments. *Conservation Biology* **13**, 1140–1150.

Brooks, T. M., Mittermeier, R. A., Mittermeier, C. G., B, d. F. G. A., Rylands, A. B., Konstant, W. R., Flick, P., Pilgrim, J., Oldfield, S., Magin, G. and Hilton-Taylor, C. (2002) Habitat loss and extinction in the hotspots of biodiversity. *Conservation Biology* **16**, 909–923.

Brooks, T. M., Collar, N. J., Green, R. E., Marsden, S. J. and Pain, D. J. (2008) The science of bird conservation. *Bird Conservation International* **18**, S2–S12.

Brooks, T. M., Wright, S. J. and Sheil, D. (2009) Evaluating the success of conservation actions in safeguarding tropical forest biodiversity. *Conservation Biology* **23**, 1448–1457.

Brooks-Moizer, F., Robertson, S. I., Edmunds, K. and Bell, D. (2009) Avian influenza H5N1 and the wild bird trade in Hanoi, Vietnam. *Ecology and Society* **14**, 28.

Brosset, A. (1990) A long-term study of the birds of M'Passa (Gabon). In: *Biogeography and Ecology of Forest Bird Communities* (eds. Keast, A.), 259–274. SPB Academic Publishing, The Hague, The Netherlands.

Brown, C. J. (1991) An investigation into the decline of the Bearded Vulture *Gypaetus barbatus* in South Africa. *Biological Conservation* **57**, 315–338.

Brown, D. R. and Sherry, T. W. (2006) Food supply controls the body condition of a migrant bird wintering in the tropics. *Oecologia* **149**, 22–42.

Brown, D. R. and Sherry, T. W. (2008) Alternative strategies of space use and response to resource change in a wintering migrant songbird. *Behavioural Ecology* **19**, 1314–1325.

Brown, J. H. and Kodric-Brown, A. (1977) Turnover rates in insular biogeography: effect of immigration on extinction. *Ecology* **58**, 445–449.

Brown, J. S. and Kotler, B. P. (2004) Hazardous duty pay and the foraging cost of predation. *Ecology Letters* **7**, 999–1014.

Brown, J. S., Kotler, B. P., Smith, R. J. and Wirtz, W. O. (1988) The effects of owl predation on the foraging behavior of heteromyid rodents. *Oecologia* **76**, 408–415.

Bruggers, R. L. and Elliott, C. C. H. (1989) *Quelea Quelea: Africa's Bird Pest*. Oxford University Press, Oxford.

Bruna, E. M. (2002) Effects of forest fragmentation on *Heliconia acuminata* seedling recruitment in central Amazonia. *Oecologia* **132**, 235–243.

Bruner, A. G., Gullison, R. E., Rice, R. E. and da Fonseca, G. A. B. (2001) Effectiveness of parks in protecting tropical biodiversity. *Science* **291**, 125–128.

Bruner, A. G., Gullison, R. E. and Balmford (2004) Financial costs and shortfalls of managing and expanding protected-area systems in developing countries. *BioScience* **54**, 1119–1126.

Bryce, S. A., Hughes, R. M. and Kaufmann, P. R. (2002) Development of a bird integrity index: using bird assemblages as indicators of riparian condition. *Environmental Management* **30**, 294–310.

Burd, M. (1994) Bateman's principle and plant reproduction: the role of pollen limitation in fruit and seed set. *Botanical Review* **60**, 81–109.

Burkey, T. V. (1993) Edge effects in seed and egg predation at two neotropical rain-forest sites. *Biological Conservation* **66**, 139–143.

Butchart, S. H. M. (2008) Red List Indices to measure the sustainability of species use and impacts of invasive alien species. *Bird Conservation International* **18**, S245–S262.

Butchart, S. H. M. and Baker, G. C. (2000) Priority sites for conservation of maleos (*Macrocephlon maleo*) in central Sulawesi. *Biological Conservation* **94**, 79–91.

Butchart, S. H. M., Stattersfield, A. J., Bennun, L. A., Shutes, S. M., Akçakaya, H. R., Baillie, J. E. M., Stuart, S. N., Hilton-Taylor, C. and Mace, G. M. (2004) Measuring global trends in the status of biodiversity: Red List indices for birds. *PLoS Biology* **2**, e383.

Butchart, S. H. M., Stattersfield, A. J. and Collar, N. J. (2006) How many bird extinctions have we prevented? *Oryx* **40**, 266–278.

Butler, J. R. A., du Toit, J. T. and Bingham, J. (2004) Free-ranging domestic dogs (*Canis familiaris*) as predators and prey in rural Zimbabwe: threats of competition and disease to large wild carnivores. *Biological Conservation* **115**, 369–378.

Butler, R. (2008) Despite financial chaos, donors pledge $100m for rainforest conservation. http://news.mongabay.com/2008/1023-fcpf.html

Buzato, S., Sazima, M. and Sazima, I. (2000) Hummingbird-pollinated floras at three Atlantic forest sites. *Biotropica* **32**, 824–841.

Cain, M. L., Milligan, B. G. and Strand, A. E. (2000) Long-distance seed dispersal in plant populations. *American Journal of Botany* **87**, 1217–1227.

Cam, E., Nichols, J. D., Hines, J. E., Sauer, J. R., Alpizar-Jara, R. and Flather, C. H. (2002) Disentangling sampling and ecological explanations underlying species–area relationships. *Ecology* **83**, 1118–1130.

Camargo, J. L. C. and Kapos, V. (1995) Complex edge effects on soil-moisture and microclimate in Central Amazonian forest. *Journal of Tropical Ecology* **11**, 205–221.

Cantaú, J. C., Sánchez, M. E., Groselet, M. and Silva, J. (2007) *The Illegal Parrot Trade in Mexico: a comprehensive assessment*. Defenders of Wildlife, Washington, DC.

Capparella, A. P., Klemens, J. A., Harper, R. G. and Frick, J. A. (2003) Lack of widespread contamination in South American resident passerines. *Bulletin of Environmental Contamination and Toxicology* **70**, 769–774.

Cardoso, M. F., Hurtt, G. C., Moore, B., Nobre, C. A. and Prins, E. M. (2003) Projecting future fire activity in Amazonia. *Global Change Biology* **9**, 656–669.

Carey, C. (2009) The impacts of climate change on the annual cycles of birds. *Philosophical Transactions of the Royal Society B* **364**, 3321–3330.

Carpenter, R. J., Read, J. and Jaffre, T. (2003) Reproductive traits of tropical rain-forest trees in New Caledonia. *Journal of Tropical Ecology* **19**, 351–365.

Carrete, M. and Tella, J. L. (2008) Wild-bird trade and exotic invasions: a new link of conservation concern. *Frontiers in Ecology and Environment* **6**, 207–211.

Carrete, M., Serrano, D., Illera, J. C., Lopez, G., Vogeli, M., Delgado, A. and Tella, J. L. (2009) Goats, birds, and emergent diseases: apparent and hidden effects of exotic species in an island environment. *Ecological Applications* **19**, 840–853.

Carriere, S. M., Andre, M., Letourmy, P., Olivier, I. and McKey, D. B. (2002) Seed rain beneath remnant trees in a slash-and-burn agricultural system in southern Cameroon. *Journal of Tropical Ecology* **18**, 353–374.

Casellini, N., Foster, K. and Hien, B. T. T. (1999) *The 'White Gold' of the Sea: a case study of sustainable harvesting of swiftlet nest in coastal Vietnam*. International Union for the Conservation of Nature and Natural Resources (IUCN), Geneva, Switzerland.

Cassey, P. (2002) Life history and ecology influences establishment success of introduced land birds. *Biological Journal of the Linnean Society* **76**, 465–480.

Cassey, P. (2003) A comparative analysis of the relative success of introduced land birds on islands. *Evolutionary Ecology Research* **5**, 1011–1021.

Castelletta, M., Sodhi, N. S. and Subaraj, R. (2000) Heavy extinctions of forest avifauna in Singapore: lessons for biodiversity conservation in Southeast Asia. *Conservation Biology* **14**, 1870–1880.

Castelletta, M., Thiollay, J. M. and Sodhi, N. S. (2005) The effects of extreme forest fragmentation on the bird community of Singapore Island. *Biological Conservation* **121**, 135–155.

Castellón, T. D. and Sieving, K. E. (2006) An experimental test of matrix permeability and corridor use by an endemic understory bird. *Conservation Biology* **20**, 135–145.

Caut, S., Angulo, E. and Courchamp, F. (2008) Dietary shift of an invasive predator: rats, seabird and sea turtles. *Journal of Applied Ecology* **45**, 428–437.

Ceballos, L., Andary, C., Delescluse, M., Gibernau, M., McKey, D. and Hossaert-McKey, M. (2002) Effects of sublethal attack by a sucking insect, *Hyalymenus tarsatus*, on *Sesbania drummondii* seeds: impact on some seed traits related to fitness. *Ecoscience* **9**, 28–36.

Chace, J. F. and Walsh, J. J. (2006) Urban effects on native avifauna: a review. *Landscape and Urban Planning* **74**, 46–69.

Chalcraft, D. R. and Resetarits, W. J. (2003) Predator identity and ecological impacts: functional redundancy or functional diversity? *Ecology* **84**, 2407–2418.

Chalfoun, A. D., Thompson, F. R. and Ratnaswamy, M. J. (2002) Nest predators and fragmentation: a review and meta-analysis. *Conservation Biology* **16**, 306–318.

Chambers, L. E., Quin, B. R., Menkhorst, P., Franklin, D. C. and Smales, I. (2008) The effects of climate on breeding in the Helmeted Honeyeater. *Emu* **108**, 15–22.

Chan, K. M. A. and Daily, G. C. (2008) The payoffs of conservation investments in tropical countryside. *Proceedings of the National Academy of Sciences of the United States of America* **105**, 19342–19347.

Chape, S., Harrison, J., Spalding, M. and Lysenko, I. (2005) Measuring the extent and effectiveness of protected areas as an indicator for meeting global biodiversity targets. *Philosophical Transactions of the Royal Society of London B* **360**, 443–455.

Chaves-Campo, J., Arevalo, J. E. and Araya, M. (2008) Altitudinal movements and conservation of Bare-necked Umbrellabirds *Cephalopterus glabricolla* of the Tilaran Mountains, Costa Rica. *Bird Conservation International* **13**, 45–58.

Chesser, R. T. (1994) Migration in South America: an overview of the austral system. *Bird Conservation International* **4**, 91–107.

Chesser, R. T. (2005) Seasonal distribution and ecology of South American austral migrant flycatchers. In: *Birds of Two Worlds: the ecology and evolution of migration* (eds. Greenberg, R. and Marra, P. P.), 168–181. The Johns Hopkins University Press, Baltimore, MD.

Chimera, C. G. and Drake, D. R. (2010) Patterns of seed dispersal and dispersal failure in a Hawaiian dry forest having only introduced birds. *Biotropica* **42**, 493–502.

Chiron, F., Shirley, S. and Kark, S. (2009) Human-related processes drive the richness of exotic birds in Europe. *Proceedings of the Royal Society of London Series B* **276**, 47–53.

Chiron, F., Shirley, S. and Kark, S. (2010) Behind the iron curtain: socio-economic and political factors shaped exotic bird introductions into Europe. *Biological Conservation* **143**, 351–356.

Chittka, L. and Waser, N. M. (1997) Why red flowers are not invisible to bees. *Israel Journal of Plant Sciences* **45**, 169–183.

Christiansen, M. B. and Pitter, E. (1997) Species loss in a forest bird community near Lagoa Santa in southeastern Brazil. *Biological Conservation* **80**, 23–32.

Christy, M. (2002) Sulawesi's disappearing flagship bird. *Species* **38**, 8–9.

Clark, A. J. and Scheuhammer, A. M. (2003) Lead poisoning in upland-foraging birds of prey in Canada. *Ecotoxicology* **12**, 23–30.

Clark, C. J., Poulsen, J. R. and Parker, V. T. (2001) The role of arboreal seed dispersal groups on the seed rain of a lowland tropical forest. *Biotropica* **33**, 606–620.

Clark, J. S., Silman, M., Kern, R., Macklin, E. and HilleRisLambers, J. (1999) Seed dispersal near and far: patterns across temperate and tropical forests. *Ecology* **80**, 1475–1494.

Clavero, M. and García-Berthou, E. (2005) Invasive species are a leading cause of animal extinctions. *Trends in Ecology and Evolution* **20**, 110.

Clavero, M., Brotons, L., Pons, P. and Sol, D. (2009) Prominent role of invasive species in avian biodiversity loss. *Biological Conservation* **142**, 2043–2049.

Cleary, D. F. R., Boyle, T. J. B., Setyawati, T., Anggraeni, C. D., van Loon, E. E. and Menken, S. B. J. (2007) Bird species and traits associated with logged and unlogged forest in Borneo. *Ecological Applications* **17**, 1184–1197.

Clough, Y., Putra, D. D., Pitopang, R. and Tscharntke, T. (2009) Local and landscape factors determine functional bird diversity in Indonesian cacao agroforestry. *Biological Conservation* **142**, 1032–1041.

Cochran, W. W. and Wikelski, M. (2005) Individual migratory tactics of New World *Catharus* thrushes. In: *Birds of Two Worlds: the ecology and evolution of migration* (eds. Greenberg, R. and Marra, P. P.), 274–189. The Johns Hopkins University Press, Baltimore, MD.

Cochrane, M. A. (2003) Fire science for rainforests. *Nature* **421**, 913–919.

Cochrane, M. A. and Laurance, W. F. (2002) Fire as a large-scale edge effect in Amazonian forests. *Journal of Tropical Ecology* **18**, 311–325.

Cochrane, M. A. and Schulze, M. D. (1999) Fire as a recurrent event in tropical forests of the eastern Amazon: effects on forest structure, biomass, and species composition. *Biotropica* **31**, 2–16.

Coleman, R. (1999) Parasitism of the herbivore *Pieris brassicae* by *Cotesia glomerata* does not benefit the host plant by reduction of herbivory. *Journal of Applied Entomology* **123**, 171–177.

Collar, N. J., Crosby, M. J. and Stattersfield, A. J. (1994) *Birds to Watch 2: the world list of threatened birds*. BirdLife International, Cambridge, UK.

Colles, A., Liow, L. H. and Prinzing, A. (2009) Are specialists at risk under environmental change? Neoecological, paleoecological and phylogenetic approaches. *Ecology Letters* **12**, 849–863.

Collias, N. E. and Collias, E. C. (1984) *Nest Building and Bird Behavior*. Princeton University Press, Princeton, NJ.

Colorado, G., Hamel, P., Rodewald, A. and Throgmartin, W. (2008) El Grupo Ceruluo: collaboration to assess nonbreeding range of Cerulean Warbler in South America. *Ornitologia Neotropical* 19, 521–529.

Colwell, R. K., Brehm, G., Cardelus, C. L., Gilman, A. C. and Longino, J. T. (2008) Global warming, elevational range shifts, and lowland biotic attrition in the wet tropics. *Science* 322, 258–261.

Compton, S. G., Craig, A. J. F. K. and Waters, I. W. R. (1996) Seed dispersal in an African fig tree: birds as high quantity, low quality dispersers? *Journal of Biogeography* 23, 553–563.

Connell, J. H. (1971) On the role of natural enemies in preventing competitive exclusion in some marine animals and in forest trees. In: *Dynamics of Populations* (eds. Boer, P. J. D. and Gradwell, G. R.), 298–312. Centre for Agricultural Publishing and Documentation, Wageningen.

Conway, C. J., Powell, G. V. N. and Nichols, J. D. (1995) Overwinter survival of Neotropical migratory birds in early successional and mature tropical forests. *Conservation Biology* 9, 855–864.

Cooney, R. (2005) Cambridge Conservation Forum discussion seminar on trade in wild birds. *Oryx* 39, 246.

Cooney, R. and Jepson, P. (2006) The international wild bird trade: what's wrong with blanket bans? *Oryx* 40, 18–23.

Cordeiro, N. J. (1992) Behavior of Blue Monkeys (*Cercopithecus mitis*) in the presence of Crowned Eagles (*Stephanoaetus coronatus*). *Folia Primatologica* 59, 203–207.

Cordeiro, N. J. and Howe, H. F. (2001) Low recruitment of trees dispersed by animals in African forest fragments. *Conservation Biology* 15, 1733–1741.

Cordeiro, N. J. and Howe, H. F. (2003) Forest fragmentation severs mutualism between seed dispersers and an endemic African tree. *Proceedings of the National Academy of Sciences of the United States of America* 100, 14052–14056.

Cordeiro, N. J., Patrick, D. A. G., Munisi, B. and Gupta, V. (2004) Role of dispersal in the invasion of an exotic tree in an East African submontane forest. *Journal of Tropical Ecology* 20, 449–457.

Corlett, R. T. (1992) The ecological transformation of Singapore: 1819–1990. *Journal of Biogeography* 19, 411–420.

Corlett, R. T. (1998) Frugivory and seed dispersal by vertebrates in the Oriental (Indomalayan) Region. *Biological Reviews* 73, 413–448.

Courchamp, F., Berec, L. and Gascoigne, J. (2008) *Allee Effects in Ecology and Conservation*. Oxford University Press, Oxford, UK.

Cox, P. A. and Elmqvist, T. (2000) Pollinator extinction in the Pacific islands. *Conservation Biology* 14, 1237–1239.

Cox, P. M., Betts, R. A., Collins, M., Harris, P. P., Huntingford, C. and Jones, C. D. (2004) Amazonian forest dieback under climate-carbon cycle projections for the 21st century. *Theoretical and Applied Climatology* 78, 137–156.

Crawford, H. S. and Jennings, D. T. (1989) Predation by birds on Spruce Budworm *Choristoneura fumiferana*: functional, numerical, and total responses. *Ecology* 70, 152–163.

Crawford, R. J. M. and Shelton, P. A. (1978) Pelagic fish and seabird interrelationships off coasts of Southwest and South Africa. *Biological Conservation* 14, 85–109.

Crick, H. Q. P. and Jones, P. J. (1992) The ecology and conservation of Palaearctic-African migrants. *Ibis* 134.

Cristinacce, A., Switzer, R. A., Cole, R. E., Jones, C. G. and Bell, D. J. (2009) Increasing use of exotic forestry tree species as refuges from nest predation by the critically endangered Mauritius fody *Foudia rubra*. *Oryx* 43, 97–103.

Croci, S., Le Quilliec, P. and Clergeau, P. (2007) Geographical range as a predictor of spatial expansion of invading birds. *Biodiversity and Conservation* 16, 511–524.

Croll, D. A., Maron, J. L., Estes, J. A., Danner, E. M. and Byrd, G. V. (2005) Introduced predators transform subarctic islands from grassland to tundra. *Science* 307, 1959–1961.

Cronk, Q. C. B. and Fuller, J. L. (1995) *Plant Invaders: the threat to natural ecosystems*. Chapman Hall, London, UK.

Crooks, K. R. and Soulé, M. E. (1999) Mesopredator release and avifaunal extinctions in a fragmented system. *Nature* 400, 563–566.

Crowley, G. and Garnett, S. (1999) Seeds of the annual grasses *Schizachyrium* spp. as a food resource for tropical granivorous birds. *Australian Journal of Ecology* 24, 208–220.

Cruden, R. W. (1972) Pollination in high-elevation ecosystems: relative effectiveness of birds and bees. *Science* 176, 1438–1440.

Cruden, R. W. and Toledo, V. M. (1977) Oriole pollination of *Erythrina breviflora*: evidence for a polytypic view of ornithophily. *Plant Systematics and Evolution* **126**,

Cummings, J. L., Mason, J. R., Otis, D. L., Davis, J. E. and Ohashi, T. J. (1994) Evaluation of methiocarb, ziram, and methyl anthranilate as bird repellents applied to *Dendrobium* orchids. *Wildlife Society Bulletin* **22**, 633–638.

Cunningham, A. A., Prakash, V., Pain, D., Ghalsasi, G. R., Welis, G. A. H., Kolte, G. N., Nighot, P., Goudar, M. S., Kshirsagar, S. and Rahmani, A. (2003) Indian vultures: victims of an infectious disease epidemic? *Animal Conservation* **6**, 189–197.

Curran, L. M., Trigg, S. N., McDonald, A. K., Astiani, D., Hardiono, Y. M., Siregar, P., Caniago, I. and Kasischke, E. (2004) Lowland forest loss in protected areas of Indonesian Borneo. *Science* **303**, 1000–1003.

da Silva, J. M. C. and Tabarelli, M. (2000) Tree species impoverishment and the future flora of the Atlantic forest of northeast Brazil. *Nature* **404**, 72–74.

Daehler, C. C. (2009) Short lag times for invasive tropical plants: evidence from experimental plantings in Hawai'i. *PloS One* **4**, e4462.

Daily, G. C., Ehrlich, P. R. and Haddad, N. M. (1993) Double keystone bird in a keystone species complex. *Proceedings of the National Academy of Sciences of the United States of America* **90**, 592–594.

Daily, G. C., Ehrlich, P. R. and Sánchez-Azofeifa, G. A. (2001) Countryside biogeography: use of human-dominated habitats by the avifauna of southern Costa Rica. *Ecological Applications* **11**, 1–13.

D'Antonio, C. M. and Vitousek, P. M. (1992) Biological invasions by exotic grasses, the grass fire cycle, and global change. *Annual Review of Ecology and Systematics* **23**, 63–87.

Darwin, C. R. (1859) *On the Origin of Species by Means of Natural Selection*. John Murray, London.

Davis, M. A., Peterson, D. W., Reich, P. B., Crozier, M., Query, T., Mitchell, E., Huntington, J. and Bazakas, P. (2000) Restoring savanna using fire: impact on the breeding bird community. *Restoration Ecology* **8**, 30–40.

Davis, N. E., O'Dowd, D. J., Green, P. T. and Mac Nally, R. (2008) Effects of an alien ant invasion on abundance, behaviour, and reproductive success of endemic island birds. *Conservation Biology* **22**, 1165–1176.

De Klerk, H. M., Fjeldsa, J., Blyth, S. and Burgess, N. D. (2004) Gaps in the protected area network for threatened Afrotropical birds. *Biological Conservation* **117**, 529–537.

Dean, W. R. J. and Milton, S. J. (2000) Directed dispersal of Opuntia species in the Karoo, South Africa: are crows the responsible agents? *Journal of Arid Environments* **45**, 305–314.

Dean, W. R. J., Milton, S. J. and Siegfried, W. R. (1990) Dispersal of seeds as nest material by birds in semiarid Karoo (South Africa) shrubland. *Ecology* **71**, 1299–1306.

Debinski, D. M. and Holt, R. D. (2000) A survey and overview of habitat fragmentation experiments. *Conservation Biology* **14**, 342–355.

Debussche, M. and Isenmann, P. (1989) Fleshy fruit characters and the choices of bird and mammal seed dispersers in a Mediterranean region. *Oikos* **56**, 327–338.

Debussche, M. and Isenmann, P. (1994) Bird-dispersed seed rain and seedling establishment in patchy Mediterranean vegetation. *Oikos* **69**, 414–426.

DeFries, R., Hansen, A., Newton, A. C. and Hansen, M. C. (2005) Increasing isolation of protected areas in tropical forests over the past twenty years. *Ecological Applications* **15**, 19–26.

Dekker, R. W. R. J. (1990) The distribution and status of nesting grounds of the maleo *Macrocephalon maleo* in Sulawesi, Indonesia. *Biological Conservation* **51**, 139–150.

del Hoyo, J., Elliott, A. and Christie, D. A. (2003) *Handbook of the Birds of the World (Vol. 8): broadbills to tapaculos*. Lynx Edicions, Barcelona.

DeVault, T. L., Rhodes, O. E. and Shivik, J. A. (2003) Scavenging by vertebrates: behavioral, ecological, and evolutionary perspectives on an important energy transfer pathway in terrestrial ecosystems. *Oikos* **102**, 225–234.

Develey, P. F. and Stouffer, P. C. (2001) Effects of roads on movements by understory birds in mixed-species flocks in central Amazonian Brazil. *Conservation Biology* **15**, 1416–1422.

Devictor, V., Julliard, R., Couvet, D. and Jiguet, F. (2008) Birds are tracking climate warming, but not fast enough. *Proceedings of the Royal Society of London Series B* **275**, 2743–2748.

Diamond, A. W. (1987) A global view of cultural and economic uses of birds. In: *The Value of Birds* (eds. Diamond, A. W. and Filion, F. L.), 99–109. International Centre for Birds of Prey, Norfolk, UK.

Diamond, J. M. (1972) Biogeographic kinetics: estimation of relaxation times for avifaunas of Southwest Pacific Islands. *Proceedings of the National Academy of Sciences of the United States of America* **69**, 3199–3203.

Didham, R. K., Tylianakis, J. M., Hutchison, M. A., Ewers, R. M. and Gemmell, N. J. (2005) Are invasive species the drivers of ecological change? *Trends in Ecology and Evolution* **20**, 470–474.

Diniz, J. A. F., Bini, L. M., Rangel, T. F., Loyola, R. D., Hof, C., Nogues-Bravo, D. and Araujo, M. B. (2009) Partitioning and mapping uncertainties in ensembles of forecasts of species turnover under climate change. *Ecography* **32**, 897–906.

Dirzo, R. and Miranda, A. (1990) Contemporary Neotropical defaunation and forest structure, function, and diversity – a sequel. *Conservation Biology* **4**, 444–447.

Dirzo, R. and Miranda, A. (1991) Altered patterns of herbivory and diversity in the forest understory: a case study of the possible consequences of contemporary defaunation. In: *Plant–Animal Interactions: evolutionary ecology in tropical and temperate regions* (eds. Price, P. W., Lewinsohn, P. W., Fernandes, G. W. and Benson, W. W.), 273–287. John Wiley & Sons, New York, NY.

Docters van Leeuwen, W. M. (1954) On the biology of some Javanese Loranthaceae and the role birds play in their life-history. *Beufortia Miscellaneous Publications* **4**, 105–207.

Doherty, P. F., Sorci, G., Royle, J. A., Hines, J. E., Nichols, J. D. and Boulinier, T. (2003) Sexual selection affects local extinction and turnover in bird communities. *Proceedings of the National Academy of Sciences of the United States of America* **100**, 5858–5862.

Dolbeer, R. A. (1990) Ornithology and integrated pest-management: Red-winged Blackbirds *Agelaius phoeniceus* and corn. *Ibis* **132**, 309–322.

Donald, P. F. (2004) Biodiversity impacts of some agricultural commodity production systems. *Conservation Biology* **18**, 17–37.

Donazar, J. A., Palacios, C. J., Gangoso, L., Ceballos, O., Gonzalez, M. J. and Hiraldo, F. (2002) Conservation status and limiting factors in the endangered population of Egyptian Vulture (*Neophron percnopterus*) in the Canary Islands. *Biological Conservation* **107**, 89–97.

Drake, J. M. and Lodge, D. M. (2006) Allee effects, propagule pressure and the probability of establishment: risk analysis for biological invasions. *Biological Invasions* **8**, 365–375.

Dranzoa, C. (1998) The avifauna 23 years after logging in Kibale National Park, Uganda. *Biodiversity and Conservation* **7**, 777–797.

du Toit, J. T., Walker, B. H. and Campbell, B. M. (2004) Conserving tropical nature: current challenges for ecologists. *Trends in Ecology and Evolution* **19**, 12–17.

Duca, C., Yokomizo, H., Marini, M. A. and Possingham, H. P. (2009) Cost-efficient conservation for the white-banged tanager (*Neothraupis fasciata*) in the Cerrado, central Brazil. *Biological Conservation* **142**, 563–574.

Dudaniec, R. Y., Kleindorfer, S. and Fessl, B. (2006) Effects of the introduced ectoparasite *Philornis downsi* on haemoglobin level and nestling survival in Darwin's Small Ground Finch (*Geospiza fuliginosa*). *Austral Ecology* **31**, 88–94.

Dugger, K. M., Faaborg, J., Arendt, W. J. and Hobson, K. A. (2004) Understanding survival and abundance of overwintering warblers: does rainfall matter? *Condor* **106**, 744–760.

Duke, N. C., Meynecke, J.-O., Dittmann, S., Ellison, A. M., Anger, K., Berger, U., Cannicci, S., Diele, K., Ewel, K. C., Field, C. D., Koedam, N., Lee, S. Y., Marchand, C., Nordhaus, I. and Dahdouh-Guebas, F. (2007) A world without mangroves. *Science* **317**, 41–42.

Duncan, R. P., Blackburn, T. M. and Sol, D. (2003) The ecology of bird introductions. *Annual Review of Ecology, Evolution and Systematics* **34**, 71–98.

Duncan, R. S. and Chapman, C. A. (1999) Seed dispersal and potential forest succession in abandoned agriculture in tropical Africa. *Ecological Applications* **9**, 998–1008.

Duncan, R. S. and Chapman, C. A. (2002) Limitations of animal seed dispersal for enhancing forest succession on degraded lands. In: *Seed Dispersal and Frugivory: ecology, evolution and conservation* (eds. Levey, D. J., Silva, W. R. and Galetti, M.), 437–450. CABI Publishing, Wallingford, UK.

Duncan, R. P., Blackburn, T. M. and Worthy, T. H. (2002) Prehistoric bird extinctions and human hunting. *Proceedings of the Royal Society of London Series B* **269**, 517–521.

Dunham, A. E. (2008) Above and below ground impacts of terrestrial mammals and birds in a tropical forest. *Oikos* **117**, 571–579.

Dunn, R. R., Harris, N. C., Colwell, R. K., Koh, L. P. and Sodhi, N. S. (2009) The sixth mass (co) extinction – are most endangered species parasites and mutualists? *Proceedings of the Royal Society of London Series B* **276**, 3037–3045.

Dunne, J. A., Williams, R. J. and Martinez, N. D. (2002) Network structure and biodiversity loss in food webs: robustness increases with connectance. *Ecology Letters* **5**, 558–567.

Durrant, K. L., Marra, P. P., Fallon, S. M., Colbeck, G. J., Gibbs, H. L., Hobson, K. A., Norris, D. R., Bernik, B., Lloyd, V. L. and Fleischer, R. C. (2008) Parasite assemblages distinguish populations of a migratory passerine on its breeding grounds. *Journal of Zoology* **274**, 318–326.

Dyer, M. I. and Ward, P. (1977) Management of pest situations. In: *Granivorous Birds in Ecosystems* (eds. Pinowski, J. and Kendeigh, S. C.), 267–300. Cambridge University Press, Cambridge, UK.

Eaton, J. (2003) Silent towers, empty skies. *Earth Island Journal* **18**, 30–33.

Ebenman, B., Law, R. and Borrvall, C. (2004) Community viability analysis: the response of ecological communities to species loss. *Ecology* **85**, 2591–2600.

Edwards, D. P., Ansell, F. A., Ahmad, A. H., Nilus, R. and Hamer, K. C. (2009) The value of rehabilitating logged rainforest for birds. *Conservation Biology* **23**, 1628–1633.

Edwards, D. P., Larsen, T. H., Docherty, T. D. S., Ansell, F. A., Hsu, W. W., Derhé, M. A., Hamer, K. C. and Wilcove, D. S. (2010) Degraded lands worth protecting: the biological importance of Southeast Asia's repeated logged forests. *Proceedings of the Royal Society of London Series B* **278**, 82–90.

Elliott, C. C. H. and Lenton, G. M. (1989) The pest status of the quelea. In: *Quelea Quelea: Africa's Bird Pest* (eds. Bruggers, R. L. and Elliott, C. C. H.), 17–34. Oxford University Press, Oxford.

Ellis, E. C., Goldewijk, K. K., Siebert, S., Lightman, D. and Ramankutty, N. (2010) Anthropogenic transformation of the biomes, 1700 to 2000. *Global Ecology and Biogeography* **19**, 589–606.

Er, K. B. H., Vardon, M. J., Tanton, M. T., Tidermann, C. R. and Webb, G. J. W. (1995) Edible Birds' Nest Swiftlets and CITES: a review of the evidence of population decline and nest harvesting effects. Centre for Resource and Environmental Studies Working Paper 1995/3, Australian National University, Canberra.

Eriksson, M. O. G. (1987) Some indicators of freshwater acidification on birds in Sweden. In: *The Value of Birds* (eds. Gilbertson, M., Elliott, J. E. and Peakall, D. B.), 183–190. International Centre for Birds of Prey, Norfolk, UK.

Estrada, A. (ed.) (1986) *Frugivores and Seed Dispersal*. Kluwer Academic Publishers, Hingham.

Estrada, A. and Coates-Estrada, R. (2005) Diversity of Neotropical migratory landbird species assemblages in forest fragments and man-made vegetation in Los Tuxtlas, Mexico. *Biodiveristy and Conservation* **14**, 1719–1734.

Estrada, A., Coates-Estrada, R. and Merritt, D. A. J. (1997) Anthropogenic landscape changes and avian diversity at Los Tuxtlas, Mexico. *Biodiversity and Conservation* **6**, 19–43.

Ewers, R. M. and Didham, R. K. (2006) Confounding factors in the detection of species responses to habitat fragmentation. *Biological Reviews* **81**, 117–142.

Ewers, R. M. and Rodrigues, A. S. L. (2008) Estimates of reserve effectiveness are confounded by leakage. *Trends in Ecology and Evolution* **23**, 113–116.

Fa, J. E., Garcia Yuste, J. E. and Castelo, R. (2000) Bushmeat markets on Bioko Island as a measure of hunting pressure. *Conservation Biology* **14**, 1602–1613.

Fa, J. E., Seymour, S., Dupain, J., Amin, R., Albrechtsen, L. and Macdonald, D. (2006) Getting to grips with the magnitude of exploitation: bushmeat in the Cross-Sanaga rivers region, Nigeria and Cameroon. *Biological Conservation* **129**, 497–510.

Faaborg, J. (2002) *Saving Migrant Birds: developing strategies for the future*. University of Texas Press, Austin, Texas, USA.

Faaborg, J. and Arendt, W. J. (1992) Long-term declines of winter resident warblers in a Puerto Rican dry forest: which species are in trouble? In: *Ecology and Conservation of Neotropical Migrant Landbirds* (eds. Hagan, J. M. III. and Johnston, D. W.), 57–63. Smithsonian Institution Press, Washington, DC.

Faaborg, J., Duffer, K. M. and Arendt, W. J. (2007) Long-term variation in the winter resident bird community of Guanica Forest, Puerto Rico: lessons for measuring and monitoring species richness. *Journal of Field Ornithology* **78**, 270–278.

Faaborg, J., Holmes, R. T., Anders, A. D., Bildstein, K. L., Dugger, K. A., Gauthreaux, S. A. J., Heglund, P., Hobson, K. A., Jahn, A. E., Johnson, D. H., Latta, S. C., Levey, D. J., Marra, P. P.,

Merkord, C. L., Nol, E., Rothstein, S. I., Sherry, T. W., Sillett, T. S., Thompson, F. R. I. and Warnock, N. (2010a) Recent advances in understanding migration systems of New World land birds. *Ecological Monographs* **80**, 3–48.

Faaborg, J., Holmes, R. T., Anders, A. D., Bildstein, K. L., Dugger, K. M., Gauthreaux, S. A. J., Heglund, P., Hobson, K. A., Jahn, A. E., Johnson, D. H., Latta, S. C., Levey, D. J., Marra, P. P., Merkord, C. L., Nol, E., Rothstein, S. I., Sherry, T. W., Sillett, T. S., Thompson, F. R. I. and Warnock, N. (2010b) Conserving migratory land birds in the New World: do we know enough? *Ecological Applications* **20**, 398–418.

Fadini, R. F., Fleury, M., Donatti, C. I. and Galetti, M. (2009) Effects of frugivore impoverishment and seed predators on the recruitment of a keystone palm. *Acta Oecologica* **35**, 188–196.

Faegri, K. (1978) *The Principles of Pollination Ecology*. Pergamon Press, New York, NY.

Fahrig, L. (2003) Effects of habitat fragmentation on biodiversity. *Annual Review of Ecology Evolution and Systematics* **34**, 487–515.

Falkowski, P., Scholes, R. J., Boyle, E., Canadell, J., Canfield, D., Elser, J., Gruber, N., Hibbard, K., Hogbeg, P., Linder, S., MacKenzie, A. F., Moore, B. III, Pedersen, T. F., Rosenthal, Y., Seitzinger, S., Smetacek, V. and Steffen, W. (2000) The global carbon cycle: a test of our knowledge of earth as a system. *Science* **290**, 291–296.

FAO (2005) *State of the World's Forests*. Food and Agricultural Organization of the United Nations, Rome.

FAO (2007) *The World's Mangroves*. Food and Agricultural Organization of the United Nations, Rome.

Farwig, N., Randrianirina, E. F., Voigt, F. A., Kraemer, M. and Böhning-Gaese, K. (2004) Pollination ecology of the dioecious tree *Commiphora guillauminii* in Madagascar. *Journal of Tropical Ecology* **20**, 307–316.

Faulquier, L., Fontaine, R., Vidal, E., Salamolard, M. and Le Corre, M. (2009) Feral cats *Felis catus* threaten the endangered endemic Barau's Petrel *Pterodroma baraui* at Reunion Island (Western Indian Ocean). *Waterbirds* **32**, 330–336.

Fayt, P., Machmer, M. M. and Steeger, C. (2005) Regulation of spruce bark beetles by woodpeckers – a literature review. *Forest Ecology and Management* **206**, 1–14.

Feeley, K. J., Gillespie, T. W., Lebbin, D. J. and Walter, H. S. (2007a) Species characteristics associated with extinction vulnerability and nestedness rankings of birds in tropical forest fragments. *Animal Conservation* **10**, 493–501.

Feeley, K. J., Wright, S. J., Supardi, M. N. N., Kassim, A. R. and Davies, S. J. (2007b) Decelerating growth in tropical forest trees. *Ecology Letters* **10**, 461–469.

Feinsinger, P. (1978) Ecological interactions between plants and hummingbirds in a successional tropical community. *Ecological Monographs* **48**, 269–287.

Feinsinger, P., Wolfe, J. A. and Swarm, L. A. (1982) Island ecology: reduced hummingbird diversity and the pollination biology of plants, Trinidad and Tobago, West Indies. *Ecology* **63**, 494–506.

Felton, A., Wood, J., Felton, A.M., Hennessey, B. and Lindenmayer, D. B. (2008) Bird community responses to reduced-impact logging in a certified forestry concession in lowland bolivia. *Biological Conservation* **141**, 545–555.

Felton, A., Fischer, J., Lindenmayer, D. B., Montague-Drake, R., Lowe, A. R., Saunders, D., Felton, A. M., Steffen, W., Munro, N. T., Youngentob, K., Gillen, J., Gibbons, P., Bruzgul, J. E., Fazey, I., Bond, S. J., Elliott, C. P., Macdonald, B. C. T., Porfirio, L. L., Westgate, M. and Worthy, M. (2009) Climate change, conservation and management: an assessment of the peer-reviewed scientific journal literature. *Biodiversity and Conservation* **18**, 2243–2253.

Ferraz, G., Russell, G. J., Stouffer, P. C., Bierregaard, R. O., Pimm, S. L. and Lovejoy, T. E. (2003) Rates of species loss from Amazonian forest fragments. *Proceedings of the National Academy of Sciences of the United States of America* **100**, 14069–14073.

Ferraz, G., Nichols, J. D., Hines, J. E., Stouffer, P. C., Bierregaard, R. O. and Lovejoy, T. E. (2007) A large-scale deforestation experiment: effects of patch area and isolation on Amazon birds. *Science* **315**, 238–241.

Filion, F. L. (1987) Birds as a socio-economic resource: a strategic concept in promoting conservation. In: *The Value of Birds* (eds. Diamond, A. W. and Filion, F. L.), 7–14. International Centre for Birds of Prey, Norfolk, UK.

Finch, D. M. and Stangel, P. W. (1993) *Status and Management of Neotropical Migratory Birds*. USDA Technical Report RM 229, USDA Forest Service, Rocky Mountain Forest and Range Experiment Station, Fort Collins, Colorado, USA.

Finke, D. L. and Denno, R. F. (2004) Predator diversity dampens trophic cascades. *Nature* **429**, 407–410.

Fischer, J. and Lindenmayer, D. B. (2007) Landscape modification and habitat fragmentation: a synthesis. *Global Ecology and Biogeography* **16**, 265–280.

Fischer, J., Stott, J. and Law, B. S. (2010) The disproportionate value of scattered trees. *Biological Conservation* **143**, 1564–1567.

Fitwater, W. D. (1994) House Sparrows. In: *Handbook of Prevention and Control Of Animal Damage* (eds. Hyngstrom, S. E., Timm, A., and Larsen, G.), E101–E108. University of Nebraska, Lincoln, NE.

Fitzpatrick, J. W. (1980) Wintering of North American tyrant flycatchers in the Neotropics. In: *Migrant Birds in the Neotropics: ecology, behavior, distribution, and conservation* (eds. Keast, A. and Morton, E. S.), 67–78. Smithsonian Institution Press, Washington, DC.

Fitzpatrick, J. W. (1981) Search strategies of tyrant flycatchers. *Animal Behaviour* **29**, 810–821.

Fjeldså, J. and Lovett, J. C. (1997) Geographical patterns of old and young species in African forest biota: the significance of specific montane areas as evolutionary centres. *Biodiversity and Conservation* **6**, 325–346.

Fjeldså, J. and Rahbek, C. (2006) Diversification of tanagers, a species rich bird group, from lowlands to montane regions of South America. *Integrative and Comparative Biology* **46**, 72–81.

Fleming, T. H., Venable, D. L. and Herrera, L. G. (1993) Opportunism vs specialization – the evolution of dispersal strategies in fleshy-fruited plants. *Vegetatio* **108**, 107–120.

Ford, H. A. (1985) Nectar-feeding birds and bird pollination: why are they so prevalent in Australia yet absent from Europe. *Proceedings of the Ecological Society of Australia* **14**, 153–158.

Ford, H. A., Paton, D. C. and Forde, N. (1979) Birds as pollinators of Australian plants. *New Zealand Journal of Botany* **17**, 509–519.

Ford, H. A., Barrett, G. W., Saunders, D. A. and Recher, H. F. (2001) Why have birds in the woodlands of Southern Australia declined? *Biological Conservation* **97**, 71–88.

Fordham, D. A. and Brook, B. W. (2010) Why tropical island endemics are acutely susceptible to global change. *Biodiversity and Conservation* **19**, 329–342.

Foster, J. T. and Robinson, S. K. (2007) Introduced birds and the fate of Hawaiian rainforests. *Conservation Biology* **21**, 1248–1257.

Foster, J. T., Tweed, E. J., Camp, R. J., Woodworth, B. L., Adler, C. D. and Telfer, T. (2004) Long-term population changes of native and introduced birds in the Alak'I swamp, Kaua'i. *Conservation Biology* **18**, 716–725.

Foster, S. A. and Janson, C. H. (1985) The relationship between seed size and establishment conditions in tropical woody plants. *Ecology* **66**, 773–780.

Franklin, D. C. (1999) Evidence of disarray amongst granivorous bird assemblages in the savannas of northern Australia, a region of sparse human settlement. *Biological Conservation* **90**, 53–68.

Frederiksen, M., Daunt, F., Harris, M. P. and Wanless, S. (2008) The demographic impact of extreme events: stochastic weather drives survival and population dynamics in a long-lived seabird. *Journal of Animal Ecology* 77, 1020–1029.

Freed, L. A., Cann, R. L., Goff, M. L., Kuntz, W. A. and Bodner, G. R. (2005) Increase in avian malaria at upper elevation in Hawai'i. *Condor* **107**, 753–764.

Freed, L. A., Medeiros, M. C. and Bodner, G. R. (2008a) Explosive increase in ecoparasites in Hawaiian forest birds. *Journal of Parasitology* **94**, 1009–1021.

Freed, L. A., Cann, R. L. and Bodner, G. R. (2008b) Incipient extinction of a major population of the Hawaii akepa owing to introduced species. *Evolutionary Ecology Research* **10**, 931–965.

Fritts, T. H. and Leasman-Tanner, D. (2001) The Brown Treesnake on Guam: how the arrival of one invasive species damaged the ecology, commerce, electrical systems, and human health on Guam: A comprehensive information source. http://www.fort.usgs.gov/resources/education/bts/bts_home.asp

Frumhoff, P. C. (1995) Conserving wildlife in tropical forests managed for timber: to provide a more viable complement to protected areas. *BioScience* **45**, 456–464.

Fry, D. M. (1995) Reproductive effects in birds exposed to pesticides and industrial chemicals. *Environmental Health Perspectives* **103**, 165–171.

Fuller, R. J., Noble, D. G., Smith, K. W. and Vanhinsbergh, D. (2005) Recent declines of populations of woodland birds in Britain: a review of possible causes. *British Birds* **98**, 116–143.

Furness, R. W. (1991) The occurrence of burrow-nesting among birds and its influence on soil fertility and stability. *Symposium of the Zoological Society of London* **63**, 53–67.

Gage, G. S., Brooke, M. D., Symonds, M. R. E. and Wege, D. C. (2005) Ecological correlates of the threat of extinction in Neotropical bird species. *Animal Conservation* 7, 161–168.

Galetti, M., Bovendorp, R. S., Fadini, R. F., Gussoni, C. O. A., Rodrigues, M., Alvarez, A. D., Guimaraes, P. R. and Alves, K. (2009) Hyper abundant mesopredators and bird extinction in an Atlantic forest island. *Zoologia* **26**, 288–298.

Galindo-Gonzalez, J., Guevara, S. and Sosa, V. J. (2000) Bat and bird-generated seed rains at isolated trees in pastures in a tropical rainforest. *Conservation Biology* **14**, 1693–1703.

Galushin, V. M. (1971) Huge urban population of birds of prey in Delhi, India. *Ibis* **113**, 522.

Ganesh, T. and Davidar, P. (2001) Dispersal modes of tree species in the wet forests of southern Western Ghats. *Current Science* **80**, 394–399.

Gangoso, L., Donazar, J. A., Scholz, S., Palacios, C. J. and Hiraldo, F. (2006) Contradictions in conservation of island ecosystems: plants, introduced herbivores and avian scavengers in the Canary Islands. *Biodiversity and Conservation* 2231–2248.

Gard, N. W. and Hooper, M. J. (1995) An assessment of potential hazards of pesticides and environmental contaminants. In: *Ecology and Management of Neotropical Birds. A synthesis and review of critical issues* (eds. Martin, T. E. and Finch, D. M.), 294–310. Oxford University Press, New York, NY.

Gardner, T., Barlow, J., Chazdon, R., Ewers, R., Harvey, C., Peres, C. A. and Sodhi, N. S. (2009) Prospects for tropical forest biodiversity in a human-modified world. *Ecology Letters* **12**, 561–582.

Garnett, S. T. and Brook, B. W. (2007) Modelling to forestall extinction of Australian tropical birds. *Journal of Ornithology* **148**, S311–S320.

Garvin, M. C., Marra, P. P. and Crain, S. K. (2004) Prevalence of Hematozoa in overwintering American Redstarts (*Setophaga ruticilla*): no evidence for local transmission. *Journal of Wildlife Diseases* **40**, 115–118.

Gascon, C. (1999) Matrix habitat and species richness in tropical forest remnants. *Biological Conservation* **91**, 223–229.

Gascon, C., Lovejoy, T. E., Bierregaard, R. O., Malcolm, J. R., Stouffer, P. C., Vasconcelos, H. L., Laurance, W. F., Zimmerman, B., Tocher, M. and Borges, S. (1999) Matrix habitat and species richness in tropical forest remnants. *Biological Conservation,* **91**, 223–229.

Gascon, C., Williamson, G. B. and da Fonseca, G. A. B. (2000) Receding forest edges and vanishing reserves. *Science* **288**, 1356–1358.

Gasner, M. R., Jankowski, J. E., Ciecka, A. L., Kyle, K. O. and Rabenold, K. N. (2010) Projecting the local impacts of climate change on a Central American montane avian community. *Biological Conservation* **143**, 1250–1258.

Gaston, K. J., Blackburn, T. M. and Goldewijk, K. K. (2003) Habitat conversion and global avian biodiversity loss. *Proceedings of the Royal Society of London Series B* **270**, 1293–1300.

Gauthreaux, S. A. J. (1999) Neotropical migrants and the Gulf of Mexico: the view from aloft. In: *Gatherings of Angels: migratory birds and their ecology* (ed. Able, K. P.), 27–50. Comstock, Ithaca, USA.

Gauthreaux, S. J. and Belser, C. G. (2005) *Radar Ornithology and the Conservation of Migratory Birds.* Bird conservation implementation and integration in the Americas: Proceedings of the Third International Partners in Flight Conference 2002 871–876. USDA Forest Service, Albany, California, USA.

Gauthreaux, S. A. J., Michi, J. E. and Belser, C. G. (2005) The temporal and spatial structure of the atmosphere and its influence on bird migration strategies. In: *Birds of Two Worlds: the ecology and evolution of migration* (eds. Greenberg, R. and Marra, P. P.), 182–193. Johns Hopkins University Press, Baltimore, MD.

Gentry, A. H. (1982) Patterns of Neotropical plant-species diversity. *Evolutionary Biology* **15**, 1–85.

Germain, R. R., Reudink, M. W., Marra, P. P. and Ratcliffe, L. M. (2010) Carotenoid-based male plumage predicts parental investment in the American Redstart. *Wilson Journal of Ornithology* **122**, 318–325.

Gibbs, H. K., Ruesch, A. S., Achard, F., Clayton, M. K., Holmgren, P., Ramankutty, N. and Foley, J. A. (2010) Tropical forests were the primary sources of new agricultural land in the 1980s and 1990s. *Proccedings of the National Academy of Sciences of the United States of Amercia* **107**, 16732–16737.

Gibson, J. P. and Wheelwright, N. T. (1995) Genetic structure in a population of a tropical *Ocotea tenera* (Lauraceae): influence of avian seed dispersal. *Oecologia* **103**, 49–54.

Giglio, L., Csiszar, I. and Justice, C. O. (2006) Global distribution and seasonality of active fires as observed with the terra and aqua moderate resolution imaging spectroradiometer (modis) sensors. *Journal of Geophysical Research-Biogeosciences* 111, G020216.

Gilardi, J. D. (2006) Captured for conservation: will cages save wild birds? A response to Cooney & Jepson. *Oryx* **40**, 24–26.

Gilbert, L. E. (1980) Food web organization and the conservation of Neotropical diversity. In: *Conservation Biology: an evolutionary-ecological perspective* (eds. Soulé, M. E. and Wilcox, B. A.), 11–33. Sinauer Associates, Sunderland, MA.

Gilbertson, M., Elliott, J. E. and Peakall, D. B. (1987) Seabirds as indicators of marine pollution. In: *The Value of Birds* (eds. Diamond, A. W. and Filion, F. L.), 231–248. ICBP, Norfolk, UK.

Gill, A. M. (1975) Fire and the Australian flora: a review. *Australian Forestry* **38**, 4–25.

Gillespie, T. W. (2000) Rarity and conservation of forest birds in the tropical dry region of Central America. *Biological Conservation* **96**, 161–168.

Gillham, M. E. (1956) The ecology of the Pembrokeshire Islands V: manuring by the colonial seabirds and mammals with a note on seed distribution by gulls. *Journal of Ecology* **44**, 428–454.

Gillham, M. E. (1960) Vegetation of New Zealand shag colonies. *Transactions of the Royal Society of New Zealand* **88**, 363–380.

Gillies, C. S. and Clair, C. C. S. (2008) Riparian corridors enhance movement of a forest specialist bird in fragmented tropical forest. *Proceedings of the National Academy of Sciences of the United States of America* **105**, 19774–19779.

Gilman, E., Brothers, N. and Kobayashi, D. R. (2005) Principles and approaches to abate seabird by-catch in longline fisheries. *Fish and Fisheries* **6**, 35–49.

Gilpin, M. E. and Diamond, J. M. (1980) Subdivision of nature reserves and the maintenance of species diversity. *Nature* **285**, 567–568.

Gilpin, M. E. and Soulé, M. E. (1986) Minimum viable populations: the processes of species extinctions. In: *Conservation Biology: the science of scarcity and diversity* (eds. Soulé, M. E.), 19–34. Sinauer Associates, Inc., Sunderland, Massachusetts.

Githiru, M., Lens, L., Bennun, L. A. and Ogol, C. (2002) Effects of site and fruit size on the composition of avian frugivore assemblages in a fragmented Afrotropical forest. *Oikos* **96**, 320–330.

Goerck, J. M. (1997) Patterns of rarity in the birds of the Atlantic forest of Brazil. *Conservation Biology* **11**, 112–118.

Gomez, L. G., Houston, D. C., Cotton, P. and Tye, A. (1994) The role of Greater Yellow-headed Vultures *Cathartes melambrotus* as scavengers in neotropical forest. *Ibis* **136**, 193–196.

Gonzalez, A. M. M., Dalsgaard, B., Ollerton, J., Timmermann, A., Olesen, J. M., Andersen, L. and Tossas, A. G. (2009) Effects of climate on pollination networks in the West Indies. *Journal of Tropical Ecology* **25**, 493–506.

Gonzalez-Solis, J., Croxall, J. P., Oro, D. and Ruiz, X. (2007) Trans-equatorial migration and mixing in the wintering areas of a pelagic seabird. *Frontiers in Ecology and the Environment* **6**, 297–301.

Gorog, A. J., Pamungkas, B. and Lee, R. J. (2005) Nesting ground abandonment by the maleo (*Macrocephalon maleo*) in North Sulawesi: identifying conservation priorities for Indonesia's endemic megapode. *Biological Conservation* **126**, 548–555.

Gottdenker, N. L., Walsh, T., Vargas, H., Merkel, J., Jimenez, G. U., Miller, R. E., Dailey, M. and Parker, P. G. (2005) Assessing the risks of introduced chickens and their pathogens to native birds in the Galapagos Archipelago. *Biological Conservation* **126**, 429–439.

Graber, J. W. and Graber, R. R. (1983) Feeding rates of warblers in spring. *Condor* **85**, 139–150.

Gradwohl, J. and Greenberg, R. (1982) The effect of a single species of avian predator on the arthropods of aerial leaf litter. *Ecology* **63**, 581–583.

Graham, C. H. (2001) Factors influencing movement patterns of Keel-billed Toucans in a fragmented tropical landscape in southern Mexico. *Conservation Biology* **15**, 1789–1798.

Graham, C. H., Moritz, C. and Williams, S. E. (2006) Habitat history improves prediction of biodiversity in rainforest fauna. *Proceedings of the National Academy of Sciences of the United States of America* **103**, 632–636.

Grainger, A. (2008) Difficulties in tracking the long-term global trend in tropical forest area. *Proceedings of the National Academy of Sciences of the United States of America* **105**, 818–823.

Grainger, A., Boucher, D. H., Frumhoff, P. C., Laurance, W. F., Lovejoy, T., McNeely, J., Niekisch, M., Raven, P., Sodhi, N. S., Venter, O. and Pimm, S. L. (2009) Biodiversity and REDD at Copenhagen. *Current Biology* **19**, R974–R976.

Gram, W. K. (1998) Winter participation by Neotropical migrant and resident birds in mixed-species flocks in northeastern Mexico. *Condor* **100**, 44–53.

Grant, P. R., Smith, J. N. M., Grant, B. R., Abbott, I. J. and Abbott, L. K. (1975) Finch numbers, owl predation and plant dispersal on Isla Daphne Major, Galapagos. *Oecologia* **19**, 239–258.

Gratz, N. G. (1999) Emerging and resurging vector-borne diseases. *Annual Review of Entomology* **44**, 51–75.

Graves, G. R. (1982) Pollination of a *Tristerix* mistletoe (Loranthaceae) by *Diglossa* (Aves, Thraupidae). *Biotropica* **14**, 316–317.

Gray, T. N. E., Chamnan, H., Borey, R., Collar, N. J. and Dolman, P. M. (2007) Habitat preferences of a globally threatened bustard provide support for community-based conservation in Cambodia. *Biological Conservation* **138**, 341–350.

Green, P. T. and Juniper, P. A. (2004) Seed mass, seedling herbivory and the reserve effect in tropical rainforest seedlings. *Functional Ecology* **18**, 539–547.

Green, R. E., Newton, I., Shultz, S., Cunningham, A. A., Gilbert, M., Pain, D. J. and Prakash, V. (2004) Diclofenac poisoning as a cause of vulture population declines across the Indian subcontinent. *Journal of Applied Ecology* **41**, 793–800.

Green, R. E., Taggart, M. A., Das, D., Pain, D. J., Kumar, C. S., Cunningham, A. A. and Cuthbert, R. (2006) Collapse of Asian vulture populations: risk of mortality from residues of the veterinary drug diclofenac in carcasses of treated cattle. *Journal of Applied Ecology* **43**, 949–956.

Green, R. J. (1993) Avian seed dispersal in and near subtropical rainforests. *Wildlife Research* **20**, 535–557.

Greenberg, R. (1979) Body size, breeding habitat, and winter exploitation systems in *Dendroica*. *Auk* **96**, 756–766.

Greenberg, R. (1980) Demographic aspects of long-distance migration. In: *Migrant Birds in the Neotropics: ecology, behavior, distribution, and conservation* (eds. Keast, A. and Morton, E. S.), 493–504. Smithsonian Institution Press, Washington, DC.

Greenberg, R. (1981) Dissimilar bill shapes in New World tropical versus temperate foliage-gleaning birds. *Oecologia* **49**, 143–147.

Greenberg, R. (1983) The role of neophobia in determining the of foraging specialization in some migrant warblers. *American Naturalist* **122**, 444–453.

Greenberg, R. (1984) The winter exploitation systems of Bay-breasted and Chestnut-sided warblers in Panama. *University of California Publications in Zoology* **116**, 1–124.

Greenberg, R. (1987) Development of dead-leaf foraging specialization in a tropica migrant warbler. *Ecology* **68**, 130–141.

Greenberg, R. (1992) Forest migrants in non-forest habitats on the Yucatan Peninsula. In: *Ecology and Conservation of Neotropical Migrant Landbirds* (eds. Hagan, J. M. III and Johnston, D. W.), 273–286. Smithsonian Press, Washington, D. C. USA

Greenberg, R. and Marra, P. P. (2005) *Birds of Two Worlds: the ecology and evolution of migration*. The Johns Hopkins University Press, Baltimore, Maryland, USA.

Greenberg, R. and Salewski, V. (2005) Ecological correlates of wintering social systems in New World and Old World migratory passerines. In: *Birds of Two Worlds: the ecology and evolution of migration* (eds. Greenberg, R. and Marra, P. P.), 336–358. The Johns Hopkins University Press, Baltimore, MD.

Greenberg, R. and Salgado-Ortiz, J. (1994) Interspecific defense of pasture trees by wintering Yellow Warblers. *Auk* **111**, 672–682.

Greenberg, R., Foster, M. S. and Marquez-Valelamar, L. (1995) The role of the White-eyed Vireo in the dispersal of Bursera fruit on the Yucatan Peninsula. *Journal of Tropical Ecology* **11**, 619–639.

Greenberg, R., Bichier, P., Cruz-Angon, A. and Reitsma, R. (1997a) Bird populations in shade and sun coffee plantations in central Guatemala. *Conservation Biology* **11**, 448–459.

Greenberg, R., Bichier, P. and Sterling, J. (1997b) Bird populations in rustic and planted shade coffee plantations of eastern Chiapas, Mexico. *Biotropica* **29**, 501–514.

Greenberg, R., Bichier, P., Cruz-Angon, A., MacVean, C., Perez, R. and Cano, E. (2000) The impact of avian insectivory on arthropods and leaf damage in some Guatemalan coffee plantations. *Ecology* **81**, 1750–1755.

Greenslade, P. (2008) Climate variability, biological control and an insect pest outbreak on Australia's Coral Sea islets: lessons for invertebrate conservation. *Journal of Insect Conservation* **12**, 333–342.

Gregory, R. D., Willis, S. G., Jiguet, F., Voříšek, P., Klvaňová, A., van Strien, A., Huntley, B., Collingham, Y. C., Couvet, D. and Green, R. E. (2009) An indicator of the impact of climatic change on European bird populations. *PLoS One* **4**, e4678.

Grimmett, R., Inskipp, C. and Inskipp, T. (1999) *A guide to the Birds of India, Pakistan, Nepal, Bangladesh, Bhhutan, Sri Lanka and the Maldives*. Princeton University Press, Princeton, New Jersey, USA.

Groom, M. J. (1992) Sand-colored Nighthawks parasitize the antipredator behavior of three nesting bird species. *Ecology* **73**, 785–793.

Groom, M. J. (2001) Consequences of subpopulation isolation for pollination, herbivory, and population growth in *Clarkia concinna concinna* (Onagraceae). *Biological Conservation* **100**, 55–63.

Gunnarsson, B. (2007) Predation on spiders: ecological mechanisms and evolutionary consequences. *Journal of Arachnology* **35**, 509–529.

Haag-Wackernagel, D. (1993) Street pigeons in Basel. *Nature* **361**, 200.

Haddad, N. M. (2003) Corridor use by diverse taxa. *Ecology* **84**, 609–615.

Haemig, P. D. (2001) Symbiotic nesting of birds with formidable animals: a review with applications to biodiversity conservation. *Biodiversity and Conservation* **10**, 527–540.

Hagan, J. M. III and Johnston, D. W. (eds.) (1992) *Ecology and Conservation of Neotropical Migrant Landbirds*. Smithsonian Institution Press, Washington, DC.

Hagemeijer, W. J. M. and Blair, M. J. (1997) *The EBCC Atlas of European Breeding Birds: their distribution and abundance*. T & A D Poyser, London, UK.

Hahn, S., Bauer, S. and Liechti, F. (2009) The natural link between Europe and Africa – 2.1 billion birds on migration. *Oikos* **118**, 624–626.

Haila, Y. (2002) A conceptual genealogy of fragmentation research: from island biogeography to landscape ecology. *Ecological Applications* **12**, 321–334.

Halaj, J. and Wise, D. H. (2001) Terrestrial trophic cascades: how much do they trickle? *American Naturalist* **157**, 262–281.

Halme, P., Hakkila, M. and Koskela, E. (2004) Do breeding Ural Owls *Strix uralensis* protect ground nests of birds? An experiment using dummy nests. *Wildlife Biology* **10**, 145–148.

Hamann, A. and Curio, E. (1999) Interactions among frugivores and fleshy fruit trees in a Philippine submontane rainforest. *Conservation Biology* **13**, 766–773.

Hamback, P. A., Oksanen, L., Ekerholm, P., Lindgren, A., Oksanen, T. and Schneider, M. (2004) Predators indirectly protect tundra plants by reducing herbivore abundance. *Oikos* **106**, 85–92.

Hamel, P., Dawson, D. K. and Keyser, P. D. (2004) How can we learn more about the Cerulean Warbler (*Dendroica cerulea*). *Auk* **121**, 7–14.

Hammond, D. S., Gourlet-Fleury, S., van der Hout, P., Ter Steege, H. and Brown, V. K. (1996) A compilation of known Guianan timber trees and the significance of their dispersal mode, seed size and taxonomic affinity to tropical rain forest management. *Forest Ecology and Management* **83**, 99–116.

Hamrick, J. L., Murawski, D. A. and Nason, J. D. (1993) The influence of seed dispersal mechanisms on the genetic structure of tropical tree populations. *Vegetatio* **108**, 281–297.

Hannah, D., Woinarski, J. C. Z., Catterall, C. P., McCosker, J. C., Thurgate, N. Y. and Fensham, R. J. (2007) Impacts of clearing, fragmentation and disturbance on the bird fauna of Eucalypt savanna woodlands in central Queensland, Australia. *Austral Ecology* **32**, 261–276.

Hansen, D. M. and Olesen, J. M. J., C. G. (2002) Trees, birds and bees in Mauritius: exploitative competition between introduced honey bees and endemic nectivorous birds? *Journal of Biogeography* **29**, 721–734.

Hansen, M. C. and DeFries, R. S. (2004) Detecting long-term global forest change using continuous fields of tree-cover maps from 8-km Advanced Very High Resolution Radiometer (AVHRR) data for the years 1982–99. *Ecosystems* **7**, 695–716.

Hansen, M. C., Stehman, S. V., Potapov, P. V., Loveland, T. R., Townshend, J. R. G., DeFries, R. S., Pittman, K. W., Arunarwati, B., Stolle, F., Steininger, M. K. and Carroll, M. (2008) Humid tropical forest clearing from 2000 to 2005 quantified by using multitemporal and multiresolution remotely sensed data. *Proceedings of the National Academy of Sciences of the United States of America* **105**, 9439–9444.

Hanski, I. (1989) Metapopulation dynamics: does it help to have more of the same? *Trends in Ecology and Evolution* **4**, 113–114.

Hardesty, J., Myers, R. L. and Fulks, W. (2005) Fire, ecosystems, and people: a preliminary assessment of fire as a global conservation issue. *The George Wright Forum* **22**, 78–87.

Harding, J. S., Hawke, D. J., Holdaway, R. N. and Winterbourn, M. J. (2004) Incorporation of marine-derived nutrients from petrel breeding colonies into stream food webs. *Freshwater Biology* **49**, 576–586.

Hardtër, R., Chow, W. Y. and Hock, O. S. (1997) Intensive plantation cropping, a source of sustainable food and energy production in the tropical rain forest areas in southeast Asia. *Forest Ecology and Management* **91**, 93–102.

Hargreaves, A. L., Johnson, S. D. and Nol, E. (2004) Do floral syndromes predict specialization in plant pollination systems? An experimental test in an "ornithophilous" African *Protea*. *Oecologia* **140**, 295–301.

Harms, K. E., Wright, S. J., Calderon, O., Hernandez, A. and Herre, E. A. (2000) Pervasive density-dependent recruitment enhances seedling diversity in a tropical forest. *Nature* **404**, 493–495.

Harper, L. H. (1989) The persistence of ant-following birds in small Amazonian forest fragments. *Acta Amazonica* **19**, 249–263.

Harper, R. G., Frick, J. A., Capparella, A. P., Borup, B., Nowak, M., Biesinger, D. and Thompson, C. F. (1996) Organochlorine pesticide contamination in neotropical migrant passerines. *Archives of Environmental Contamination and Toxicology* **31**, 386–390.

Harper, K. A., Macdonald, S. E., Burton, P. J., Chen, J. Q., Brosofske, K. D., Saunders, S. C., Euskirchen, E. S., Roberts, D., Jaiteh, M. S. and Esseen, P. A. (2005) Edge influence on forest structure and composition in fragmented landscapes. *Conservation Biology* **19**, 768–782.

Harris, G. M. and Pimm, S. L. (2004) Bird species' tolerance of secondary forest habitats and its effects on extinction. *Conservation Biology* **18**, 1607–1616.

Harris, G. M. and Pimm, S. L. (2008) Range size and extinction risk in forest birds. *Conservation Biology* **22**, 163–171.

Harvell, C. D., Mitchell, C. E., Ward, J. R., Altizer, S., Dobson, A. P., Ostfeld, R. S. and Samuel, M. D. (2002) Ecology – climate warming and disease risks for terrestrial and marine biota. *Science* **296**, 2158–2162.

Harvey, C. A. (2000) Windbreaks enhance seed dispersal into agricultural landscapes in Monteverde, Costa Rica. *Ecological Applications* **10**, 155–173.

Hawes, J., Barlow, J., Gardner, T. A. and Peres, C. A. (2008) The value of forest strips for understorey birds in an Amazonian plantation landscape. *Biological Conservation* **141**, 2262–2278.

Hawke, D. J., Holdaway, R. N., Causer, J. E. and Ogden, S. (1999) Soil indicators of pre-European seabird breeding in New Zealand at sites identified by predator deposits. *Australian Journal of Soil Research* **37**, 103–113.

Hayes, K. R. and Barry, S. C. (2008) Are there consistent predictors of invasion success? *Biological Invasions* **10**, 483–506.

Haynes-Sutton, A. M. (1987) The value of seabirds as a socio-economic resource in Jamaica. In: *The Value of Birds* (eds. Diamond, A. W. and Filion, F. L.), 77–81. ICBP, Norfolk, UK.

Heather, B. D. and Robertson, H. A. (1996) *The Field Guide to the Birds of New Zealand*. Penguin Books Ltd, Viking, New Zealand.

Hegde, S. G., Shaanker, R. U. and Ganeshaiah, K. N. (1991) Evolution of seed size in the bird-dispersed tree *Santalum album* – a tradeoff between seedling establishment and dispersal efficiency. *Evolutionary Trends in Plants* **5**, 131–135.

Heine, J. C. and Speir, T. W. (1989) Ornithogenic soils of the Cape Bird Adelie Penguin rookeries, Antarctica. *Polar Biology* **10**, 89–100.

Heldbjerg, H. and Fox, T. (2008) Long-term population declines in Danish trans-Saharan migrant birds. *Bird Study* **55**, 267–279.

Henle, K., Davies, K. F., Kleyer, M., Margules, C. and Settele, J. (2004) Predictors of species sensitivity to fragmentation. *Biodiversity and Conservation* **13**, 207–251.

Herrera, C. M. and Pellmyr, O. (2002) *Plant–Animal Interactions: An Evolutionary Approach*. Blackwell Science, Malden, MA.

Hertel, F. (1994) Diversity in body size and feeding morphology within past and present vulture assemblages. *Ecology* **75**, 1074–1084.

Hewitt, G. M. (1996) Some genetic consequences of ice ages, and their role in divergence and speciation. *Biological Journal of the Linnean Society* **58**, 247–276.

Hewitt, N. and Kellman, M. (2002) Tree seed dispersal among forest fragments: II. Dispersal abilities and biogeographical controls. *Journal of Biogeography* **29**, 351–363.

Heynen, I. (1999) Ecuadorian Hillstar *Oreotrochilus chimborazo*. In: *Handbook of the Birds of the World (Vol. 5): Barn-owls to Hummingbirds* (eds. Hoyo, J. D., Elliott, A. and Sargatal, J.), 623. Lynx Edicions, Barcelona.

Hilty, S. L. (1980) Flowering and fruiting periodicity in a premontane rain forest in Pacific Colombia. *Biotropica* **12**, 292–306.

Hjerpe, J., Hedenas, H. and Elmqvist, T. (2001) Tropical rain forest recovery from cyclone damage and fire in Samoa. *Biotropica* **33**, 249–259.

Hobbs, J. J. (2004) Problems in the harvest of edible birds' nests in Sarawak and Sabah, Malaysian Borneo. *Biodiversity and Conservation* **13**, 2209–2226.

Hobson, K. A. (2005) Flying fingerprints: making connections with stable isotopes and trace elements. In: *Birds of Two Worlds: the ecology and evolution of migration* (eds. Greenberg, R. and Marra, P. P.), 235–248. The Johns Hopkins University Press, Baltimore, MD.

Hobson, K. A. and Wassenaar, L. I. (1997) Linking breeding and wintering grounds of Neotropical migrant songbirds using stable isotope analysis of feathers. *Oecologia* **109**, 132–141.

Hobson, K. A., Wassenaar, L. I., Mila, F., Lovette, I., Dingle, C. and Smith, T. B. (2003) Stable isotopes as indicators of altitudinal distributions and movements in an Ecuadorean hummingbird community. *Oecologia* **136**, 302–308.

Holbrook, K. M. and Loiselle, B. A. (2009) Dispersal in a Neotropical tree, *Virola flexuosa* (Myristicaceae): does hunting of large vertebrates limit seed removal? *Ecology and Society* **90**, 1449–1455.

Holbrook, K. M. and Smith, T. B. (2000) Seed dispersal and movement patterns in two species of *Ceratogymna* hornbills in a West African tropical lowland forest. *Oecologia* **125**, 249–257.

Holbrook, K. M., Smith, T. B. and Hardesty, B. D. (2002) Implications of long-distance movements of frugivorous rain forest hornbills. *Ecography* **25**, 745–749.

Holdsworth, A. R. and Uhl, C. (1997) Fire in Amazonian selectively logged rain forest and the potential for fire reduction. *Ecological Applications* **7**, 713–725.

Hole, D. G., Willis, S. G., Pain, D. J., Fishpool, L. D., Butchart, S. H. M., Collingham, Y. C., Rahbek, C. and Huntley, B. (2009) Projected impacts of climate change on a continent-wide protected area network. *Ecology Letters* **12**, 420–431.

Holl, K. D., Loik, M. E., Lin, E. H. V. and Samuels, I. A. (2000) Tropical montane forest restoration in Costa Rica: overcoming barriers to dispersal and establishment. *Restoration Ecology* **8**, 339–349.

Holmes, R. T. (1990) Ecological and evolutionary impacts of bird predation on forest insects: an overview. In: *Avian Foraging: Theory, Methodology, and Applications* (eds. Morrison, M. L., Ralph, C. J., Verner, J. and J. R. Jehl, J.), 6–13. Allen Press, Lawrence, KS.

Holmes, R. T. (2007) Understanding population change in migratory songbirds: long-term and experimental studies of Neotropical migrants in breeding and winter areas. *Ibis* **149**, 2–13.

Holmes, R. T., Schultz, J. C. and Nothnagle, P. (1979) Bird predation on forest insects – an exclosure experiment. *Science* **206**, 462–463.

Homberger, D. G. and Beissinger, S. R. (2006) Bird trade: Conservation strategy or extinction catalyst? *Acta Zoologica Sinica* **52**(Suppl), 46–47.

Hooks, C. R. R., Pandey, R. R. and Johnson, M. W. (2003) Impact of avian and arthropod predation on lepidopteran caterpillar densities and plant productivity in an ephemeral agroecosystem. *Ecological Entomology* **28**, 522–532.

Hoover, J. P. (2003) Decision rules for site fidelity in a migratory bird, the Prothonotary Warbler. *Ecology* **84**, 416–430.

Houghton, J. (2004) *Global Warming: the complete briefing*. Cambridge University Press, Cambridge, UK.

Houston, D. C. (1974) Role of griffon vultures *Gyps africanus* and *Gyps ruppellii* as scavengers. *Journal of Zoology* **172**, 35–46.

Houston, D. C. (1979) The adaptation of scavengers. In: *Serengeti, Dynamics of an Ecosystem* (eds. Sinclair, A. R. E. and Griffiths, N.), 263–286. University of Chicago Press, Chicago.

Houston, D. C. (1983) The adaptive radiation of the griffon vultures. In: *Vulture Biology and Management* (eds. Wilbur, S. R. and Jackson, J. A.), 360–363. University of California Press, Berkeley.

Houston, D. C. (1986) Scavenging efficiency of Turkey Vultures in tropical forest. *Condor* **88**, 318–323.

Houston, D. C. (1987) The effect of reduced mammal numbers on *Cathartes* vultures in Neotropical forests. *Biological Conservation* **41**, 91–98.

Houston, D. C. (1988) Competition for food between Neotropical vultures in forest. *Ibis* **130**, 402–417.

Houston, D. C. (1994) Family Cathartidae (New World vultures). In: *Handbook of the Birds of the World (Vol. 2): New World Vultures to Guineafowl* (eds. Hoyo, J. D., Elliott, A. and Sargatal, J.), 24–41. Lynx Ediciones, Barcelona.

Houston, D. C. and Cooper, J. E. (1975) The digestive tract of the Whiteback Griffon Vulture and its role in disease transmission among wild ungulates. *Journal of Wildlife Diseases* **11**, 306–313.

Howard, P. C., Viskanic, P., Davenport, T. R. B., Kigenyi, F. W., Baltzer, M., Dickinson, C. J., Lwanga, J. S., Matthews, R. A. and Balmford, A. (1998) Complementarity and the use of indicator groups for reserve selection in Uganda. *Nature* **394**, 472–475.

Howe, H. F. (1984) Implications of seed dispersal by animals for tropical reserve management. *Biological Conservation* **30**, 261–282.

Howe, H. F. (1989) Scatter- and clump-dispersal and seedling demography: hypothesis and implications. *Oecologia* **79**, 417–426.

Howe, H. F. (1993) Aspects of variation in a Neotropical seed dispersal system. *Vegetatio* **108**, 149–162.

Howe, H. F. and De Steven, D. (1979) Fruit production, migrant bird visitation, and seed dispersal of Guarea glabra in Panama. *Oecologia* **39**, 185–196.

Howe, H. F. and Miriti, M. N. (2000) No question: seed dispersal matters. *Trends in Ecology and Evolution* **15**, 434–436.

Howe, H. F. and Miriti, M. N. (2004) When seed dispersal matters. *BioScience* **54**, 651–660.

Howe, H. F. and Smallwood, J. (1982) Ecology of seed dispersal. *Annual Review of Ecology and Systematics* **13**, 201–228.

Howe, H. F. and Vandekerckhove, G. A. (1981) Removal of Wild Nutmeg (*Virola surinamensis*) crops by birds. *Ecology* **62**, 1093–1106.

Howe, H. F., Schupp, E. W. and Westley, L. C. (1985) Early consequences of seed dispersal for a Neotropical tree (*Virola surinamensis*). *Ecology* **66**, 781–791.

Huang, H.-W. (2009) *Bycatch of Taiwanese Tuna Longline Fisheries in the Pacific Ocean*. Western and Central Pacific Fisheries Commission, Port Via, Vanuatu.

Huber, S. K. (2008) Effects of the introduced parasite *Philornis downsi* on nestling growth and mortality in the medium ground finch (*Geospiza fortis*). *Biological Conservation* **141**, 601–609.

Hughes, B. J., Martin, G. R. and Reynolds, S. J. (2008) Cats and seabirds: effects of feral domestic cat *Felis silvestris catus* eradication on the population of Sooty Terns *Onychoprion fuscata* on Ascension Island, South Atlantic. *Ibis* **150**, 122–131.

Hughes, J. B., Daily, G. C. and Ehrlich, P. R. (1997) Population diversity, its extent and extinction. *Science* **278**, 689–692.

Hughes, J. B., Daily, G. C. and Ehrlich, P. R. (2002) Conservation of tropical forest birds in countryside habitats. *Ecology Letters* **5**, 121–129.

Huntley, B., Collingham, Y. C., Green, R. E., Hilton, G. M., Rahbek, C. and Willis, S. G. (2006) Potential impacts of climatic change upon geographical distributions of birds. *Ibis* **148**, 8–28.

Huong, S. L. and Sodhi, N. S. (1997) Status of the Oriental Magpie Robin *Copsychus saularis* in Singapore. *Malayan Nature Journal* **50**, 347–354.

Huston, M. A. (1997) Hidden treatments in ecological experiments: re-evaluating the ecosystem function of biodiversity. *Oecologia* **110**, 449–460.

Hutchins, H. E., Hutchins, S. A. and Liu, B. W. (1996) The role of birds and mammals in Korean Pine (*Pinus koraiensis*) regeneration dynamics. *Oecologia* **107**, 120–130.

Hutchinson, G. E. (1957) Concluding Remarks. Cold Spring Harbor Symposia on Quantitative Biology, 415–427. Cold Spring Harbor, NY.

Hutto, R. L. (1989) The effect of habitat alteration on migratory land birds in a West Mexican tropical deciduous forest: A conservation perspective. *Conservation Biology* **3**, 138–148.

Hutto, R. L. 1992. Habitat distributions of migratory landbirds in western Mexico. In: *Ecology and Conservation of Neotropical Migrant Landbirds* (eds. Hagan, J. M. III and Johnston, D. W.), 211–239. Smithsonian Institution Press, Washington, DC.

Hutto, R. L. (1995) Composition of bird communities following stand-replacement fires in northern rocky-mountain (USA) conifer forests. *Conservation Biology* **9**, 1041–1058.

Hutto, R. L. (2008) The ecological importance of severe wildfires: some like it hot. *Ecological Applications* **18**, 1827–1834.

Iason, G. R., Duck, C. D. and Clutton-Brock, T. H. (1986) Grazing and reproductive success of Red Deer – the effect of local enrichment by gull colonies. *Journal of Animal Ecology* **55**, 507–515.

Ims, R. A. and Andreassen, H. P. (2000) Spatial synchronization of vole population dynamics by predatory birds. *Nature* **408**, 194–196.

Ingle, N. R. (2003) Seed dispersal by wind, birds, and bats between Philippine montane rainforest and successional vegetation. *Oecologia* **134**, 251–261.

Iñigo-Elias, E. E. and Ramos, M. A. (1991) The psittacine trade in Mexico. In: *Neotropical Wildlife Use and Conservation* (eds. Robinson, J. G. and Redford, K. H.), 380–392. University of Chicago Press, Chicago, IL.

IPCC (Intergovernmental Panel on Climate Change) (2007) *Fourth Assessment Report: climate change 2007, the physical science basis*. Cambridge University Press, Cambridge, UK.

Isaac, J. L., De Gabriel, J. L. and Goodman, B. A. (2008) Microclimate of daytime den sites in a tropical possum: Implications for the conservation of tropical arboreal marsupials. *Animal Conservation* **11**, 281–287.

Ishtiaq, F., Beadell, J. S., Baker, A. J., Rahmani, A. R., Jhala, Y. V. and Fleischer, R. C. (2006) Prevalence and evolutionary relationships of haematozoan parasites in native versus introduced populations of common myna *Acridotheres tristis. Proceedings of the Royal Society of London Series B* **273**, 587–594.

IUCN (2010a) Summary Tables. IUCN Red List of Threatened Species. Version 2010.4. http:/ www.iucnredlist.org/about/summary-statistics#Tables_3_4.

IUCN (2010b) IUCN Red List of Threatened Species. Version 2010.4. http://www.iucnredlist.org.

Ives, A. R. and Dobson, A. P. (1987) Antipredator behavior and the population-dynamics of simple predator-prey systems. *American Naturalist* **130**, 431–447.

Izhaki, I. and Safriel, U. N. (1989) Why are there so few exclusively frugivorous birds? Experiments on fruit digestibility. *Oikos* **54**, 23–32.

Jackson, P. S. W., Cronk, Q. C. B. and Parnell, J. A. N. (1988) Notes on the regeneration of two rare Mauritian endemic trees. *Tropical Ecology* **298**, 98–106.

Jacquemot, A. and Filion, F. L. (1987) The economic significance of birds in Canada. In: *The Value of Birds* (eds. Diamond, A. W. and Filion, F. L.), 15–21. ICBP, Norfolk, UK.

Jaeger, M. E. and Elliott, C. C. H. (1989) Quelea as a resource. In: *Quelea Quelea: Africa's bird pest* (eds. Bruggers, R. L. and Elliott, C. C. H.), 327–338. Oxford University Press, Oxford, UK.

Jahn, A. E., Levey, D. J. and Smith, K. G. (2004) Reflections across hemispheres: a system-wide approach to New World bird migration. *Auk* **121**, 1005–1013.

Jahn, A. E., Levy, D. J., Hostetler, J. A. and Mamani, A. M. (2010) Determinants of partial bird migration in the Amazon Basin. *Journal of Animal Ecology* **79**, 983–992.

Jaksic, F. M. (2004) El Niño effects on avian ecology: lessons learned from the southeastern Pacific. *Ornitologia Neotropical* **15**, 61–72.

Jaksic, F. M., Feinsinger, P. and Jimenez, J. E. (1996) Ecological redundancy and long-term dynamics of vertebrate predators in semiarid Chile. *Conservation Biology* **10**, 252–262.

James, A., Gaston, K. J. and Balmford, A. (2001) Can we afford to conserve biodiversity. *BioScience* **51**, 43–52.

James, F. C. and McCulloch, C. E. (1995) The strength of inferences about causes of trends in populations. In: *Ecology and Management of Neotropical birds. A synthesis and review of critical issues* (eds. Martin, T. E. and Finch, D. M.), 40–54. Oxford University Press, New York, NY.

James, F. C., McCulloch, C. E. and Wiedenfeld, D. A. (1996) New approaches to the analysis of population trends in land birds. *Ecology* **77**, 13–27.

Jankowski, J. E., Robinson, S. K. and Levey, D. J. (2010) Squeezed at the top: interspecific aggression may constrain elevational ranges in tropical birds. *Ecology* **91**, 1877–1884.

Jansson, R. (2003) Global patterns in endemism explained by past climatic change. *Proceedings of the Royal Society of London Series B* **270**, 583–590.

Jantti, A., Aho, T., Hakkarainen, H., Kuitunen, M. and Suhonen, J. (2001) Prey depletion by the foraging of the Eurasian Treecreeper, *Certhia familiaris*, on tree-trunk arthropods. *Oecologia* **128**, 488–491.

Janzen, D. H. (1967) Why mountain passes are higher in the tropics. *American Naturalist* **101**, 233–249.

Janzen, D. H. (1970) Herbivores and the number of tree species in tropical forests. *American Naturalist* **104**, 501–528.

Janzen, D. H. (1972) Escape in space by *Sterculia apetala* seeds from bug *Dysdercus fasciatus* in a Costa Rican deciduous forest. *Ecology* **53**, 350–361.

Jarvinen, A. (1994) Global warming and egg size of birds. *Ecography* **17**, 108–110.

Jenkins, C. N. and Joppa, L. (2009) Expansion of the global terrestrial protected area system. *Biological Conservation* **142**, 2166–2174.

Jenkins, C. N., Alves, M. A. S. and Pimm, S. L. (2010) Avian conservation priorities in a top ranked biodiversity hotspot. *Biological Conservation* **143**, 992–998.

Jenkins, M. (2003) Prospects for biodiversity. *Science* **302**, 1175–1177.

Jennings, M. C. (1992) The House Crow *Corvus splendens* in Aden (Yemen) and an attempt at its control. *Sandgrouse* **14**, 27–33.

Jensch, D. and Ellenberg, H. (1999) The hornbill *Tockus semifasciatus* as a seed-disperser and ecological indicator, and forest rehabilitation in eastern Ivory Coast. *Revue d Ecologie-La Terre Et La Vie* **54**, 333–350.

Jepson, P. and Ladle, R. J. (2005) Bird-keeping in Indonesia: conservation impacts and the potential for substitution-based conservation responses. *Oryx* **39**, 442–448.

Jepson, P. and Ladle, R. J. (2009) Governing bird-keeping in Java and Bali: evidence from a household survey. *Oryx* **43**, 364–374.

Jepson, P., Prana, M., Sujatnika and Amama, F. (2008) Developing a certification system for captive-bred birds in Indonesia. *TRAFFIC Bulletin* **22**, 7–9.

Jeschke, J. M. (2008) Across islands and continents, mammals are more successful invaders than birds. *Diversity and Distributions* **14**, 913–916.

Jeschke, J. M. and Strayer, D. L. (2008) Are threat status and invasion success two sides of the same coin? *Ecography* **31**, 124–130.

Jetz, W., Rahbek, C. and Colwell, R. K. (2004) The coincidence of rarity and richness and the potential signature of history in centers of endemism. *Ecology Letters* **7**, 1180–1191.

Jetz, W., Wilcove, D. S. and Dobson, A. P. (2007) Projected impacts of climate and land-use change on the global diversity of birds. *PLoS Biology* **5**, e157.

Jetz, W., Sekercioglu, C. H. and Böhning-Gaese, K. (2008a) The worldwide variation in avian clutch size across species and space. *PLoS Biology* **6**, 2650–2657.

Jetz, W., Sekercioglu, C. H. and Watson, J. E. M. (2008b) Ecological correlates and conservation implications of overestimating species geographic ranges. *Conservation Biology* **22**, 110–119.

Jha, P. K. (2003) Health and social benefits from improving community hygiene and sanitation: an Indian experience. *International Journal of Environmental Health Research* **13**, S133–S140.

Jiguet, F., Julliard, R., Thomas, C. D., Dehorter, O., Newson, S. E. and Couvet, D. (2006) Thermal range predicts bird population resilience to extreme high temperatures. *Ecology Letters* **9**, 1321–1330.

Johns, A. D. (1986) Effects of selective logging on the ecological organization of a peninsular Malaysian rainforest avifauna. *Forktail* **1**, 65–79.

Johns, A. D. (1989) Recovery of a Peninsular Malaysian rainforest avifauna following selective timber logging: the first twelve years. *Forktail* **4**, 89–105.

Johns, A. G. (1996) Bird population persistence in Sabahan logging concessions. *Biological Conservation* **75**, 3–10.

Johnson, E. A., Miyanishi, K. and Bridge, S. R. J. (2001) Wildfire regime in the boreal forest and the idea of suppression and fuel buildup. *Conservation Biology* **15**, 1554–1557.

Johnson, M. D., Sherry, T. W., Holmes, R. T. and Marra, P. P. (2006) Assessing habitat quality for a migratory songbird wintering in natural and agricultural habitats. *Conservation Biology* **20**, 1433–1444.

Johnson, M. D., Kellermann, J. L. and Stercho, A. M. (2010) Pest reduction services by birds in shade and sun coffee in Jamaica. *Animal Conservation* **13**, 140–147.

Johnston, R. F. and Janiga, M. (1995) *Feral Pigeons*. Oxford University Press, New York, NY.

Jones, C. G., Lawton, J. H. and Shachak, M. (1994) Organisms as ecosystem engineers. *Oikos* **69**, 373–386.

Jones, C. G., Heck, W., Lewis, R. E., Mungroo, Y., Slade, G. and Cade, T. (2008a) The restoration of the Mauritius Kestrel *Falco punctatus* population. *Ibis* **137**, S173–S180.

Jones, E. R., Wishnie, M. H., Deago, J., Sautu, A. and Cerezo, A. (2004) Facilitating natural regeneration in *Saccharum spontaneum* (L.) grasslands within the Panama Canal watershed: effects of tree species and tree structure on vegetation recruitment patterns. *Forest Ecology and Management* **191**, 171–183.

Jones, F. A., Peterson, C. J. and Haines, B. L. (2003) Seed predation in Neotropical pre-montane pastures: site, distance, and species effects. *Biotropica* **35**, 219–225.

Jones, H. P., Tershy, B. R., Zavaleta, E. S., Croll, D. A., Keitt, B. S., Finkelstein, M. E. and Howald, G. R. (2008b) Severity of the effects of invasive rats on seabirds: a global review. *Conservation Biology* **22**, 16–26.

Jones, J. and Francis, C. M. (2003) The effects of light characteristics on avian mortality at lighthouses. *Journal of Avian Biology* **34**, 328–333.

Jones, J., Norris, D. R., Given, M. K., Barg, J. J., Kyser, T. K. and Robertson, R. J. (2008c) Migratory connectivity and rate of population decline in a vulnerable songbird. *Condor* **110**, 538–544.

Jones, M. J., Sullivan, M. S., Marsden, S. J. and Linsley, M. D. (2001) Correlates of extinction risk of birds from two Indonesian Islands. *Biological Journal of the Linnean Society* **73**, 65–79.

Jonzen, N., Linden, A., Ergon, T., Knudsen, E., Vik, J. O., Rubolini, D., Piacentini, D., Brinch, C., Spina, F., Karlsson, L., Stervander, M., Andersson, A., Wladenstrom, J., Lehikoinen, A., Edvardsen, E., Solvang, R. and Stenseth, N.-C. (2006) Rapid advance of spring arrival dates in long-distance migratory birds. *Science* **312**, 1959–1961.

Jonzen, N., Ergon, T., Linden, A. and Stenseth, N. C. (2007) Bird migration and climate. *Climate Research* **35**, 1–180.

Joppa, L. N., Loarie, S. R. and Pimm, S. L. (2008) On the protection of "protected areas". *Proceedings of the National Academy of Sciences of the United States of America* **105**, 6673–6678.

Jordano, P. (1987) Patterns of mutualistic interactions in pollination and seed dispersal: connectance, dependence asymmetries, and coevolution. *American Naturalist* **129**, 657–677.

Jordano, P. (2000) Fruits and frugivory. In: *Seeds: the ecology of regeneration in plant communities* (eds. Fenner, M.), 125–165. CAB International, New York, NY.

Jordano, P. and Godoy, J. A. (2000) RAPD variation and population genetic structure in *Prunus mahaleb* (Rosaceae), an animal-dispersed tree. *Molecular Ecology* **9**, 1293–1305.

Joseph, L. (1997) Towards a broader view of Neotropical migrants: consequences of a reexamination of austral migration. *Ornitologia Neotropical* **8**, 31–36.

Junk, W. J. (2002) Long-term environmental trends and the future of tropical wetlands. *Environmental Conservation* **29**, 414–435.

Kang, N., Hails, C. J. and Sigurdsson, J. B. (1991) Nest construction and egg-laying in edible nest-swiftlets Aerodramus spp and the implications for harvesting. *Ibis* **133**, 170–177.

Kaplan, R. (1997) *The Ends of the Earth: from Togo to Turkmenistan, from Iran to Cambodia, a Journey to the Frontiers of Anarchy*. Vintage Books, New York, NY.

Karels, T. J., Dobson, F. S., Trevino, H. S. and Skibiel, A. L. (2008) The biogeography of avian extinctions on oceanic islands. *Journal of Biogeography* **35**, 1106–1111.

Karesh, W. B., Cook, R. A., Bennett, E. L. and Newcomb, J. (2005) Wildlife trade and global disease emergence. *Emerging Infectious Diseases* **11**, 1000–1002.

Karesh, W. B., Cook, R. A., Gilbert, M. and Newcomb, J. (2007) Implications of wildlife trade on the movement of avian influenza and other infectious diseases. *Journal of Wildlife Diseases* **43**, S55–S59.

Karr, J. R. (1976) On the relative abundance of migrants from the north temperate zone in tropical habitats. *Wilson Bulletin* **88**, 433–458.

Karr, J. R. (1982) Population variability and extinction in the avifauna of a tropical land bridge island. *Ecology and Society* **63**, 1975–1978.

Karr, J. R. (1990) Avian survival rates and the extinction process on Barro Colorado island, Panama. *Conservation Biology* **4**, 391–397.

Kattan, G. H. (1992) Rarity and vulnerability: the birds of the Cordillera Central of Colombia. *Conservation Biology* **6**, 64–70.

Kattan, G. H., Alvarez-López, H. and Giraldo, M. (1994) Forest fragmentation and bird extinctions: San Antonio eighty years later. *Conservation Biology* **8**, 138–146.

Keane, A., Brooke, M. D. and McGowan, P. J. K. (2005) Correlates of extinction risk and hunting pressure in gamebirds (Galliformes). *Biological Conservation* **126**, 216–233.

Keast, A. and Morton, E. S. (1980) *Migrant Birds in the Neotropics: ecology, behavior, distribution, and conservation*. Smithsonian Institution Press, Washington, DC.

Keighery, G. J. (1980) Bird pollination in south Western Australia: a checklist. *Entwicklungsgeschichte Und Systematik Der Pflanzen* **135**, 171–176.

Keith, J. O. and Bruggers, R. L. (1998) Review of hazards to raptors from pest control in Sahelian Africa. *Journal of Raptor Research* **32**, 151–158.

Keitt, B. S., Wilcox, C., Tershy, B. R., Croll, D. A. and Donlan, C. J. (2002) The effect of feral cats on the population viability of black-vented shearwaters (*Puffinus opisthomelas*) on Natividad Island, Mexico. *Animal Conservation* **5**, 217–223.

Kellermann, J. L., Johnson, M. D., Stercho, A. M. and Hackett, S. C. (2008) Ecological and economic services provided by birds on Jamaican Blue Mountain coffee farms. *Conservation Biology* **22**, 1177–1185.

Kelly, D., Ladley, J. J. and Robertson, A. W. (2004) Is dispersal easier than pollination? Two tests in new Zealand Loranthaceae. *New Zealand Journal of Botany* **42**, 89–103.

Kelly, D., Ladley, J. J., Robertson, A. W., Anderson, S. H., Wotton, D. M. and Wiser, S. K. (2010) Mutualisms with the wreckage of an avifauna: the status of bird pollination and fruit-dispersal in New Zealand. *New Zealand Journal of Ecology* **34**, 66–85.

Kelly, J. F., Altudorei, V., Sharp, Z. D. and Finch, D. H. (2002) Insights into Wilson's Warbler migration from analysis of hydrogen stable-isotope ratios. *Oecologia* **130**, 216–221.

Kemp, A. (2001) Family Bucerotidae (Hornbills). In: *Handbook of the Birds of the World, Vol. 6, mousebirds to hornbills* (eds. Hoyo, J. D., Elliot, A. and Sargatal, J.), 436–523. Lynx Edicions, Barcelona, Spain.

Kennedy, P. G., Hausmann, N. J., Wenk, E. H. and Dawson, T. E. (2004) The importance of seed reserves for seedling performance: an integrated approach using morphological, physiological, and stable isotope techniques. *Oecologia* **141**, 547–554.

Kerbes, R. H., Kotanen, P. M. and Jefferies, R. L. (1990) Destruction of wetland habitats by Lesser Snow Geese: a keystone species on the west coast of Hudson Bay. *Journal of Applied Ecology* **271**, 242–258.

Kerr, A. M. (1993) Low frequency of stabilimenta in orb webs of Argiope appensa (Araneae: Araneidae) from Guam: an indirect effect of an introduced avian predator? *Pacific Science* **47**, 328–337.

Kerr, J. T. and Burkey, T. V. (2002) Endemism, diversity, and the threat to tropical moist forest extinctions. *Biodiversity and Conservation* **11**, 695–704.

Kilpatrick, A. M., LaDeau, S. L. and Marra, P. M. (2007) Ecology of West Nile Virus transmission and its impact on birds in the western hemisphere. *Auk* **124**, 1121–1136.

Kingsford, R. T., Watson, J. E. M., Lundquist, C. J., Venter, O., Hughes, L., Johnston, E. L., Atherton, J., Gawel, M., Keith, D. A., Mackey, B. G., Morley, C., Possingham, H. P., Raynor, B., Recher, H. F. and Wilson, K. A. (2009) Major conservation policy issues for biodiversity in Oceania. *Conservation Biology* **23**, 834–840.

Kinnaird, M. F. (1998) Evidence for effective seed dispersal by the Sulawesi Red-knobbed Hornbill, *Aceros cassidix*. *Biotropica* **30**, 50–55.

Kinnaird, M. F. and O'Brien, T. G. (1998) Ecological effects of wildfire on lowland rainforest in Sumatra. *Conservation Biology* **12**, 954–956.

Kinzelbach, R. K. (1995) Avifauna and climatic changes in the 16th century – new sources and findings in historical ornithology. *Naturwissenschaften* **82**, 499–508.

Kirika, J. M., Farwig, N. and Böhning-Gaese, K. (2008) Effects of local disturbance of tropical forests on frugivores and seed removal of a small-seeded Afrotropical tree. *Conservation Biology* **22**, 318–328.

Kirk, D. A., Evenden, M. D. and Mineau, P. (1996) Past and current attempts to evaluate the role of birds as predators of insect pests in temperate agriculture. *Current Ornithology* **13**, 175–269.

Kitamura, S., Yumoto, T., Poonswad, P., Chuailua, P., Plongmai, K., Maruhashi, T. and Noma, N. (2002) Interactions between fleshy fruits and frugivores in a tropical seasonal forest in Thailand. *Oecologia* **133**, 559–572.

Kitamura, S., Suzuki, S., Yumoto, T., Poonswad, P., Chuailua, P., Plongmai, K., Noma, N., Maruhashi, T. and Suckasam, C. (2004) Dispersal of *Aglaia spectabilis*, a large-seeded tree species in a moist evergreen forest in Thailand. *Journal of Tropical Ecology* **20**, 421–427.

Kleindorfer, S. and Dudaniec, R. Y. (2006) Increased prevalence of avian poxvirus in Darwin's finches and its effect on male pairing success. *Journal of Avian Biology* **37**, 69–76.

Klink, C. A. and Machado, R. B. (2005) Conservation of the Brazilian Cerrado. *Conservation Biology* **19**, 707–713.

Knick, S. T., Dobkin, D. S., Rotenberry, J. T., Schroeder, M. A., Haegen, W. M. V. and Riper, C. v. (2003) Teetering on the edge or too late? Conservation and research issues for avifauna of sagebrush habitats. *Condor* **105**, 611–634.

Knutson, M. G., Niemi, G. J., Newton, W. E. and Friberg, M. A. (2004) Avian nest success in Midwestern forests fragmented by agriculture. *Condor* **106**, 116–130.

Kodandapani, N., Cochrane, M. A. and Sukumar, R. (2008) A comparative analysis of spatial, temporal, and ecological characteristics of forest fires in seasonally dry tropical ecosystems in the Western Ghats, India. *Forest Ecology and Management* **256**, 607–617.

Koh, L. P. (2008) Birds defend oil palms from herbivorous insects. *Ecological Applications* **18**, 821–825.

Koh, L. P. and Gardner, T. A. (2010) Conservation in human-modified landscapes. In: *Conservation Biology for All* (eds. Sodhi, N. S. and Ehrlich, P. R.), 236–261. Oxford University Press, Oxford, UK.

Koh, L. P. and Ghazoul, J. (2010) A matrix-calibrated species–area model for predicting biodiversity losses due to land-use change. *Conservation Biology* **24**, 994–1001.

Koh, L. P. and Wilcove, D. S. (2008) Is oil palm agriculture really destroying tropical biodiversity? *Conservation Letters* **1**, 60–64.

Koh, L. P., Dunn, R. R., Sodhi, N. S., Colwell, R. K., Proctor, H. C. and Smith, V. S. (2004) Species coextinctions and the biodiversity crisis. *Science* **305**, 1632–1634.

Kolar, C. S. and Lodge, D. M. (2001) Progress in invasion biology: predicting invaders. *Trends in Ecology and Evolution* **16**, 199–204.

Komar, O. (2006) Ecology and conservation of birds in coffee plantations: a critical review. *Bird Conservation International* **16**, 1–23.

Komdeur, J. (1994) Conserving the Seychelles warbler *Acrocephalus sechellensis* by translocation from cousin island to the islands of Aride and Cousine. *Biological Conservation* **67**, 143–152.

Komdeur, J. (1996) Breeding of the Seychelles Magpie Robin *Copsychus sechellarum* and implications for its conservation. *Ibis* **138**, 485–498.

Kong, Y. C., Keung, W. M., Yip, T. T., Ko, K. M., Tsao, S. W. and Ng, M. H. (1987) Evidence that epidermal growth factor is present in swiftlet's (*Collocalia*) nest. *Comparative Biochemistry and Physiology* **87**, 221–226.

Korpimaki, E. and Norrdahl, K. (1991) Do breeding nomadic avian predators dampen population fluctuations of small mammals? *Oikos* **62**, 195–208.

Korpimaki, E. and Norrdahl, K. (1998) Experimental reduction of predators reverses the crash phase of small-rodent cycles. *Ecology* **79**, 2448–2455.

Kotler, B. P., Brown, J. S., Smith, R. J. and Wirtz, W. O. (1988) The effects of morphology and body size on rates of owl predation on desert rodents. *Oikos* **53**, 145–152.

Krook, K., Bond, W. J. and Hockey, P. A. R. (2007) The effect of grassland shifts on the avifauna of a South African savanna. *Ostrich* **78**, 271–279.

Kruger, S. C. and Lawes, M. J. (1997) Edge effects at an induced forest-grassland boundary: forest birds in the ongoye forest reserve, kwazulu-natal. *South African Journal of Zoology* **32**, 82–91.

Kull, C. A. and Laris, P. (2009) Fire ecology and fire politics in Mali and Madagascar. In: *Tropical Fire Ecology: Climate Change, Land Use and Ecosystem Dynamics* (ed. Cochrane, M. A.), 171–226. Springer-Praxis, Heidelberg, Germany.

La Sorte, F. A. and Thompson, F. R. III. (2007) Poleward shifts in winter ranges of North American birds. *Ecology* **88**, 1803–1812.

Lacher, T. E. and Goldstein, M. I. (1997) Tropical ecotoxicology: status and needs. *Environmental Toxicology and Chemistry* **16**, 100–111.

LaDeau, S. L., Kilpatrick, A. M. and Marra, P. P. (2007) West Nile Virus emergence and large-scale declines of North American bird populations. *Nature* **447**, 710–713.

Ladley, J. J. and Kelly, D. (1995) Explosive New Zealand mistletoe. *Nature* **378**, 766.

Ladley, J. J. and Kelly, D. (1996) Dispersal, germination and survival of New Zealand mistletoes (Loranthaceae): dependence on birds. *New Zealand Journal of Ecology* **20**, 69–79.

Lagos, V. O., Contreras, L. C., Meserve, P. L., Gutierrez, J. R. and Jaksic, F. M. (1995) Effects of predation risk on space use by small mammals: a field experiment with a Neotropical rodent. *Oikos* 74, 259–264.

Lambert, F. R. (1992) The consequences of selective logging for Bornean lowland forest birds. *Philosophical Transactions of the Royal Society of London B* 335, 443–457.

Lambertucci, S. A., Trejo, A., Di Martino, S., Sanchez-Zapata, J. A., Donazar, J. A. and Hiraldo, F. (2009) Spatial and temporal patterns in the diet of the Andean condor: ecological replacement of native fauna by exotic species. *Animal Conservation* 12, 338–345.

Lasso, E. and Naranjo, M. E. (2003) Effect of pollinators and nectar robbers on nectar production and pollen deposition in *Hamelia patens* (Rubiaceae). *Biotropica* 35, 57–66.

Latta, S. C. (2003) Effects of scaley-leg mite infestations on body condition and site fidelity of migratory warblers in the Dominican Republic. *Auk* **120**, 730–743.

Latta, S. C. and Baltz, M. E. (1997) Population limitation in Neotropical migratory birds: comments on Rappole and MacDonald (1994). *Auk* **114**, 754–762.

Latta, S. C. and Brown, C. (1999) Autumn stopover ecology of the Blackpoll Warbler (*Dendroica striata*) in thorn scrub forest of the Dominican Republic. *Canadian Journal of Zoology* 77, 1147–1156.

Latta, S. C. and Faaborg, J. (2001) Winter site fidelity of Prairie Warblers in the Dominican Republic. *Condor* **103**, 455–468.

Latta, S. C. and Faaborg, J. (2002) Demographic and population responses of Cape May Warblers wintering in multiple habitats. *Ecology* **83**, 2502–2515.

Latta, S. C. and Faaborg, J. (2009) Benefits of studies of overwintering birds for understanding resident bird ecology and promoting development of conservation capacity. *Conservation Biology* **23**, 286–293.

Latta, S. C., Rimmer, C. C. and MacFarland, K. P. (2003) Winter bird communities in four different habitats along an elevational gradient on Hispaniola. *Condor* **105**, 179–197.

Laundre, J. W., Hernandez, L. and Altendorf, K. B. (2001) Wolves, elk, and bison: Reestablishing the "landscape of fear" in Yellowstone National Park, USA. *Canadian Journal of Zoology* **79**, 1401–1409.

Laurance, S. G. W. and Gomez, M. S. (2005) Clearing width and movements of understory rainforest birds. *Biotropica* 37, 149–152.

Laurance, S. G. W., Stouffer, P. C. and Laurance, W. E. (2004) Effects of road clearings on movement patterns of understory rainforest birds in central Amazonia. *Conservation Biology* **18**, 1099–1109.

Laurance, W. F. (1999) Reflections on the tropical deforestation crisis. *Biological Conservation* **91**, 109–117.

Laurance, W. F. (2004) Forest–climate interactions in fragmented tropical landscapes. *Philosophical Transactions of the Royal Society B* **359**, 345–352.

Laurance, W. F. (2006) A change in climate. *Tropinet* **17**, 1–3.

Laurance, W. F. (2008a) Theory meets reality in fragmented forests. *Animal Conservation* **11**, 364–365.

Laurance, W. F. (2008b) Theory meets reality: how habitat fragmentation research has transcended island biogeographic theory. *Biological Conservation* **141**, 1731–1744.

Laurance, W. F. and Bierregaard, R. O., Jr. (1997) *Tropical Forest Remnants: ecology, management, and conservation of fragmented communities*. University of Chicago Press, Chicago.

Laurance, W. F. and Curran, T. J. (2008) Impacts of wind disturbance on fragmented tropical forests: a review and synthesis. *Austral Ecology* **33**, 399–408.

Laurance, W. F. and Useche, D. C. (2009) Environmental synergisms and extinction of tropical species. *Conservation Biology* **23**, 1427–1437.

Laurance, W. F., Vasconcelos, H. L. and Lovejoy, T. E. (2000) Forest loss and fragmentation in the Amazon: implications for wildlife conservation. *Oryx* **34**, 39–45.

Laurance, W. F., Goosem, M. and Laurance, S. G. W. (2009) Impacts of roads and linear clearings on tropical forests. *Trends in Ecology and Evolution* **24**, 659–669.

Lawler, J. J., Shafer, S. L., White, D., Kareiva, P., Maurer, E. P., Blaustein, A. R. and Bartlein, P. J. (2009) Projected climate-induced faunal change in the Western Hemisphere. *Ecology* **90**, 588–597.

Laybourne, R. (1974) Collision between a vulture and an aircraft at an altitude of 37000 feet. *Wilson Bulletin* **86**, 461–462.

Le Bohec, C., Durant, J. M., Gauthier-Clerc, M., Stenseth, N. C., Park, Y. H., Pradel, R., Gremillet, D., Gendner, J. P. and Le Maho, Y. (2008) King penguin population threatened by Southern Ocean warming. *Proceedings of the National Academy of Sciences of the United States of America* **105**, 2493–2497.

Leck, C. F. (1979) Avian extinctions in an isolated tropical wet-forest preserve, Ecuador. *Auk* **96**, 343–352.

Lee, R. J., Gorog, A. J., Dwiyarhreni, A., Siwu, S., Riley, J., Alexander, H., Paoli, G. D. and Ramono, W. (2005a) Wildlife trade and implications for law enforcement in Indonesia: a case study from North Sulawesi. *Biological Conservation* **123**, 477–488.

Lee, T. M. and Jetz, W. (2008) Future battlegrounds for conservation under global change. *Proceedings of the Royal Society of London Series B* **275**, 1261–1270.

Lee, T. M., Soh, M. C. K., Sodhi, N. S., Koh, L. P. and Lim, S. L. H. (2005b) Effects of habitat disturbance on mixed species bird flocks in a tropical submontane forest. *Biological Conservation* **122**, 193–204.

Lee, T. M., Sodhi, N. S. and Prawiradilaga, D. M. (2007) The importance of protected areas for the forest and endemic avifauna of Sulawesi (Indonesia). *Ecological Applications* **17**, 1727–1741.

Lee, T. M., Sodhi, N. S. and Prawiradilaga, D. M. (2008a) Birds, local people and protected areas in Sulawesi, Indonesia. In: *Biodiversity and Human Livelihoods in Protected Areas* (eds. Sodhi, N. S., Acciaioli, G., Erb, M. and Tan, A. K.-J.), 78–94. Cambridge University Press, Cambridge, UK.

Lee, Y. F., Kuo, Y. M., Lin, Y. H., Chu, W. C., Wu, S. H., Wang, H. H. and Chao, J. T. (2008b) Spatiotemporal variation in avian diversity and the short-term effects of typhoons in tropical reef-karst forests on Taiwan. *Zoological Science* **25**, 593–603.

Lees, A. C. and Peres, C. A. (2006) Rapid avifaunal collapse along the Amazonian deforestation frontier. *Biological Conservation* **133**, 198–211.

Lees, A. C. and Peres, C. A. (2008a) Conservation value of remnant riparian forest corridors of varying quality for Amazonian birds and mammals. *Conservation Biology* **22**, 439–449.

Lees, A. C. and Peres, C. A. (2008b) Avian life-history determinants of local extinction risk in a hyper-fragmented neotropical forest landscape. *Animal Conservation* **11**, 128–137.

Lees, A. C. and Peres, C. A. (2009) Gap-crossing movements predict species occupancy in Amazonian forest fragments. *Oikos* **118**, 280–290.

Lefebvre, G., Poulin, B. and McNeil, R. (1994) Temporal dynamics of mangrove bird communities in Venezuela with special reference to migrant warblers. *Auk* **111**, 233–236.

Lens, L., van Dongen, S., Wilder, C. M., Brooks, T. M. and Matthysen, E. (1999) Fluctuating asymmetry increases with habitat disturbance in seven bird species of a fragmented afrotropical forest. *Proceedings of the Royal Society of London Series B* **266**, 1241–1246.

Lens, L., van Dongen, S., Norris, K., Githiru, M. and Matthysen, E. (2002a) Avian persistence in fragmented rainforest. *Science* **298**, 1236–1238.

Lens, L., van Dongen, S. and Matthysen, E. (2002b) Fluctuating asymmetry as an early warning system in the critically endangered Taita thrush. *Conservation Biology* **16**, 479–487.

Leopold, A. (1939) A biotic view of land. *Journal of Forestry* **37**, 727–730.

Leven, M. R. and Corlett, R. T. (2004) Invasive birds in Hong Kong, China. *Ornithological Science* **3**, 43–55.

Lever, C. (1987) *Introduced Birds of the World*. Longman Scientific and Technical, Harlow, UK.

Levey, D. J. (1987) Seed size and fruit-handling techniques of avian frugivores. *American Naturalist* **129**, 471–485.

Levey, D. J. (1994) Why we should adopt a broader view of Neotropical migrants. *Auk* **111**, 233–236.

Levey, D. J., Moermond, T. C. and Denslow, J. S. (1994) Frugivory: an overview. In: *La Selva: ecology and natural history of a Neotropical rain forest* (eds. Mcdade, L. A., Bawa, K. S., Hespenheide, H. A. and Hartshorn, G. S.). University of Chicago Press, Chicago.

Levins, R. (1970) Extinctions. In: *Some Mathematical Questions in Biology: lectures on mathematics in the life sciences* (ed. Gesternhaber, M.), 77–107. American Mathematical Society, Providence, Rhode Island.

Li, W., Fu, R. and Dickinson, R. E. (2006) Rainfall and its seasonality over the Amazon in the 21st century as assessed by the coupled models for the IPCC AR4. *Journal of Geophysical Research* **111**, D02111.

Liberatori, F. and Penteriani, V. (2001) A long-term analysis of the declining population of the Egyptian Vulture in the Italian peninsula: distribution, habitat preference, productivity, and conservation implications. *Biological Conservation* **101**, 381–389.

Lichtenberg, J. S. and Lichtenberg, D. A. (2002) Weak trophic interactions among birds, insects, and white oak saplings (*Quercus alba*). *American Midland Naturalist* **148**, 338–349.

Lim, H. C. and Sodhi, N. S. (2004) Responses of avian guilds to urbanization in a tropical city. *Landscape and Urban Planning* **66**, 199–215.

Lim, H. C., Sodhi, N. S., Brook, B. W. and Soh, M. C. K. (2003) Undesirable aliens: factors determining the distribution of three invasive bird species in Singapore. *Journal of Tropical Ecology* **19**, 685–695.

Lim, K. S. (1992) *Vanishing Birds of Singapore*. The Nature Society, Singapore.

Lindell, C. A., Riffell, S. K., Kaiser, S. A., Battin, A. L., Smith, M. L. and Sisk, T. D. (2007) Edge responses of tropical and temperate birds. *Wilson Journal of Ornithology* **119**, 205–220.

Lindenmayer, D. B. and Franklin, J. F. (2002) *Conserving Forest Biodiversity: a comprehensive multi-scaled approach*. Island Press, Washington, DC.

Lindenmayer, D. B., Fischer, J. and Cunningham, R. B. (2005) Native vegetation cover thresholds associated with species responses. *Biological Conservation* **124**, 311–316.

Linhart, Y. B., Feinsinger, P., Beach, J. H., Busby, W. H., Murray, K. G., Pounds, W. Z., Kinsman, S., Guindon, C. A. and Kooiman, M. (1987) Disturbance and predictability of flowering patterns in bird-pollinated cloud forest plants. *Ecology* **68**, 1696–1710.

Loarie, S. R., Duffy, P. B., Hamilton, H., Asner, G. P., Field, C. B. and Ackerly, D. D. (2009) The velocity of climate change. *Nature* **462**, 1052–1055.

Lockwood, J. L. (1999) Using taxonomy to predict success among introduced avifauna: relative importance of transport and establishment. *Conservation Biology* **13**, 560–567.

Loiselle, B. A. (1990) Seeds in droppings of tropical fruit-eating birds: importance of considering seed composition. *Oecologia* **82**, 494–500.

Loiselle, B. A. and Blake, J. G. (1991) Temporal variation in birds and fruit along an elevational gradient in Costa Rica. *Ecology* **72**, 180–193.

Loiselle, B. A. and Blake, J. G. (1999) Dispersal of Melastome seeds by fruit-eating birds of tropical forest understory. *Ecology* **80**, 330–336.

Loiselle, B. A. and Blake, J. G. (2002) Potential consequences of extinction of frugivorous birds for shrubs of a tropical forest. In: *Seed Dispersal and Frugivory: Ecology, Evolution, and Conservation* (eds. Levey, D. J., Silva, W. R. and Galetti, M.), 397–406. CABI International, New York, NY.

Loiselle, B. A. and Hoppes, W. G. (1983) Nest predation in insular and mainland lowland rainforest in Panama. *Condor* **85**, 93–95.

Long, J. L. (1981) *Introduced Birds of the World: the worldwide history, distribution and influence of birds introduced to new environments*. Universe Books, New York, NY.

Lucas, R. M., Honzak, M., Curran, P. J., Foody, G. M., Milne, R., Brown, T. and Amaral, S. (2000) Mapping the regional extent of tropical forest regeneration stages in the Brazilian Legal Amazon using NOAA AVHRR data. *International Journal of Remote Sensing* **21**, 2855–2881.

Luck, G. W. and Daily, G. C. (2003) Tropical countryside bird assemblages: richness, composition, and foraging differ by landscape context. *Ecological Applications* **13**, 235–247.

Lui, C. Y., Amidon, G. L., Berardi, R. R., Fleisher, D., Youngberg, C. and Dressman, J. B. (1986) Comparison of gastrointestinal pH in dogs and humans – implications on the use of the beagle dog as a model for oral absorption in humans. *Journal of Pharmaceutical Sciences* **75**, 271–274.

Lundberg, J. and Moberg, F. (2003) Mobile link organisms and ecosystem functioning: implications for ecosystem resilience and management. *Ecosystems* **6**, 87–98.

Luteyn, J. L. (2002) Diversity, adaptation, and endemism in Neotropical Ericaceae: biogeographical patterns in the Vaccinieae. *Botanical Review* **68**, 55–87.

Luther, D. A. and Greenberg, R. (2009) Mangroves: a global perspective on the evolution and conservation of their terrestrial vertebrates. *BioScience* **59**, 602–612.

Lwanga, J. S. (2003) Forest succession in Kibale National Park, Uganda: implications for forest restoration and management. *African Journal of Ecology* **41**, 9–22.

Lynch, J. F. (1991) Effects of Hurricane Gilbert on birds in a dry tropical forest in the Yucatan Peninsula. *Biotropica* **23**, 488–496.

Lynch, J. F. (1992) Distribution of overwintering neotropical migrants in the Yucatan Peninsula. II. Use of native and human-modified vegetation. In: *Ecology and Conservation of Neotropical Migrant Landbirds* (eds. Hagan, J. M. III. and Johnston, D. W.), 178–195. Smithsonian Institution Press, Washington, DC.

Mabey, S. E. and Morton, E. S. (1992) Demography and territorial behavior of wintering Kentucky Warblers in Panama. In: *Ecology and Conservation of Neotropical Migrant Landbirds* (eds. Hagan, J. M. III. and Johnston, D. W.), 329–336. Smithsonian Institution Press, Washington, DC.

Mac Nally, R., Bennett, A. F., Thomson, J. R., Radford, J. Q., Unmack, G., Horrocks, G. and Vesk, P. A. (2009) Collapse of an avifauna: climate change appears to exacerbate habitat loss and degradation. *Diversity and Distributions* **15**, 720–730.

MacArthur, R. H. and Wilson, E. O. (1967) *The Theory of Island Biogeography*. Princeton University Press, Princeton, NJ.

Mack, A. L. (1993) The sizes of vertebrate-dispersed fruits – a Neotropical-Paleotropical comparison. *American Naturalist* **142**, 840–856.

Mack, A. L. (1995) Distance and nonrandomness of seed dispersal by the Dwarf Cassowary *Casuarius bennetti*. *Ecography* **18**, 286–295.

Mack, A. L. (1998) An advantage of large seed size: tolerating rather than succumbing to seed predators. *Biotropica* **30**, 604–608.

Mack, A. L., Ickes, K., Jessen, J. H., Kennedy, B. and Sinclair, J. R. (1999) Ecology of *Aglaia mackiana* (Meliaceae) seedlings in a New Guinea rain forest. *Biotropica* **31**, 111–120.

Mack, R. N., Simberloff, D., Lonsdale, W. M., Evans, H., Clout, M. and Bazzaz, F. A. (2000) Biotic invasions: causes, epidemiology, global consequences, and control. *Ecological Applications* **10**, 689–710.

Magige, F. J., Holmern, T., Stokke, S., Mlingwa, C. and Røskaft, E. (2009) Does illegal hunting affect density and behaviour of African grassland birds? A case study on ostrich (*Struthio camelus*). *Biodiversity and Conservation* **18**, 1361–1373.

Makana, J. R. and Thomas, S. C. (2004) Dispersal limits natural recruitment of African mahoganies. *Oikos* **106**, 67–72.

Malanson, G. P. and Armstrong, M. P. (1996) Dispersal probability and forest diversity in a fragmented landscape. *Ecological Modelling* **87**, 91–102.

Malhi, Y., Roberts, J. T., Betts, R. A., Killeen, T. J., Li, W. H. and Nobre, C. A. (2008) Climate change, deforestation, and the fate of the Amazon. *Science* **319**, 169–172.

Mancini, P. L., Bugoni, L., Neves, T., Monteiro, D. S. and Estima, S. C. (2009) The effect of light toriline on seabird by-catch and fish catch rates in the pelagic longline fishery off Southern Brazil. *Collection Volume of Scientific Papers by the International Commission for the Conservation of Aquatic Tunas* **64**, 2499–2507.

Mandon-Dalger, I. C. P., Tassin, J., Riviere, J. N. and Gatti, S. (2004) Relationships between alien plants and an alien bird species on Reunion Island. *Journal of Tropical Ecology* **20**, 635–642.

Manne, L. L., Brooks, T. M. and Pimm, S. L. (1999) Relative risk of extinction of passerine birds on continents and islands. *Nature* **399**, 258–261.

Mantyla, E., Klemola, T. and Haukioja, E. (2004) Attraction of Willow Warblers to sawfly-damaged Mountain Birches: novel function of inducible plant defences? *Ecology Letters* **7**, 915–918.

Margules, C. R. and Pressey, R. L. (2000) Systematic conservation planning. *Nature* **405**, 243–253.

Marini, M. A. (2001) Effects of forest fragmentation on birds of the cerrado region, Brazil. *Bird Conservation International* **11**, 13–25.

Marini, M. A., Barbet-Massin, M., Lopes, L. E. and Jiguet, F. (2009) Predicted climate-driven bird distribution changes and forecasted conservation conflicts in a Neotropical savanna. *Conservation Biology* **23**, 1558–1567.

Markwell, T. J. and Daugherty, C. H. (2002) Invertebrate and lizard abundance is greater on seabird-inhabited islands than on seabird-free islands in the Marlborough Sounds, New Zealand. *Ecoscience* **9**, 293–299.

Maron, J. L., Estes, J. A., Croll, D. A., Danner, E. M., Elmendorf, S. C. and Buckelew, S. L. (2006) An introduced predator alters Aleutian island plant communities by thwarting nutrient subsidies. *Ecological Monographs* **76**, 3–24.

Marquis, R. J. and Whelan, C. J. (1994) Insectivorous birds increase growth of White Oak through consumption of leaf-chewing insects. *Ecology* **75**, 2007–2014.

Marra, P. P. (2000) The role of behavioral dominance in structuring patterns of habitat occupancy in a migratory bird during the nonbreeding season. *Behavioral Ecology* **11**, 299–308.

Marra, P. P. and Holmes, R. T. (2001) Consequences of dominance-based habitat segregation in American Redstarts during the nonbreeding season. *Auk* **118**, 92–104.

Marra, P. P., Hobson, K. A. and Holmes, R. T. (1998a) Linking winter and summer events in a migratory bird using stable isotopes. *Science* **282**, 1884–1886.
Marra, P. P., Hobson, K. A. and Holmes, R. T. (1998b) Corticosterone levels as indicators of habitat quality: effects of habitat segregation in a migratory bird during the nonbreeding period. *Oecologia* **116**, 284–292.
Marsden, S. J. (1998) Changes in bird abundance following selective logging on Seram, Indonesia. *Conservation Biology* **12**, 605–611.
Marsden, S. J., Whiffin, M., Galetti, M. and Fielding, A. H. (2005) How well will Brazil's system of Atlantic forest reserves maintain viable bird populations? *Biodiversity and Conservation* **12**, 2835–2853.
Martensen, A. C., Pimentel, R. G. and Metzger, J. P. (2008) Relative effects of fragment size and connectivity on bird community in the Atlantic Rain Forest: implications for conservation. *Biological Conservation* **141**, 2184–2192.
Martin, C. M. and Martin, L. R. (1982) Pigeon control: an integrated approach. *Proceedings of Vertebrate Pest Conference* **10**, 190–192.
Martin, J. B. I. and Fitzgerald, L. (2005) A taste for novelty in invading house sparrows, *Passer domesticus*. *Behavioral Ecology* **16**, 702–707.
Martin, T. E. and Finch, D. M. (1995) *Ecology and Management of Neotropical Migratory Birds: a synthesis and review of critical issues*. Oxford University Press, New York, NY.
Martin, T. G., Chades, I., Arcese, P., Marra. P. P., Possingham, H. P. and Norris, D. R. (2007) Optimal conservation of migratory species. *PLoS One* **2**, e751.
Martinez-Morales, M. A., Cruz, P. C. and Cuaron, A. D. (2009) Predicted population trends for Cozumel Curassows (*Crax rubra griscomi*): empirical evidence and predictive models in the face of climate change. *Journal of Field Ornithology* **80**, 317–327.
Mas, A. H. and Dietsch, T. V. (2004) Linking shade-coffee certification to biodiversity conservation: butterflies and birds in Chiapas, Mexico. *Ecological Applications* **14**, 642–654.
Mason, D. (1996) Responses of Venezuelan understory birds to selective logging, enrichment strips, and vine cutting. *Biotropica* **28**, 296–309.
Matlock, R. B. J., Rogers, D., Edwards, P. J. and Martin, S. G. (2002) Avian communities in forest fragments and reforestation areas associated with banana plantations in Costa Rica. *Agriculture, Ecosystems and Environment* **91**, 199–215.
Matthews, E. (2001) *Understanding the FRA 2000*. World Resources Institute, Washington, DC.
Mazia, C. N., Kitzberger, T. and Chaneton, E. J. (2004) Interannual changes in folivory and bird insectivory along a natural productivity gradient in northern Patagonian forests. *Ecography* **27**, 29–40
Mazia, C. N., Chaneton, E. J., Kitzberger, T. and Garibaldi, L. A. (2009) Variable strength of top-down effects in Nothofagus forests: bird predation and insect herbivory during an ENSO event. *Austral Ecology* **34**, 359–367.
McConkey, K. R. and Drake, D. R. (2002) Extinct pigeons and declining bat populations: are large seeds still being dispersed in the tropical Pacific? In: *Seed Dispersal and Frugivory: ecology, evolution and conservation* (eds. Levey, D. J., Silva, W. R. and Galetti, M.), pp. 381–395. CABI Publishing, Wallingford, UK.
McConkey, K. R., Drake, D. R., Meehan, H. J. and Parsons, N. (2003) Husking stations provide evidence of seed predation by introduced rodents in Tongan rain forests. *Biological Conservation* **109**, 221–225.
McCulloch, M. N., Tucker, G. M. and Baillie, S. R. (1992) The hunting of migratory birds in Europe: a ringing recovery analysis. *Ibis* **134**(Suppl.), 55–65.
McElligott, A. G., Maggini, I., Hunziker, L. and Konig, B. (2004) Interactions between Red-billed Oxpeckers and Black Rhinos in captivity. *Zoo Biology* **23**, 347–354.
McKechnie, A. E. (2008) Phenotypic flexibility in basal metabolic rate and the changing view of avian physiological diversity: a review. *Journal of Comparative Physiology B – Biochemical Systemic and Environmental Physiology* **178**, 235–247.
McKey, D. (1975) The ecology of co-evolved seed dispersal systems. In: *Coevolution of Animals and Plants* (eds. Gilbert, L. E. and Raven, P. H.), 157–191. University of Texas Press, Austin, TX.

McKinney, M. L. (2006) Urbanization as a major cause of biotic homogenization. *Biological Conservation* **127**, 247–260.

McLain, D. K., Moulton, M. P. and Sanderson, J. G. (1999) Sexual selection and extinction: the fate of plumage-dimorphic and plumage-monomorphic birds introduced onto islands. *Evolutionary Ecology Research* **1**, 549–565.

McNab, B. K. (2009) Ecological factors affect the level and scaling of avian BMR. *Comparative Biochemistry and Physiology a – Molecular & Integrative Physiology* **152**, 22–45.

McWilliam, A. N. and Cheke, R. A. (2004) A review of the impacts of control operations against the Red-billed Quelea (*Quelea quelea*) on non-target organisms. *Environmental Conservation* **31**, 130–137.

Medeiros, A. C., Loope, L. L., Conant, P. and McElvaney, S. (1997) Status, ecology, and management of the invasive plant, *Miconia calvescens* DC (Melastomataceae) in the Hawaiian Islands. *Bishop Museum Occassional Papers* **48**, 23–36.

Medina, F. M. and Nogales, M. (2009) A review on the impacts of feral cats (*Felis silvestris catus*) in the Canary Islands: implications for the conservation of its endangered fauna. *Biodiversity and Conservation* **18**, 829–846.

Meehan, H. J., McConkey, K. R. and Drake, D. R. (2002) Potential disruptions to seed dispersal mutualisms in Tonga, Western Polynesia. *Journal of Biogeography* **29**, 695–712.

Meinzingen, W. W., Bashir, E. S. A., Parker, J. D., Heckel, J.-U. and Elliott, C. C. H. (1989) Lethal control of quelea. In: *Quelea Quelea: Africa's Bird Pest* (eds. Elliott, C. C. H. and Allan, R. G.), pp. 293–316. Oxford University Press, Oxford, UK.

Melo-Vasquez, I., Ochoa-Quintero, J. M., Lopez-Arevalo, H. F. and Velasquez-Sandino, P. (2008) Potential habitat loss and subsistence hunting of Blue Billed Curassow (*Crax alberti*), a Colombian critically endangered endemic bird. *Caldasia* **30**, 161–177.

Menon, S., Islam, M. Z. U. and Peterson, A. T. (2009) Projected climate change effects on nuthatch distribution and diversity across Asia. *Raffles Bulletin of Zoology* **57**, 569–575.

Mesquita, R. C. G., Delamonica, P. and Laurance, W. F. (1999) Effect of surrounding vegetation on edge-related tree mortality in Amazonian forest fragments. *Biological Conservation* **91**, 129–134.

Mettke-Hofmann, C. and Greenberg, R. (2005) Behavioral and cognitive adaptations to long distance migration. In: *Birds of Two Worlds* (eds. Greenberg, R. and Marra, P. P.), 114–123. Johns Hopkins University Press, Baltimore, MD.

Metzger, J. P. (2009) Time-lag in biological responses to landscape changes in a highly dynamic Atlantic forest region. *Biological Conservation* **142**, 1166–1177.

Meyer, G. A. and Witmer, M. C. (1998) Influence of seed processing by frugivorous birds on germination success of three North American shrubs. *American Midland Naturalist* **140**, 129–139.

Mila, B., Smith, T. B. and Wayne, R. K. (2006) Postglacial expansion drives the evolution of long-distance migration in a songbird. *Evolution* **60**, 2403–2409.

Millennium Ecosystem Assessment (MEA) (2005) *Ecosystems and Human Well-being: synthesis.* Washington, DC.

Miller, G. H., Fogel, M. L., Magee, J. W., Gagan, M. K., Clarke, S. J. and Johnson, B. J. (2005) Megafaunal extinction in Pleistocene Australia and a human role in ecosystem collapse. *Science* 287–290.

Mills, M. S. L. (2004) Bird community responses to savanna fires: should managers be concerned? *South African Journal of Wildlife Research* **34**, 1–11.

Milner-Gulland, E. J. and Bennett, E. L. (2003) Wild meat: the bigger picture. *Trends in Ecology and Evolution* **18**, 351–357.

Milton, S. J., Dean, W. R. J., Kerley, G. I. H., Hoffman, H. T. and Whitford, W. G. (1998) Dispersal of seeds as nest material by the Cactus Wren. *Southwestern Naturalist* **43**, 449–452.

Minnich, R. A. (2001) An integrated model of two fire regimes. *Conservation Biology* **15**, 1549–1553.

Mitani, J. C., Sanders, W. J., Lwanga, J. S. and Windfelder, T. L. (2001) Predatory behavior of Crowned Hawk-eagles (*Stephanoaetus coronatus*) in Kibale National Park, Uganda. *Behavioral Ecology and Sociobiology* **49**, 187–195.

Mitchell, C. L., Boinski, S. and Vanschaik, C. P. (1991) Competitive regimes and female bonding in two species of squirrel monkeys (*Saimiri oerstedi* and *S. sciureus*). *Behavioral Ecology and Sociobiology* **28**, 55–60.

Mitra, S. S. and Sheldon, F. H. (1993) Use of an exotic tree plantation by Bornean lowland forest birds. *Auk* **110**, 529–540.

Moat, J. and Smith, P. (2007) *Atlas of the Vegetation of Madagascar, Atlas de la Végétation de Madagascar*. Kew Publishing, Royal Botanical Gardens, Kew, UK.

Møller, A. P. (1995) Patterns of fluctuating asymmetry in sexual ornaments of birds from marginal and central populations. *American Naturalist* **145**, 316–327.

Møller, A. P. (2009) Basal metabolic rate and risk-taking behaviour in birds. *Journal of Evolutionary Biology* **22**, 2420–2429.

Mols, C. M. M. and Visser, M. E. (2002) Great Tits can reduce caterpillar damage in apple orchards. *Journal of Applied Ecology* **39**, 888–899.

Mönkkönen, M. and Reunanen, P. (1999) On critical thresholds in landscape connectivity: a management perspective. *Oikos* **84**, 302–305.

Mönkkönen, M., Helle, P. and Welsh, D. (1992) Perspectives on Palaearctic and Nearctic bird migration: comparisons and overview of life-history and ecology of migrant passerines. *Ibis* **134**(Suppl.), 7–13.

Montaldo, N. H. (2000) Reproductive success of bird-dispersed plants in a subtropical forest relict in Argentina. *Revista Chilena De Historia Natural* 73, 511–524.

Montgomery, B. R., Kelly, D. and Ladley, J. J. (2001) Pollinator limitation of seed set in *Fuchsia perscandens* (Onagraceae) on Banks Peninsula, South Island, New Zealand. *New Zealand Journal of Botany* **39**, 559–565.

Mooney, K. A., Gruner, D. S., Barber, N. A., van Bael, S. A., Philpott, S. M. and Greenberg, R. (2010) Interactions among predators and the cascading effects of vertebrate insectivores on arthropod communities and plants. *Proceedings of the National Academy of Sciences of the United States of America* **107**, 7335–7340.

Mooney, N. (1998) Status and conservation of raptors in Australia's tropics. *Journal of Raptor Research* **32**, 64–73.

Moore, F. R. (1999) Neotropical migrants and the Gulf of Mexico: the Cheniers of Louisiana and stopover ecology. In: *Gatherings of Angels: migratory birds and their ecology* (ed. Able, K. P.), 51–62. Comstock, Ithaca, NY.

Moore, F. R. (2000) *Stopover Ecology of Nearctic-Neotropical Landbird Migrants: habitat relations and conservation implications*. Cooper Ornithological Society, Lawrence, Kansas, KS.

Moore, J. L., Balmford, A., Brook, T., Burgess, N. D., Hansen, L. A., Rahbek, C. and Williams, P. H. (2003) Performance of Sub-Saharan vertebrates as indicator groups for identifying priority areas for conservation. *Conservation Biology* **17**, 207–218.

Moore, F. R., Smith, R. J. and Sandberg, R. (2005) Stopover ecology of intercontinental migrants: en route problems and consequences for reproductive performance. In: *Birds of Two Worlds: the ecology and evolution of migration* (eds. Greenberg, R. and Marra, P. P.), 251–261. The Johns Hopkins University Press, Baltimore, MD.

Moore, R. P., Robinson, W. D., Lovette, I. J. and Robinson, T. R. (2008) Experimental evidence for extreme dispersal limitation in tropical forest birds. *Ecology Letters* **11**, 960–968.

Mora, J. A. (1997) Transboundary pollution: persistent organochlorine pesticides in migrant birds of the southwestern United States and Mexico. *Environmental Toxicology and Chemistry* **25**, 3–11.

Mordecai, R. S., Cooper, R. J. and Justicia, R. (2009) A threshold response to habitat disturbance by forest birds in the Choco Andean corridor, Northwest Ecuador. *Biodiversity and Conservation* **18**, 2421–2431.

Morris, W. F., Pfister, C. A., Tuljapurkar, S., Haridas, C. V., Boggs, C. L., Boyce, M. S., Bruna, E. M., Church, D. R., Coulson, T., Doak, D. F., Forsyth, S., Gaillard, J. M., Horvitz, C. C., Kalisz, S., Kendall, B. E., Knight, T. M., Lee, C. T. and Menges, E. S. (2008) Longevity can buffer plant and animal populations against changing climatic variability. *Ecology* **89**, 19–25.

Mortensen, H. S., Dupont, Y. L. and Olesen, J. M. (2008) A snake in paradise: disturbance of plant reproduction following extirpation of bird flower-visitors on Guam. *Biological Conservation* **141**, 2146–2154.

Morton, E. S. (1971) Food and migration habits of Eastern Kingbirds in Panama. *Auk* 88, 925–926.

Morton, E. S. (1979) Effective pollination of *Erythrina fusca* by the Orchard Oriole (*Icterus spurious*): coevolved behavioral manipulation? *Annals of the Missouri Botanical Garden* **66**, 482–489.

Morton, E. S. (1980) Adaptations to seasonal changes by migrant land birds in the Panama Canal Zone. In: *Migrant Birds in the Neotropics: ecology, behavior, distribution, and conservation* (eds. Keast, A and Morton, E. S.), 437–453. Smithsonian Institution Press, Washington, DC.

Morton, E. S. (1977) Intratropical migration in the Yellow-green Vireo and Piratic Flycatcher. *Auk* **94** 97–106.

Morton, E. S. (1990) Habitat segregation by sex in the Hooded Warbler: experiments on proximate causation and discussion of its evolution. *American Naturalist* **135**, 319–333.

Morton, E. S. and Greenberg, R. (1989) The outlook for migratory songbirds: "future shock" for birders. *American Birds* **43**, 178–183.

Moulton, M. P., Sanderson, J. G. and Labisky, R. F. (2001) Patterns of success in game bird (Aves: Galliformes) introductions to the Hawaiian islands and New Zealand. *Evolutionary Ecology Research* **3**, 507–519.

Mourato, S., Ozdemiroglu, E. and Foster, V. (2000) Evaluating health and environmental impacts of pesticide use: implications for the design of ecolabels and pesticide taxes. *Environmental Science & Technology* **34**, 1456–1461.

Mundy, P., Butchart, D., Ledger, J. and Piper, S. E. (1992) *The Vultures of Africa*. The Penrose Press, Johannesburg.

Munn, C. A. (1986) Birds that cry wolf. *Nature* **319**, 143–145.

Munn, C. A. and Terborgh, J. W. (1979) Multi-species territoriality in neotropical foraging flocks. *Condor* **81**, 338–347.

Murakami, M. and Nakano, S. (2000) Species-specific bird functions in a forest-canopy food web. *Proceedings of the Royal Society of London Series B* **267**, 1597–1601.

Murakami, M. and Nakano, S. (2002) Indirect effect of aquatic insect emergence on a terrestrial insect population through by birds predation. *Ecology Letters* **5**, 333–337.

Murcia, C. (1995) Edge effects in fragmented forests – implications for conservation. *Trends in Ecology and Evolution* **10**, 58–62.

Murphy, D. J. and Kelly, D. (2001) Scarce or distracted? Bellbird (*Anthornis melanura*) foraging and diet in an area of inadequate mistletoe pollination. *New Zealand Journal of Ecology* **25**, 69–81.

Murphy, G. I. (1981) Guano and the anchovetta fishery. *Research Management and Environmental Uncertainty* **11**, 81–106.

Murphy, S. R., Reid, N., Yan, Z. and Venables, W. N. (1993) Differential passage time of mistletoe fruits through the gut of honeyeaters and flowerpeckers: effects on seedling establishment. *Oecologia* **93**, 171–176.

Murray, K. G. (1988) Avian seed dispersal of three Neotropical gap-dependent plants. *Ecological Monographs* **58**, 271–298.

Murton, R., Thearle, R. and Thompson, J. (1972) Ecological studies of the feral pigeon *Columba livia* var. I. Population, breeding biology and methods of control. *Journal of Applied Ecology* **9**, 835–874.

Myers, N. (1996) Environmental services of biodiversity. *Proceedings of the National Academy of Sciences of the United States of America* **93**, 2764–2769.

Myers, N., Mittermeier, R. A., Mittermeier, C. G., da Fonseca, G. A. B. and Kent, J. (2000) Biodiversity hotspots for conservation priorities. *Nature* **403**, 853–858.

Myers, R. L. (2006) *Living with Fire – sustaining ecosystems and livelihoods through integrated fire management*. The Nature Conservancy, Arlington, VA.

Nabhan, G. P. and Buchmann, S. L. (1997) Services provided by pollinators. In: *Nature's Services* (ed. Daily, G. C.), 133–150. Island Press, Washington DC.

Nagendra, H. (2008) Do parks work? Impact of protected areas on land cover clearing. *Ambio* **37**, 330–337.

Narang, M. L., Rana, R. S. and Prabhakar, M. (2000) Avian species involved in pollination and seed dispersal of some forestry species in Himachal Pradesh. *Journal of the Bombay Natural History Society* **97**, 215–222.

Nathan, R. and Muller-Landau, H. C. (2000) Spatial patterns of seed dispersal, their determinants and consequences for recruitment. *Trends in Ecology and Evolution* **15**, 278–285.

Naughton-Treves, L., Holland, M. and Brandon, K. (2005) The role of protected areas in conserving biodiversity and sustaining livelihoods. *Annual Review of Environment and Resources* **30**, 219–252.

Naylor, R. and Ehrlich, P. R. (1997) Natural pest control services and agriculture. In: *Nature's Services* (ed. Daily, G. C.), 151–174. Island Press, Washington, DC.

Negro, J. J., Grande, J. M., Tella, J. L., Garrido, J., Hornero, D., Donazar, J. A., Sanchez-Zapata, J. A., Benitez, J. R. and Barcell, M. (2002) Coprophagy: an unusual source of essential carotenoids. *Nature* **416**, 807–808.

Nelson, J. T., Woodworth, B. L., Fancy, S. G., Lindsey, G. D. and Tweed, E. J. (2002) Effectiveness of rodent control and monitoring techniques for a montane rainforest. *Wildlife Society Bulletin* **30**, 82–92.

Nepstad, D. C., Verissimo, A., Alencar, A., Nobre, C., Lima, E., Lefebvre, P., Schlesinger, P., Potter, C., Moutinho, P., Mendoza, E., Cochrane, M. and Brooks, V. (1999) Large-scale impoverishment of Amazonian forests by logging and fire. *Nature* **398**, 505–508.

Nepstad, D., Carvalho, G., Barros, A. C., Alencar, A., Capobianco, J. P., Bishop, J., Moutinho, P., Lefebvre, P. and Silva, L. S. (2001) Road paving, fire regime feedbacks, and the future of Amazon forests. *Forest Ecology and Management* **154**, 395–407.

Nepstad, D., Schwartzman, S., Bamberger, B., Santilli, M., Ray, D., Schlesinger, P., Lefebvre, P., Alencar, A., Prinz, E., Fiske, G. and Rolla, A. (2006) Inhibition of Amazon deforestation and fire by parks and indigenous lands. *Conservation Biology* **20**, 65–73.

Ness, J. H. and Bronstein, J. L. (2004) The effects of invasive ants on perspective ant mutualists. *Biological Invasions* **6**, 445–461.

Newmark, W. D. (1991) Tropical forest fragmentation and the local extinction of understory birds in the Eastern Usambara Mountains, Tanzania. *Conservation Biology* **5**, 67–78.

Newton, I. (1979) *Population Ecology of Raptors*. T. & A.D. Poyser, Birkhamsted, UK.

Newton, I. (1995) Relationships between breeding and wintering ranges in Palaearctic-African migrants. *Ibis* **137**, 241–249.

Newton, I. (1998) *Population Limitation in Birds*. Academic Press, London, UK.

Newton, I. (2004) The recent declines of farmland bird populations in Britain: an appraisal of causal factors and conservation actions. *Ibis* **146**, 579–600.

Niesten, E. T., Rice, R. E., Ratay, S. M. and Paratore, K. (2004) *Commodities and Conservation: the need for greater habitat protection in the tropics*. Centre for Applied Biodiversity Science, Conservation International, Washington, DC.

Nijman, V. (2010) An overview of international wildlife trade from Southeast Asia. *Biodiversity and Conservation* **19**, 1101–1114.

Niven, D. K., Butcher, G. S. and Bancroft, G. T. (2009) *Birds and Climate Change: ecological disruption in motion*, National Audubon Society, New York, NY.

Njabo, K. Y. and Sorenson, M. D. (2009) Origin of Bannerman's Turaco Tauraco bannermani in relation to historical climate change and the distribution of West African montane forests. *Ostrich* **80**, 1–7.

Noble, J. C. (1975) Effects of Emus (*Dromaius novaehollandiae* Latham) on distribution of Nitre Bush (*Nitraria billardieri* Dc). *Journal of Ecology* **63**, 979–984.

Nogales, M., Medina, F. M., Quilis, V. and Gonzalez-Rodriguez, M. (2001) Ecological and bio-geographical implications of Yellow-Legged Gulls (*Larus cachinnans* Pallas) as seed dispersers of *Rubia fruticosa* Ait. (Rubiaceae) in the Canary Islands. *Journal of Biogeography* **28**, 1137–1145.

Nogales, M., Quilis, V., Medina, F. M., Mora, J. L. and Trigo, L. S. (2002) Are predatory birds effective secondary seed dispersers? *Biological Journal of the Linnean Society* **75**, 345–352.

Norrdahl, K., Suhonen, J., Hemminki, O. and Korpimaki, E. (1995) Predator presence may benefit – kestrels protect curlew nests against nest predators. *Oecologia* **101**, 105–109.

Norrdahl, K., Klemola, T., Korpimaki, E. and Koivula, M. (2002) Strong seasonality may attenuate trophic cascades: vertebrate predator exclusion in boreal grassland. *Oikos* **99**, 419–430.

Norris, D. R., Marra, P. P., Kyser, T. K., Sherry, T. W. and Ratcliffe, L. M. (2004) Tropical winter habitat limits reproductive success on the temperate breeding grounds in a migratory bird. *Proceedings of the Royal Society of London Series B* **271**, 59–64.

Norris, K. and Harper, N. (2004) Extinction processes in hot spots of avian biodiversity and the targeting of pre-emptive conservation action. *Proceedings of the Royal Society of London Series B* **271**, 123–130.

Norton, D. A. (1991) *Trilepidea adamsii* – an obituary for a species. *Conservation Biology* **5**, 52–57.

Norton, D. A., Hobbs, R. J. and Atkins, L. (1995) Fragmentation, disturbance, and plant distribution: mistletoes in woodland remnants in the Western Australian wheatbelt. *Conservation Biology* **9**, 426–438.

Norton, D. A., Delange, P. J., Garnock-Jones, P. J. and Given, D. R. (1997) The role of seabirds and seals in the survival of coastal plants: lessons from New Zealand *Lepidium* (Brassicaceae). *Biodiversity and Conservation* **6**, 765–785.

Nygåard, T. and Gjershaug, J. O. (2001) The effects of low levels of pollutants on the reproduction of Golden Eagles in Western Norway. *Ecotoxicology* **10**, 285–290.

Nyström, M. and Folke, C. (2001) Spatial resilience of coral reefs. *Ecosystems* **4**, 406–417.

Oaks, J. L., Gilbert, M., Virani, M. Z., Watson, R. T., Meteyer, C. U., Rideout, B. A., Shivaprasad, H. L., Ahmed, S., Chaudhry, M. J. I., Arshad, M., Mahmood, S., Ali, A. and Khan, A. A. (2004) Diclofenac residues as the cause of vulture population decline in Pakistan. *Nature* **427**, 630–633.

O'Connor, J. A., Sulloway, F. J., Robertson, J. and Kleindorfer, S. (2010) *Philornis downsi* parasitism is the primary cause of nestling mortality in the critically endangered Darwin's medium tree finch (*Camarhynchus pauper*). *Biodiversity and Conservation* **19**, 853–866.

Ohlemuller, R., Anderson, B. J., Araujo, M. B., Butchart, S. H. M., Kudrna, O., Ridgely, R. S. and Thomas, C. D. (2008) The coincidence of climatic and species rarity: high risk to small-range species from climate change. *Biology Letters* **4**, 568–572.

Oliver, J. D. and Legovic, T. (1988) Okefenokee marshland before, during and after nutrient enrichment by a bird rookery. *Ecological Modelling* **43**, 195–223.

Oliver, J. D. and Schoenberg, S. A. (1989) Residual influence of macronutrient enrichment on the aquatic food web of an Okefenokee swamp (Georgia, USA) abandoned bird rookery. *Oikos* **55**, 175–182.

Olupot, W., McNeilage, A. J. and Plumptre, A. J. (2009) *An Analysis of Socioeconomics of Bushmeat Hunting at Major Hunting Sites in Uganda*. Working Paper 38, Wildlife Conservation Society, Kampala, Uganda.

Oniki, Y. (1979) Is nesting success of birds low in the tropics? *Biotropica* **11**, 60–69.

O'Reilly, L., Ogada, D., Palmer, T. M. and Keesing, F. (2006) Effects of fire on bird diversity and abundance in an East African savanna. *African Journal of Ecology* **44**, 165–170.

Orgeas, J., Vidal, E. and Ponel, P. (2003) Colonial seabirds change beetle assemblages on a Mediterranean island. *Ecoscience* **10**, 38–44.

Orme, C. D. L., Davies, R. G., Burgess, M., Eigenbrod, F., Pickup, N., Olson, V. A., Webster, A. J., Ding, T., Rasmussen, P. C., Ridgely, R. S., Stattersfield, A. J., Bennett, P. M., Blackburn, T. M., Gaston, K. J. and Owens, I. P. F. (2005) Global hotspots of species richness are not congruent with endemism or threat. *Nature* **436**, 1016–1019.

Otterstrom, S. M., Schwartz, M. W. and Velazquez-Rocha, I. (2006) Responses to fire in selected tropical dry forest trees. *Biotropica* **38**, 592–598.

Otvos, I. S. (1979) The effects of insectivorous bird activities in forest ecosystems: an evaluation. In: *The Role of Insectivorous Birds in Forest Ecosystems* (eds. Dickson, J. G., Conner, E. N., Fleet, R. R., Kroll, J. C. and Jackson, J. A.), 341–374. Academic Press, New York, NY.

Owiunji, I. and Plumptre, A. J. (1998) Bird communities in logged and unlogged compartments in Budongo Forest, Uganda. *Forest Ecology and Management* **108**, 115–126.

Oyama, M. D. and Nobre, C. A. (2003) A new climate-vegetation equilibrium state for Tropical South America. *Geophysical Research Letters* **30**, 2199–2203.

Paaby, P., Clark, D. B. and Gonzalez, H. (1991) Training rural residents as naturalist guides: evaluation of a pilot project in Costa Rica. *Conservation Biology* **5**, 542–546.

Pacheco, L. F. and Simonetti, J. A. (2000) Genetic structure of a mimosoid tree deprived of its seed disperser, the spider monkey. *Conservation Biology* **14**, 1766–1775.

Packer, A. and Clay, K. (2000) Soil pathogens and spatial patterns of seedling mortality in a temperate tree. *Nature* **404**, 278–281.

Padian, K. and Chiappe, L. M. (1998) The origin and early evolution of birds. *Biological Reviews* **73**, 1–42.

Pagiola, S. (2008) Payments for environmental services in Costa Rica. *Ecological Economics* **65**, 712–724.

Pain, D. J., Cunningham, A. A., Donald, P. F., Duckworth, J. W., Houston, D. C., Katzner, T., Parry-Jones, J., Poole, C., Prakash, V., Round, P. and Timmins, R. (2003) Causes and effects of temporospatial declines of *Gyps* vultures in Asia. *Conservation Biology* **17**, 661–671.

Pain, D. J., Martins, T. L. F., Boussekey, M., Diaz, S. H., Downs, C. T., Ekstrom, J. M. M., Garnett, S. T., McNiven, D., Gilardi, J. D., Primot, P., Rouys, S., Saoumoé, M., C., S., Tamungang, S. A., Theuerkauf, J., Villafuerte, D., Verfailles, L., Widmann, P. and Widmann, I. D. (2006) Impact of protection and nest take and nesting success of parrots in Africa, Asia and Australasia. *Animal Conservation* **9**, 322–330.

Paine, R. T. and Schindler, D. E. (2002) Ecological pork: novel resources and the trophic reorganization of an ecosystem. *Proceedings of the National Academy of Sciences of the United States of America* **99**, 554–555.

Paine, R. T., Wootton, J. T. and Boersma, P. D. (1990) Direct and indirect effects of Peregrine Falcon predation on seabird abundance. *Auk* **107**, 1–9.

Palomo, G., Iribarne, O. and Martinez, M. M. (1999) The effect of migratory seabirds guano on the soft bottom community of a SW Atlantic coastal lagoon. *Bulletin of Marine Science* **65**, 119–128.

Palmer, M., Bernhardt, E., Chornesky, E., Collins, S., Dobson, A., Duke, C., Gold, B., Jacobson, R., Kingsland, S., Kranz, R., Mappin, M., Martinez, M. L., Micheli, F., Morse, J., Pace, M., Pascual, M., Palumbi, S., Reichman, O. J., Simons, A., Townsend, A. and Turner, M. (2004) Ecology for a crowded planet. *Science* **304**, 1251–1252.

Parasharya, B. M., Dodia, J. F., Mathew, K. L. and Yadav, D. N. (1994) Natural regulation of white grub (*Holotrichia* sp, Scarabidae) by birds in agroecosystem. *Journal of Biosciences* **19**, 381–389.

Parmesan, C. (2006) Ecological and evolutionary responses to recent climate change. *Annual Review of Ecology Evolution and Systematics* **37**, 637–669.

Parra, V., Vargas, C. F. and Eguiarte, L. E. (1993) Reproductive biology, pollen and seed dispersal, and neighborhood size in the hummingbird-pollinated *Echeveria gibbiflora* (Crassulaceae). *American Journal of Botany* **80**, 153–159.

Parrish, J. K., Marvier, M. and Paine, R. T. (2001) Direct and indirect effects: interactions between Bald Eagles and Common Murres. *Ecological Applications* **11**, 1858–1869.

Parry-Jones, J. (2001) The Parsi project and Indian griffon vultures. In: *4th Eurasian Congress on Raptors* (eds. Katzner, T. and Parry-Jones, J.), 17–18. Raptor Research Foundation, Seville, Spain.

Paton, D. C. (2000) Disruption of bird-plant pollination systems in southern Australia. *Conservation Biology* **14**, 1232–1234.

Patten, M. A. and Bolger, D. T. (2003) Variation in top-down control of avian reproductive success across a fragmentation gradient. *Oikos* **101**, 479–488.

Pauchard, A., Aguayo, M., Peña, E. and Urrutia, R. (2006) Multiple effects of urbanization on the biodiversity of developing countries: the case of a fast-growing metropolitan area (Concepción, Chile). *Biological Conservation* **127**, 272–281.

Pautasso, M., Böhning-Gaese, K., Clergeau, P., Cueto, V. R., Dinetti, M., Fernández-Juricic, E., Kaisanlahti-Jokimäki, M.-L., Jokimäki, J., McKinney, M. L., Sodhi, N. S., Storch, D., Tomialojc, L., Weisberg, P. J., Woinarski, J., Fuller, R. A. and Cantarello, E. (2010) Global macroecology of bird assemblages in urbanized and semi-natural ecosystems. *Global Ecology and Biogeography*, in press.

Peach, W., Baillie, S. and Underhill, L. G. (1991) Survival of British Sedge Warblers *Acrocephalus schoenobaenus* in relation to West African rainfall. *Ibis* **133**, 300–305.

Peakall, D. B. and Boyd, H. (1987) Birds as bio-indicators of environmental conditions. In: *The Value of Birds* (eds. Diamond, A. W. and Filion, F. L.), 113–118. International Council for Bird Preservation, Norfolk, UK.

Pearson, D. L., Anderson, C. D., Mitchell, B. R., Rosenberg, M. S., Navarrete, R. and Coopmans, P. (2010) Testing hypotheses of bird extinctions at Rio Palenque, Ecuador, with informal species lists. *Conservation Biology* **24**, 500–510.

Pearson, O. P. (1966) Prey of carnivores during one cycle of mouse abundance. *Journal of Animal Ecology* **35**, 217–233.

Pedlowski, M. A., Matricardi, E. A. T., Skole, D., Cameron, S. R., Chomentowski, W., Fernandes, C. and Lisboa, A. (2005) Conservation units: a new deforestation frontier in the Amazonian state of Rondonia, Brazil. *Environmental Conservation* **32**, 149–155.

Peh, K. S. H. (2007) Potential effects of climate change on elevational distributions of tropical birds in Southeast Asia. *Condor* **109**, 437–441.

Peh, K. S.-H. (2010) Invasive species in Southeast Asia: the knowledge so far. *Biodiversity and Conservation* **19**, 1083–1099.

Peh, K. S.-H. and Sodhi, N. S. (2002) Characteristics of nocturnal roosts of house crows in Singapore. *Journal of Wildlife Management* **66**, 1128–1133.

Peh, K. S. H., de Jong, J., Sodhi, N. S., Lim, S. L.-H. and Yap, C. A. M. (2005) Lowland rainforest avifauna and human disturbance: persistence of primary forest birds in selectively logged forests and mixed-rural habitats of southern Peninsular Malaysia. *Biological Conservation* **123**, 489–505.

Peh, K. S. H., Sodhi, N. S., de Jong, J., Sekercioglu, C. H., Yap, C. A.-M. and Lim, S. L.-H. (2006) Conservation value of degraded habitats for forest birds in southern Peninsular Malaysia. *Diversity and Distributions* **12**, 572–581.

Pejchar, L. and Jeffrey, J. (2004) Sap-feeding behavior and tree selection in the endangered Akiapolaau (*Hemignathus munroi*) in Hawaii. *Auk* **121**, 548–556.

Pejchar, L., Pringle, R. M., Ranganathan, J., Zook, J. R., Duran, G., Oviedo, F. and Daily, G. C. (2008) Birds as agents of seed dispersal in a human-dominated landscape in southern Costa Rica. *Biological Conservation* **141**, 536–544.

Pell, A. S. and Tidemann, C. R. (1997) The impact of two exotic hollow-nesting birds on two native parrots in savannah and woodland in eastern Australia. *Biological Conservation* **79**, 145–153.

Peres, C. A. (1999) Ground fires as agents of mortality in a Central Amazonian forest. *Journal of Tropical Ecology* **15**, 535–541.

Peres, C. A. (2001) Synergistic effects of subsistence hunting and habitat fragmentation on amazonian forest vertebrates. *Conservation Biology* **15**, 1490–1505.

Peres, C. A. (2005) Why we need megareserves in Amazonia. *Conservation Biology* **19**, 728–733.

Peres, C. A. and Palacois, E. (2007) Basin-wide effects of game harvest on vertebrate population densities in Amazonian forests: implications for animal-mediated seed dispersal. *Biotropica* **39**, 304–315.

Peres, C. A. and van Roosmalen, M. G. M. (1996) Avian dispersal of "mimetic seeds" of *Ormosia lignivalvis* by terrestrial granivores: deception or mutualism? *Oikos* **75**, 249–258.

Peres, C. A. and van Roosmalen, M. (2002) Primate frugivory in two species-rich Neotropical forests: implications for the demography of large-seeded plants in overhunted areas. In: *Seed Dispersal and Frugivory: Ecology, Evolution and Conservation* (eds. Levey, D. J., Silva, W. R. and Galetti, M.), 83–98. CABI Publishing, Wallingford, UK.

Peres, C. A., Gardner, T. A., Barlow, J., Zuanon, J., Michalski, F., Lees, A. C., Vieira, I. C. G., Moreira, F. M. S. and Feeley, K. J. (2010) Biodiversity conservation in human-modified Amazonian forest landscapes. *Biological Conservation* **143**, 2314–2327.

Perfecto, I., Mas, A., Dietsch, T. and Vandermeer, J. (2003) Conservation of biodiversity in coffee agroecosystems: a tri-taxa comparison in southern Mexico. *Biodiversity and Conservation* **12**, 1239–1252.

Perfecto, I., Vandermeer, J. H., Bautista, G. L., Nunez, G. I., Greenberg, R., Bichier, P. and Langridge, S. (2004) Greater predation in shaded coffee farms: the role of resident Neotropical birds. *Ecology* **85**, 2677–2681.

Peterjohn, B. G., Sauer, J. R. and Robbins, C. S. (1995) Population trends from the North American breeding bird survey. In: *Ecology and Management of Neotropical birds. A synthesis and review of critical issues* (eds. Martin, T. E. and Finch, D. M.), 3–39. Oxford University Press, New York, NY.

Peters, M. K. and Okalo, B. (2009) Severe declines of ant-following birds in African rainforest fragments are facilitated by a subtle change in army ant communities. *Biological Conservation* **142**, 2050–2058.

Peters, R. H. (1983) *The Ecological Implications of Body Size*. Cambridge University Press, New York, NY.

Peterson, A. T., Sanchez-Cordero, V., Soberon, J., Bartley, J., Buddemeier, R. W. and Navarro-Siguenza, A. G. (2001) Effects of global climate change on geographic distributions of Mexican Cracidae. *Ecological Modelling* **144**, 21–30.

Peterson, A. T., Ortega-Huerta, M. A., Bartley, J., Sanchez-Cordero, V., Soberon, J., Buddemeier, R. H. and Stockwell, D. R. B. (2002) Future projections for Mexican faunas under global climate change scenarios. *Nature* **416**, 626–629.

Petit, D. R., Lynch, J. F., Hutto, R. L., Blake, J. G. and Waide, R. B. (1995) Habitat use and conservation in the Neotropics. In: *Ecology and Management of Neotropical Birds. A synthesis and review of critical issues* (eds. Martin, T. E. and Finch, D. M.), 145–197. Oxford University Press, New York, NY.

Petit, L. J. and Petit, D. R. (2003) Evaluating the importance of human-modified lands for Neotropical bird conservation. *Conservation Biology* **19**, 687–694.

Petit, L. J., Petit, D. R., Christian, D. G. and Powell, H. D. W. (1999) Bird communities of natural and modified habitats in Panama. *Ecography* **22**, 292–304.

Phat, N. K., Knorr, W., and Kim, S. (2004). Appropriate measures for conservation of terrestrial carbon stocks – analysis of trends of forest management in Southeast Asia. *Forest Ecology and Management* **191**, 283–299.

Phillips, O. L., Malhi, Y., Higuchi, N., Laurance, W. F., Nunez, P. V., Vasquez, R. M., Laurance, S. G., Ferreira, L. V., Stern, M., Brown, S. and Grace, J. B. (1998) Changes in the carbon balance of tropical forests: evidence from long-term plots. *Science* **282**, 439–442.

Philpott, S. M., Greenberg, R., Bichier, P. and Perfecto, I. (2004) Impacts of major predators on tropical agroforest arthropods: comparisons within and across taxa. *Oecologia* **140**, 140–149.

Pianka, E. R. (1966) Latitudinal gradients in species diversity: a review of concepts. *American Naturalist* **100**, 33–45.

Pimentel, D., Zuniga, R. and Morrison, D. (2005) Update on the environmental and economic costs associated with alien-invasive species in the United States. *Ecological Economics* **52**, 273–288.

Pimm, S., Moulton, M. P. and Justice, L. J. (1995a) Bird extinctions in the central Pacific. In: *Extinction Rates* (eds. Lawton, J. H. and May, R. M.), 75–87. Oxford University Press, Oxford, UK.

Pimm, S., Raven, R., Peterson, A., Sekercioglu, C. and Ehrlich, P. R. (2006) Human impacts on the rates of recent, present, and future bird extinctions. *Proceedings of the National Academy of Sciences of the United States of America* **103**, 10941–10946.

Pimm, S. L. (1996) Lessons from a kill. *Biodiversity and Conservation* **5**, 1059–1067.

Pimm, S. L. and Jenkins, C. N. (2010) Extinctions and the practice of preventing them. In: *Conservation Biology for All* (eds. Sodhi, N. S. and Ehrlich, P. R.), 181–198. Oxford University Press, Oxford, UK.

Pimm, S. L., Jones, H. L. and Diamond, J. (1988) On the risk of extinction. *American Naturalist* **132**, 757–785.

Pimm, S. L., Russell, G. J., Gittleman, J. L. and Brooks, T. M. (1995b) The future of biodiversity. *Science* **269**, 347–350.

Pitelka, F. A., Tomich, P. Q. and Treichel, G. W. (1955) Ecological relations of jaegers and owls as lemming predators near Barrow, Alaska. *Ecological Monographs* **25**, 85–118.

Pizo, M. A. (1997) Seed dispersal and predation in two populations of *Cabralea canjerana* (Meliaceae) in the Atlantic Forest of southeastern Brazil. *Journal of Tropical Ecology* **13**, 559–578.

Pizo, M. A. (2002) The seed dispersers and fruit syndromes of *Myrtaceae* in the Brazilian Atlantic forest. In: *Seed Dispersal and Frugivory: ecology, evolution and conservation* (eds. Levey, D. J., Silva, W. R. and Galetti, M.), 129–144. CABI Publishing, Wallingford, UK.

Polis, G. A. and Hurd, S. D. (1996) Linking marine and terrestrial food webs: allochthonous input from the ocean supports high secondary productivity on small islands and coastal land communities. *American Naturalist* **147**, 396–423.

Pomeroy, D. E. (1975) Birds as scavengers of refuse in Uganda. *Ibis* **117**, 69–81.

Pons, P. and Wendenburg, C. (2005) The impact of fire and forest conversion into savanna on the bird communities of West Madagascan dry forests. *Animal Conservation* **8**, 183–193.

Poonswad, P., Sukkasem, C., Phataramata, S., Hayeemuida, S., Plongmai, K., Chuailua, P., Thiensongrusame, P. and Jirawatkavi, N. (2005) Comparison of cavity modification and community involvement as strategies for hornbill conservation in Thailand. *Biological Conservation* **122**, 385–393.

Posa, M. R. C. and Sodhi, N. S. (2006) Effects of anthropogenic land use on forest birds and butterflies in Subic Bay, Philippines. *Biological Conservation* **129**, 256–270.

Posa, M. R. C., Diesmos, A. C., Sodhi, N. S. and Brooks, T. M. (2008) Hope for threatened tropical biodiversity: lessons from the Philippines. *BioScience* **58**, 231–240.

Possingham, H. P. (1992) Habitat selection by two species of nectarivore – habitat quality isolines. *Ecology* **73**, 1903–1912.

Post, E., Peterson, R. O., Stenseth, N. C. and McLaren, B. E. (1999) Ecosystem consequences of wolf behavioural response to climate. *Nature* **401**, 905–907.

Potts, S. G., Biesmeijer, J. C., Kremen, C., Neumann, P., Schweiger, O. and Kunin, W. E. (2010) Global pollinator declines: trends, impacts and drivers. *Trends in Ecology and Evolution* **25**, 345–353.

Poulin, B. and Lefebvre, G. (1996) Dietary relationships of migrant and resident birds from a humid forest in Central Panama. *Auk* **113**, 277–287.

Pounds, J. A., Fogden, M. P. L. and Campbell, J. H. (1999) Biological response to climate change on a tropical mountain. *Nature* **398**, 611–615.

Pounds, J. A., Bustamante, M. R., Coloma, L. A., Consuegra, J. A., Fogden, M. P. L., Foster, P. N., La Marca, E., Masters, K. L., Merino-Viteri, A., Puschendorf, R., Ron, S. R., Sanchez-Azofeifa, G. A., Still, C. J. and Young, B. E. (2006) Widespread amphibian extinctions from epidemic disease driven by global warming. *Nature* **439**, 161–167.

Powell, G. V. N. (1979) Structure and dynamics of interspecific flocks in a Neotropical mid-elevation forest. *Auk* **96**, 375–390.

Powell, G. V. N. (1985) Sociobiology and adaptive significance of interspecific foraging flocks in the neotropics. *Ornithological Monograph* **36**, 713–732.

Powell, G. V. N. and Bjork, R. (1994a) Implications of altitudinal migration for conservation strategies to protect tropical diversity: a case study of the resplendent quetzal *Pharomachrus mocinno* at Monteverde, Costa Rica. *Bird Conservation International* **4**, 161–174.

Powell, G. V. N. and Bjork, R. (1994b) Habitat linkages and the conservation of tropical biodiversity as indicated by seasonal migrations of three-wattled bellbirds. *Conservation Biology* **18**, 500–509.

Powell, G. V. N., Fourqurean, J. W., Kenworthy, W. J. and Zieman, J. C. (1991) Bird colonies cause seagrass enrichment in a subtropical estuary – observational and experimental evidence. *Estuarine Coastal and Shelf Science* **32**, 567–579.

Power, M. E. (1984) Depth distributions of armored catfish – predator-induced resource avoidance. *Ecology* **65**, 523–528.

Powlesland, R. G., Butler, D. J. and Westbrooke, I. M. (2008) Was tropical cyclone Heta or hunting by people responsible for decline of the Lupe (*Ducula pacifica*) (Aves: Columbidae) population on Niue during 1994–2004? *Pacific Science* **62**, 461–471.

Prakash, V. (1999) Status of vultures in Keoladeo National Park, Bharatpur, Rajasthan, with special reference to population crash in *Gyps* species. *Journal of the Bombay Natural History Society* **96**, 365–378.

Prakash, V., Pain, D. J., Cunningham, A. A., Donald, P. F., Prakash, N., Verma, A., Gragi, R., Sivakumar, S. and Rahmani, A. R. (2003) Catastrophic collapse of Indian white-backed *Gyps bengalensis* and long-billed *Gyps indicus* vulture populations. *Biological Conservation* **109**, 381–390.

Prawiradilaga, D. M. (1997) The maleo *Macrocephalon maleo* on Buton. *Bulletin of the British Ornithologists' Club* **117**, 237.

Preston, K., Rotenberry, J. T., Redak, R. A. and Allen, M. F. (2008) Habitat shifts of endangered species under altered climate conditions: importance of biotic interactions. *Global Change Biology* **14**, 2501–2515.

Price, J. P. and Wagner, W. L. (2004) Speciation in Hawaiian angiosperm lineages: cause, consequence and mode. *Evolution* **58**, 2185–2200.

Price, O. F. (2004) Indirect evidence that frugivorous birds track fluctuating fruit resources among rainforest patches in the Northern Territory, Australia. *Austral Ecology* **29**, 137–144.

Primack, R. B., Miller-Rushing, A. J. and Dharaneeswaran, K. (2009) Changes in the flora of Thoreau's Concord. *Biological Conservation* **142**, 500–508.

Proctor, M., Yeo, P. and Lack, A. (1996) *The Natural History of Pollination*. Timber Press, Portland, OR.

Prugh, L. R., Hodges, K. E., Sinclair, A. R. E. and Brashares, J. S. (2008) Effect of habitat area and isolation on fragmented animal populations. *Proceedings of National Academy of Sciences of the United States of America* **105**, 20770–20775.

Pulliam, H. R. (1988) Sources, sinks, and population regulation. *American Naturalist* **132**, 652–661.

Putz, F. E., Blate, G. M., Redford, K. H., Fimbel, R. and Robinson, J. (2001) Tropical forest management and conservation of biodiversity: an overview. *Conservation Biology* **15**, 7–20.

Quammen, D. (1997) *The Song of the Dodo: island biogeography in an age of extinctions*. Touchstone Books, New York, NY.

Quinn, J. L., Prop, J., Kokorev, Y. and Black, J. M. (2003) Predator protection or similar habitat selection in Red-breasted Goose nesting associations: extremes along a continuum. *Animal Behaviour* **65**, 297–307.

Rabenold, K. N. (1978) Foraging strategies, diversity and seasonality in bird communities of Appalachian spruce-fir forests. *Ecological Monographs* **48**, 397–424.

Ralph, C. J. and Rich, T. D. (2005) Bird Conservation Implementation and Integration in the Americas. Proceedings of the Third International Partners in Flight Conference: March 20–24, 2002: Asilomar, California. General Technical Report PSW-GTR-191, Pacific Southwest Research Station, Forest Service, United States Department of Agriculture, Albany, California, USA.

Raman, T. R. S. (2001) Effect of slash-and-burn shifting cultivation on rainforest birds in Mizoram, northeast India. *Conservation Biology* **15**, 685–698.

Raman, T. R. S., Rawat, G. S. and Johnsingh, A. J. T. (1998) Recovery of tropical rainforest avifauna in relation to vegetation succession following shifting cultivation in Mizoram, north-east India. *Journal of Applied Ecology* **35**, 214–231.

Ramirez-Bastida, P., Navarro-Siguenza, A. G. and Peterson, A. T. (2008) Aquatic bird distributions in Mexico: designing conservation approaches quantitatively. *Biodiversity and Conservation* **17**, 2525–2558.

Ranganathan, J. and Daily, G. C. (2010) Countryside biogeography: harmonizing biodiversity and agriculture. In: *Conservation Biology for All* (eds. Sodhi, N. S. and Ehrlich, P. R.), 251–253. Oxford University Press, Oxford, UK.

Ranganathan, J., Daniels, R. J. R., Chandran, M. D. S., Ehrlich, P. R. and Daily, G. C. (2008) Sustaining biodiversity in ancient tropical countryside. *Proceedings of the National Academy of Sciences of the United States of America* **105**, 17852–17854.

Rappole, J. H. (1995) *The Ecology of Migrant Birds: a Neotropical perspective*, Smithsonian Institution Press, Washington, DC.

Rappole, J. H. and Hubalek, Z. (2003) Migratory birds and West Nile Virus. *Journal of Applied Microbiology* **94**, 47–58.

Rappole, J. H. and MacDonald, M. V. (1994) Cause and effect in population declines of migratory birds. *Auk* **111**, 652–660.

Rappole, J. H. and Morton, E. S. (1985) Effects of habitat alteration on a tropical forest bird community. *Ornithological Monographs* **36**, 1013–1021.

Rappole, J. H., Ramos, M. A. and Winker, K. (1989) Wintering Wood Thrush movements and mortality in southern Veracruz. *Auk* **106**, 402–410.

Rappole, J. H., King, D. I. and Barrow, W. C. J. (1999) Winter ecology of the endangered Golden-cheeked Warbler. *Condor* **101**, 762–770.

Ratcliffe, D. (1993) *The Peregrine Falcon*. Princeton University Press, Princeton, NJ.

Rathcke, B. J. (2000) Hurricane causes resource and pollination limitation of fruit set in a bird-pollinated shrub. *Ecology* **81**, 1951–1958.

Raven, P. H. (1972) Why are bird-visited flowers predominantly red? *Evolution* **26**, 674–674.

Ray, D., Nepstad, D. and Moutinho, P. (2005) Micrometeorological and canopy controls of fire susceptibility in a forested Amazon landscape. *Ecological Applications* **15**, 1644–1678.

Redford, K. H. (1992) The empty forest. *BioScience* **42**, 412–422.

Reed, J. M. (1999) The role of behaviour in recent avian extinctions and endangerments. *Conservation Biology* **13**, 232–241.

Regal, P. J. (1977) Ecology and evolution of flowering plant dominance. *Science* **196**, 622–629.

Reid, N. (1991) Coevolution of mistletoes and frugivorous birds. *Australian Journal of Ecology* **16**, 457–469.

Reino, L., Moya-Larano, J. and Heitor, A. C. (2009) Using survival regression to study patterns of expansion of invasive species: will the common waxbill expand with global warming? *Ecography* **32**, 237–246.

Remsen, J. V. Jr. and Parker, T. A. III. (1983) The contribution of river-created habitats to bird species richness in Amazonia. *Biotropica* **15**, 223–231.

Remsen, J. V. Jr. and Parker, T. A. III. (1984) Arboreal dead-leaf-searching birds of the Neotropics. *Condor* **86**, 36–41.

Remsen, J. V. Jr. and Robinson, S. K. (1990) A classification scheme for foraging behavior of birds in terrestrial habitats. *Studies in Avian Biology* **13**, 144–160.

Renjifo, L. M. (1999) Composition changes in a subandean avifauna after long-term forest fragmentation. *Conservation Biology* **13**, 1124–1139.

Renjifo, L. M. (2001) Effect of natural and anthropogenic landscape matrices on the abundance of subandean bird species. *Ecological Applications* **11**, 14–31.

Renne, I. J., Barrow, W. C., Randall, L. A. J. and Bridges, W. C. (2002) Generalized avian dispersal syndrome contributes to Chinese Tallow Tree (*Sapium sebiferum*, Euphorbiaceae) invasiveness. *Diversity and Distributions* 8, 285–295.

Renner, S. S. (1989) A survey of reproductive biology in neotropical Melastomataceae and Memecylaceae. *Annals of the Missouri Botanical Garden* 76, 496–518.

Renner, S. S. (2005) Rewardless flowers in the angiosperms and the role of insect cognition in their evolution. In: *Plant-Pollinator Interactions: from specialization to generalization* (eds. Waser, N. M. and Ollerton, J.), 123–144. University of Chicago Press, Chicago, IL.

Restrepo, C., Sargent, S., Levey, D. J. and Watson, D. M. (2002) The role of vertebrates in the diversification of New World mistletoes. In: *Seed Dispersal and Frugivory: ecology, evolution and conservation* (eds. Levey, D. J., Silva, W. R. and Galetti, M.), 83–98. CABI Publishing, Wallingford, UK.

Reudink, M. W., Marra, P. P., Kyser, T. K., Boag, P. T., Langin, K. M. and Ratcliffe, L. M. (2009a) Non-breeding season events influence polygyny and extra-pair paternity in a long-distance migratory bird. *Proceedings of the Royal Society of London Series B* 276, 1619–1626.

Reudink, M. W., Studds, C. E., Kyser, T. K., Marra, P. P. and Ratcliffe, L. M. (2009b) Plumage brightness predicts non-breeding season territory quality in a long-distance migratory songbird. *Journal of Avian Biology* 40, 34–41.

Rhymer, J. M. and Simberloff, D. (1996) Extinction by hybridization and introgression. *Annual Review of Ecology and Systematics* 27, 83–109.

Ribeiro, M. C., Metzger, J. P., Martensen, A. C., Ponzoni, F. J. and Hirota, M. M. (2009) The Brazilian Atlantic Forest: how much is left, and how is the remaining forest distributed? Implications for conservation. *Biological Conservation* 142, 1141–1153.

Ribon, R., Simon, J. E. and De Mattos, G. T. (2003) Bird extinctions in Atlantic forest fragments of the Viçosa Region, Southeastern Brazil. *Conservation Biology* 17, 1827–1839.

Rich, T. D., Beardmore, C. J., Berlanga, H., Blancher, P. J., Bradstreet, M. S. W., Butcher, G. S., Demarest, D. W., Dunn, E. H., Hunter, W. C., Iñigo-Elias, E. E., Kennedy, J. A., Martell, A. M., Panjabi, A. O., Pashley, D. N., Rosenberg, K. V., Rustay, C. M., Wendt, J. S. and Will, T. C. (2004) *Partners in Flight North American Landbird Conservation Plan*. Cornell Laboratory of Ornithology, Ithaca, New York, NY.

Ricklefs, R. E., Fallon, S. M., Latta, S. C., Swanson, B. L. and Bermingham, E. (2005) Migrants and their parasites: a bridge between two worlds. In: *Birds of Two Worlds: the ecology and evolution of migration* (eds. Greenberg, R. and Marra, P. P.), 210–221. The Johns Hopkins University Press, Baltimore, MD.

Robbins, C. S. (1979) *Effects of Forest Fragmentation oBird Populations*. Management of north central and northeastern forests for nongame birds USDA Forest General Technical Report NC-51, 198–212. USDA Forest Service North Central Experiment Station, St. Paul, Minnesota, USA.

Robbins, C. S., Sauer, J. R., Greenberg, R. S. and Droege, S. (1989) Population declines in North American birds that migrate to the Neotropics. *Proceedings of the National Academy of Sciences of the United States of America* 86, 7658–7662.

Robbins, C. S., Dowell, B. A., Dawson, D. K., Colon, J. A., Estrada, R., Sutton, A., Sutton, R. and Weyer, D. (1992a) Comparison of Neotropical migrant landbird populations wintering in tropical forest, isolated fragments, and agriculture habitats. In: *Ecology and Conservation of Neotropical Migrant Landbirds* (eds. Hagan, J. M. III. and Johnston, D. W.), 549–562. Smithsonian Institution Press, Washington, DC.

Robbins, C. S., Fitzpatrick, J. W. and Hamel, P. B. (1992b) A warbler in trouble: *Dendroica cerulea*. In: *Ecology and Conservation of Neotropical Migrant Landbirds* (eds. Hagan, J. M. III. and Johnston, D. W.), 408–418. Smithsonian Institution Press, Washington, DC.

Robertson, A. W., Kelly, D., Ladley, J. J. and Sparrow, A. D. (1999) Effects of pollinator loss on endemic New Zealand mistletoes (Loranthaceae). *Conservation Biology* 13, 499–508.

Robinson, G. R. and Handel, S. N. (1993) Forest restoration on a closed landfill: rapid addition of new species by bird dispersal. *Conservation Biology* 7, 271–278.

Robinson, J. M. (2001) The dynamics of avicultural markets. *Environmental Conservation* 28, 76–85.

Robinson, S. K. (1985) Coloniality in the Yellow-rumped Cacique as a defense against nest predators. *Auk* 102, 506–519.

Robinson, S. K. (1994) Habitat selection and foraging ecology of raptors in Amazonian Peru. *Biotropica* **26**, 443–458.

Robinson, S. K. (1997) The case of the missing songbirds. *Consequences* **3**, 3–16.

Robinson, S. K., Terborgh, J. and Fitzpatrick, J. W. (1988) Habitat selection and relative abundance of migrants in southeastern Peru. *Proceedings of the International Ornithological Congress* **19**, 2298–2307.

Robinson, S. K., Terborgh, J. and Munn, C. A. (1990) Lowland tropical forest bird communities of a site in western Amazonia. In: *Biogeography and Ecology of Forest Bird Communities* (ed. Keast, A), 229–274. SPB Academic Publishing, The Hague, The Netherlands.

Robinson, S. K., Fitzpatrick, J. W. and Terborgh, J. (1995) Distribution and habitat use of neotropical migrant landbirds in the Amazon basin and Andes. *Bird Conservation International* **5**, 305–323.

Robinson, W. D. (1999) Long-term changes in the avifauna of Barro Colorado Island, Panama, a tropical forest isolate. *Conservation Biology* **13**, 85–97.

Robinson, W. D., Robinson, T. R., Robinson, S. K. and Brawn, J. D. (2000) Nesting success of understory forest birds in Central Panama. *Journal of Avian Biology* **31**, 151–164.

Rodrigues, A. S. L., Andelman, S. J., Bakarr, M. I., Boitani, L., Brooks, T. M., M., C. R., Fishpool, L. D. C., da Fonseca, G. A. B., Gaston, K. J., Hoffmann, M., Long, J. S., Marquet, P. A., Pilgrim, J. D., Pressey, R. L., Schipper, J., Sechrest, W., Stuart, S. N., Underhill, L. G., Waller, R. W., Watts, M. E. J. and Yan, X. (2004) Effectiveness of the global protected area network in representing species diversity. *Nature* **428**, 640–643.

Rodriguez, C., Torres, R. and Drummond, H. (2008) Eradicating introduced mammals from a forested tropical island. *Biological Conservation* **130**, 98–105.

Rodriguez-Girones, M. A. and Santamaria, L. (2004) Why are so many bird flowers red? *PLoS Biology* **2**, 1515–1519.

Roemer, G. W., Donlan, C. J. and Courchamp, F. (2002) Golden Eagles, feral pigs, and insular carnivores: how exotic species turn native predators into prey. *Proceedings of the National Academy of Sciences of the United States of America* **99**, 791–796.

Rompré, G., Robinson, W. D., Desrochers, A. and Angehr, G. (2009) Predicting declines in avian species richness under nonrandom patterns of habitat loss in a Neotropical landscape. *Ecological Application* **19**, 1614–1627.

Rosenberg, K. V. (1990) Dead leaf foraging specialization in tropical forest birds: measuring resource availability and use. *Studies in Avian Biology* **13**, 360–368.

Rosenzweig, M. L. (2003) *Win-win Ecology: how the earth's species can survive in the midst of human enterprise*. Oxford University Press, New York, NY.

Rossiter, N. A., Setterfield, S. A., Douglas, M. M. and Hutley, L. B. (2003) Testing the grass-fire cycle: alien grass invasion in the tropical savannas of northern Australia. *Diversity and Distributions* **9**, 169–176.

Rost, J., Pons, P. and Bas, J. M. (2009) Can salvage logging affect seed dispersal by birds into burned forests? *Acta Oecologica* **35**, 763–768.

Rubenstein, D. R., Chamberlain, C. P., Holmes, R. T., Ayres, M. P., Waldbauer, J. R., Graves, G. R. and Tuross, N. C. (2002) Linking breeding and wintering ranges of a migratory songbird using stable isotopes. *Science* **295**, 1062–1065.

Ruiz-Gutierrez, V., Gavin, T. A. and Dhondt, A. A. (2008) Habitat fragmentation lowers survival of a tropical forest bird. *Ecological Applications* **18**, 838–846.

Ruiz-Gutierrez, V., Zipkin, E. F. and Dhondt, A. A. (2010) Occupancy dynamics in a tropical bird community: unexpectedly high forest use by birds classified as non-forest species. *Journal of Applied Ecology* **47**, 621–630.

Runge, M. C. and Marra, P. P. (2005) Modeling seasonal interactions in the population dynamics of migratory birds. In: *Birds of Two Worlds: the ecology and evolution of migration* (eds. Greenberg, R. and Marra, P. P.), 375–389. The Johns Hopkins University Press, Baltimore, MD.

Ruxton, G. D. and Houston, D. C. (2004) Obligate vertebrate scavengers must be large soaring fliers. *Journal of Theoretical Biology* **228**, 431–436.

Ryall, C. (1992) Predation and harassment of native bird species by the Indian House Crow *Corvus splendens*, in Mombassa, Kenya. *Scopus* **16**, 1–8.

Sabat, P., Cavieres, G., Veloso, C., Canals, M. and Bozinovic, F. (2009) Intraspecific basal metabolic rate varies with trophic level in rufous-collared sparrows. *Comparative Biochemistry and Physiology A-Molecular & Integrative Physiology* **154**, 502–507.

Sachs, J. D., Baillie, J. E. M., Sutherland, W. J., Armsworth, P., Ash, N., Bateson, P., Beddington, J., Blackburn, T., Collen, B., Gardiner, B., Gaston, K. J., Godfray, H. C. J., Green, R., Harvey, P., Homewood, K., House, B., Hutton, J., Knapp, S., Kumpel, N., MacDonald, D., Mace, G. M., Mallett, J., Matthews, A., May, R., Petchey, O., Purvis, A., Raffaelli, D., Rodrigues, A., Roe, D., Safi, K., Toulmin, C., Turner, K., Walpole, M., Watson, R. and Jones, K. E. (2009) Biodiversity conservation and the Millennium Development Goals. *Science* **325**, 1502–1503.

Saether, B. E., Engen, S., Møller, A. P., Visser, M. E., Matthysen, E., Fiedler, W., Lambrechts, M. M., Becker, P. H., Brommer, J. E., Dickinson, J., Du Feu, C., Gehlbach, F. R., Merila, J., Rendell, W., Robertson, R. J., Thomson, D. and Torok, J. (2005) Time to extinction of bird populations. *Ecology* **86**, 693–700.

Safford, R. J. and Jones, C. G. (1997) Did organochlorine pesticide use cause declines in Mauritian forest birds? *Biodiversity and Conservation* **6**, 1445–1451.

Safford, R. J. and Jones, C. G. (1998) Strategies of land-bird conservation on Mauritius. *Conservation Biology* **12**, 169–176.

Sakai, A. K., Wagner, W. L. and Mehrhoff, L. A. (2002) Patterns of endangerment in the Hawaiian flora. *Systematic Biology* **51**, 276–302.

Sánchez-Azofeifa, G. A., Quesada-Mateo, C., Gonzalez-Quesada, P., Dayanandan, S. and Bawa, K. S. (1999) Protected areas and conservation of biodiversity in the tropics. *Conservation Biology* **13**, 407–411.

Sanchez-Pinero, F. and Polis, G. A. (2000) Bottom-up dynamics of allochthonous input: direct and indirect effects of seabirds on islands. *Ecology* **81**, 3117–3132.

Sanderson, F. J., Donald, P. F., Cain, D. J., Burfield, I. J. and nan Bommel, F. P. J. (2006) Long-term declines in Afro-Palearctic migrant birds. *Biological Conservation* **131**, 93–105.

Sanford, R. L., Saldarriaga, J., Clark, K. E., Uhl, C. and Herrera, R. (1985) Amazon rain-forest fires. *Science* **227**, 53–55.

Santos, T. and Telleria, J. (1994) Influence of forest fragmentation on seed consumption and dispersal of Spanish juniper *Juniperus thurifera*. *Biological Conservation* **70**, 129–134.

Santos, T., Telleria, J. L. and Virgos, E. (1999) Dispersal of Spanish juniper *Juniperus thurifera* by birds and mammals in a fragmented landscape. *Ecography* **22**, 193–204.

Sanz, J. J. (2001) Experimentally increased insectivorous bird density results in a reduction of caterpillar density and leaf damage to Pyrenean Oak. *Ecological Research* **16**, 387–394.

Sarasola, J. H., Negro, J. J., Hobson, K. A., Bortolotti, G. R. and Bildstein, K. L. (2008) Can a 'wintering area effect' explain population status of Swainson's hawks? A stable isotope approach *Diversity and Distributions* **14**, 686–691.

Sargent, O. H. (1918) Fragments on the flower biology of Westralian plants. *Annals of Botany* **23**, 265–274.

Saunders, D. A., Hobbs, R. J. and Margules, C. R. (1991) Biological consequences of ecosystem fragmentation – a review. *Conservation Biology* **5**, 18–32.

Sazima, I., Buzato, S. and Sazima, M. (1995) The Saw-billed Hermit *Ramphodon naevius* and its flowers in southeastern Brazil. *Journal für Ornithologie* **136**, 195–206.

Schaller, G. B. (1968) *The Serengeti Lion*. University of Chicago Press, Chicago, IL.

Scharlemann, J. P., Green, R. E. and Balmford, A. (2004) Land-use trends in Endemic Bird Areas: global expansion of agriculture in areas of high conservation value. *Global Change Biology* **10**, 2046–2051.

Scheuerlein, A. and Gwinner, E. (2006) Reduced nestling growth of East African Stonechats *Saxicola torquata axillaris* in the presence of a predator. *Ibis* **148**, 468–476.

Schmitt, C. B., Burgess, N. D., Coad, L. and Belokurov, A. (2009) Global analysis of the protection status of world's forests. *Biological Conservation* **142**, 2122–2130.

Schmitz, O. J., Hamback, P. A. and Beckerman, A. P. (2000) Trophic cascades in terrestrial systems: a review of the effects of carnivore removals on plants. *American Naturalist* **155**, 141–153.

Schneider, N. and Griesser, M. (2009) Influence and value of different water regimes on avian species richness in arid inland Australia. *Biodiversity and Conservation* **18**, 457–471.

Schneider, S. H. (1989) *Global Warming*. Sierra Club, San Francisco, CA.

Schneider, S. H. and Londer, R. (1984) *The Coevolution of Climate and Life*. Sierra Club Books, San Francisco, CA.

Schuchmann, K. L. (1999) Family Trochilidae (Hummingbirds). In: *Handbook of the Birds of the World (Vol. 5): Barn-owls to hummingbirds* (eds. del Hoyo, J., Elliott, A. and Sargatal, J.), 468–535. Lynx Edicions, Barcelona.

Schupp, E. W., Milleron, T. and Russo, S. E. (2002) Dissemination limitation and the origin and maintenance of species-rich tropical forests. In: *Seed Dispersal and Frugivory: Ecology, Evolution and Conservation* (eds. Levey, D. J., Silva, W. R. and Galetti, M.), 19–33. CABI Publishing, Wallingford, UK.

Schüz, E. and König, K. (1983) Old World vultures and man. In: *Vulture Biology and Management* (eds. Wilbur, S. R. and Jackson, J. H.), 461–469. University of California Press, Berkeley, CA.

Schwager, M., Covas, R., Blaum, N. and Jeltsch, F. (2008) Limitations of population models in predicting climate change effects: a simulation study of sociable weavers in southern Africa. *Oikos* **117**, 1417–1427.

Schwartz, M. W., Iverson, L. R., Prasad, A. M., Matthews, S. N. and O'Connor, R. J. (2006) Predicting extinctions as a result of climate change. *Ecology* **87**, 1611–1615.

Scott, A. C. and Glasspool, I. J. (2006) The diversification of Paleozoic fire systems and fluctuations in atmospheric oxygen concentration. *Proceedings of the National Academy of Sciences of the United States of America* **103**, 10861–10865.

Seabloom, E. W., Dobson, A. P. and Stoms, D. M. (2002) Extinction rates under nonrandom patterns of habitat loss. *Proceedings of the National Academy of Sciences of the United States of America* **99**, 11229–11234.

Seaman, B. S. and Schulze, C. H. (2010) The importance of gallery forests in the tropical lowlands of Costa Rica for understorey forest birds. *Biological Conservation* **143**, 391–398.

Seavy, N. E. (2006) Physiological correlates of habitat association in East African sunbirds (Nectariniidae). *Journal of Zoology* **270**, 290–297.

Seitre, R. and Seitre, J. (1992) Causes of land-bird extinction in French Polynesia. *Oryx* **26**, 215–222.

Sekercioglu, C. H. (2002a) Effects of forestry practices on the vegetation structure and bird community of Kibale National Forest, Uganda. *Biological Conservation* **107**, 229–240.

Sekercioglu, C. H. (2002b) Impacts of birdwatching on human and avian communities. *Environmental Conservation* **29**, 282–289.

Sekercioglu, C. H. (2002c) Forest fragmentation hits insectivorous birds hard. *The Scientific World* **1**, 62–64.

Sekercioglu, C. H. (2006a) Ecological significance of bird populations. In: *Handbook of the Birds of the World* (eds. Hoyo, J. D., Elliott, A. and Christie, D. A.), 15–51. Lynx Edicions, Barcelona.

Sekercioglu, C. H. (2006b) Increasing awareness of avian ecological function. *Trends in Ecology and Evolution* **21**, 464–471.

Sekercioglu, C. H. (2007) Conservation ecology: area trumps mobility in fragment bird extinctions. *Current Biology* **17**, R283–R286.

Sekercioglu, C. H. (2009) Tropical ecology: riparian corridors connect fragmented forest bird populations. *Current Biology* **19**, R210–R213.

Sekercioglu, C. H. (2010a) Ecosystem functions and services. In: *Conservation Biology for All* (eds. Sodhi, N. S. and Ehrlich, P. R.), 45–72. Oxford University Press, Oxford.

Sekercioglu, C. H. (2010b) Partial migration in tropical birds: the frontier of movement ecology. *Journal of Animal Ecology* **79**, 933–936.

Sekercioglu, C. H. (2011) Functional extinctions of bird pollinators cause plant declines. *Science* in press.

Sekercioglu, C. H. and Riley, A. (2005) A brief survey of the birds in Kumbira Forest, Gabela, Angola. *Ostrich* **76**, 111–117.

Sekercioglu, C. H., Ehrlich, P. R., Daily, G. C., Aygen, D., Goehring, D. and Sandi, R. F. (2002) Disappearance of insectivorous birds from tropical forest fragments. *Proceedings of the National Academy of Sciences of the United States of America* **99**, 263–267.

Sekercioglu, C. H., Daily, G. C. and Ehrlich, P. R. (2004) Ecosystem consequences of bird declines. *Proceedings of the National Academy of Sciences of the United States of America* **101**, 1842–1847.

Sekercioglu, C. H., Loarie, S. R., Oviedo Brenes, F., Ehrlich, P. R. and Daily, G. C. (2007) Persistence of forest birds in the Costa Rican agricultural countryside. *Conservation Biology* **21**, 482–494.

Sekercioglu, C. H., Schneider, S. H., Fay, J. P. and Loarie, S. R. (2008) Climate change, elevational range shifts, and bird extinctions. *Conservation Biology* **22**, 140–150.

Senior, K. (2006) Battle over wild bird trade ban. *Frontiers in Ecology and Environment* **4**, 509.

Severinghaus, L. L. and Chi, L. (1999) Prayer animal release in Taiwan. *Biological Conservation* **89**, 301–304.

Sezen, U. U., Chazdon, R. L. and Holsinger, K. E. (2005) Genetic consequences of tropical second-growth forest regeneration. *Science* **307**, 891.

Shahabuddin, G. (1997) Preliminary observations on the role of coffee plantations as avifaunal refuges in the Palni Hills of the Western Ghats. *Journal of the Bombay Natural History Society* **94**, 10–21.

Shanahan, M., So, S., Compton, S. G. and Corlett, R. (2001) Fig-eating by vertebrate frugivores: a global review. *Biological Reviews* **76**, 529–572.

Shapcott, A. (1999) Vagility and the monsoon rain forest archipelago of northern Australia: patterns of genetic diversity in *Syzygium nervosum* (Myrtaceae). *Biotropica* **31**, 579–590.

Sharam, G. J., Sinclair, A. R. E. and Turkington, R. (2009) Serengeti birds maintain forests by inhibiting seed predators. *Scyence* **325**(5936), 51.

Sheldon, F. H., Styring, A. and Hosner, P. A. (2010) Bird species richness in a Bornean exotic tree plantation: a long-term perspective. *Biological Conservation* **143**, 399–407.

Shepherd, C. R. (2006) The bird trade in Medan, north Sumatra: an overview. *Birding Asia* **5**, 16–24.

Sherry, T. W. (1984) Comparative dietary ecology of sympatric, insectivorous Neotropical flycatchers (Tyrannidae). *Ecological Monographs* **54**, 313–338.

Sherry, T. W. and Holmes, R. T. (1995) Summer versus winter limitations of populations: what are the issues and what is the evidence? In: *Ecology and Management of Neotropical birds. a synthesis and review of critical issues* (eds. Martin, T. E. and Finch, D. M.), 85–120. Oxford University Press, New York, NY.

Sherry, T. W. and Holmes, R. T. (1996) Winter habitat quality, population limitation, and conservation of neotropical-nearctic migrant birds. *Ecology* **77**, 36–48.

Sherry, T. W., Johnson, M. D. and Strong, A. M. (2005) Does winter food limit populations of migratory birds? In: *Birds of Two Worlds: the ecology and evolution of migration* (eds. Greenberg, R. and Marra, P. P.), 414–425. The Johns Hopkins University Press, Baltimore, MD.

Shieh, B.-S., Lin, Y.-H., Lee, T.-W., Chang, C.-C. and Cheng, K.-T. (2006) Pet trade as sources of introduced bird species in Taiwan. *Taiwania* **51**, 81–86.

Shiels, A. B. and Walker, L. R. (2003) Bird perches increase forest seeds on Puerto Rican landslides. *Restoration Ecology* **11**, 457–465.

Shlisky, A., Alencar, A. A. C., Nolasco, M. M. and Curran, L. M. (2009) Global fire regime conditions, threats, and opportunities for fire management in the tropics. In: *Tropical Fire Ecology: climate change, land use, and ecosystem dynamics* (eds. Cochrane, M. A.). Springer-Praxis, Heidelberg, Germany.

Shoo, L. P., Williams, S. E. and Hero, J. M. (2005a) Climate warming and the rainforest birds of the Australian Wet Tropics: using abundance data as a sensitive predictor of change in total population size. *Biological Conservation* **125**, 335–343.

Shoo, L. P., Williams, S. E. and Hero, J. M. (2005b) Potential decoupling of trends in distribution area and population size of species with climate change. *Global Change Biology* **11**, 1469–1476.

Shoo, L. P., Williams, S. E. and Hero, J. M. (2006) Detecting climate change induced range shifts: where and how should we be looking? *Austral Ecology* **31**, 22–29.

Shultz, S., Noe, R., McGraw, W. S. and Dunbar, R. I. M. (2004) A community-level evaluation of the impact of prey behavioural and ecological characteristics on predator diet composition. *Proceedings of the Royal Society of London Series B* **271**, 725–732.

Shwartz, A., Strubbe, D., Butler, C. J., Matthysen, E. and Kark, S. (2009) The effect of enemy-release and climate conditions on invasive birds: a regional test using the rose-ringed parakeet (*Psittacula krameri*) as a case study. *Diversity and Distributions* **15**, 310–318.

Siegert, F., Ruecker, G., Hinrichs, A. and Hoffmann, A. A. (2001) Increased damage from fires in logged forests during droughts caused by El Nino. *Nature* **414**, 437–440.

Sieving, K. E. (1992) Nest predation and differential insular extinction among selected forest birds of Central Panama. *Ecology* **73**, 2310–2328.

Sigel, B. J., Sherry, T. W. and Young, B. E. (2006) Avian community response to lowland tropical rainforest isolation: 40 years of change at La Selva Biological Station, Costa Rica. *Conservation Biology* **20**, 111–121.

Sigel, B. J., Robinson, W. D. and Sherry, T. W. (2010) Comparing bird community response to forest fragmentation in two lowland Central American reserves. *Biological Conservation* **143**, 340–350.

Sih, A., Crowley, P., McPeek, M., Petranka, J. and Strohmeier, K. (1985) Predation, competition, and prey communities: a review of field experiments. *Annual Review of Ecology and Systematics* **16**, 269–311.

Silander, J. A. (1978) Density-dependent control of reproductive success in *Cassia biflora*. *Biotropica* **10**, 292–296.

Sillett, T. S. and Holmes, R. T. (2002) Variation in survivorship of a migratory songbird throughout its annual cycle. *Journal of Animal Ecology* **71**, 296–308.

Sillett, T. S. and Holmes, R. T. (2005) Long-term demographic trends, limiting factors, and the strength of density dependence in a breeding population of a migratory songbird. In: *Birds of Two Worlds: the ecology and evolution of migration* (eds. Greenberg, R. and Marra, P. P.), 426–436. The Johns Hopkins University Press, Baltimore, MD.

Sillett, T. S., Holmes, R. T. and Sherry, T. W. (2000) Impacts of a global climate cycle on population dynamics of a migratory songbird. *Science* **288**, 2040–2042.

Silman, M. R. (1996) Regeneration from Seed in a Neotropical Rain Forest. PhD dissertation, Duke University, Durham, NC.

Silva, J. M. C. D., Uhl, C. and Murray, G. (1996) Plant succession, landscape management, and the ecology of frugivorous birds in abandoned Amazonian pastures. *Conservation Biology* **10**, 491–503.

Simberloff, D. (1992a) Do species–area curves predict extinction in fragmented forest? In: *Tropical Deforestation and Species Extinction* (eds. Whitmore, T. C. and Sayer, J. A.), 75–89. Chapman and Hall, London.

Simberloff, D. (1992b) Species-area relationships, fragmentation, and extinction in tropical forests. *Malayan Nature Journal* **45**, 398–413.

Simberloff, D. (1992c) Extinction, survival, and effects of birds introduced to the Mascarenes. *Acta Oecologica* **6**, 663–678.

Simberloff, D. (2001) Biological invasions – how are they affecting us, and what can we do about them? *Western North American Naturalist* **61**, 308–315.

Simberloff, D. S. and Abele, L. G. (1976) Island biogeography and conservation – strategy and limitations. *Science* **193**, 1032–1032.

Simberloff, D. and Cox, J. (1987) Consequences and costs of conservation corridors. *Conservation Biology* **1**, 63–71.

Simberloff, D. and Von Holle, B. (1999) Positive interactions of non-indigenous species: invasional meltdown? *Biological Invasions* **1**, 21–32.

Sinclair, A. R. E., Mduma, S. A. R. and Arcese, P. (2002) Protected areas as biodiversity benchmarks for human impact: agriculture and the Serengeti avifauna. *Proceedings of the Royal Society of London Series B* **269**, 2401–2405.

Sipura, M. (1999) Tritrophic interactions: willows, herbivorous insects and insectivorous birds. *Oecologia* **121**, 537–545.

Sirami, C., Seymour, C., Midgley, G. and Barnard, P., 2009. The impact of shrub encroachment on savanna bird diversity from local to regional scale. *Diversity and Distributions* **15**, 948–957.

Skorupa, J. P. (1989) Crowned Eagles *Stephanoaetus coronatus* in rainforest – observations on breeding chronology and diet at a nest in Uganda. *Ibis* **131**, 294–298.

Skowno, A. L. and Bond, W. J. (2003) Bird community composition in an actively managed savanna reserve, importance of vegetation structure and vegetation composition. *Biodiversity and Conservation* **12**, 2279–2294.

Slik, J. W. F. and Balen, S. V. (2006) Bird community changes in response to single and repeated fires in a lowland tropical rainforest of eastern Borneo. *Biodiversity and Conservation* **15**, 4425–4451.

Slik, J. W. F. and Eichhorn, K. A. O. (2003) Fire survival of lowland tropical rain forest trees in relation to stem diameter and topographic position. *Oecologia* **137**, 446–455.

Smith, A. P., Horning, N. and Moore, D. (1997) Regional biodiversity planning and lemur conservation with GIS in western Madagascar. *Conservation Biology* **11**, 498–512.

Smith, D. (2009) Vultures Face Extinction as Gamblers Seek Visions of the Future. Guardian, Johannesburg. http://www.guardian.co.uk/world/2009/dec/30/vulture-extinct-gambling

Smith, J. A. M., Reitsma, L. R., Rockwood, L. L. and Marra, P. P. (2008) Roosting behavior of a Neotropical Migrant songbird, the Northern Waterthrush *Seiurus noveboracensis* during the nonbreeding season. *Journal of Avian Biology* **39**, 460–465.

Smith, J. F. (2001) High species diversity in fleshy-fruited tropical understory plants. *American Naturalist* **157**, 646–653.

Smith, K. F., Sox, D. F. and Lafferty, K. D. (2006) Evidence for the role of infectious disease in species extinction and endangerment. *Conservation Biology* **20**, 1349–1357.

Smith, P., Gregory, P. J., van Vuuren, D., Obersteiner, M., Havlik, P., Rounsevell, M., Woods, J., Stehfast, E. and Bellarby, J. (2010) Competition for land. *Philosophical Transactions of the Royal Society of London B* **365**, 2941–2957.

Smith, T. B., Freed, L. A., Lepson, J. K. and Carothers, J. H. (1995) Evolutionary consequences of extinctions in populations of a Hawaiian honeycreeper. *Conservation Biology* **9**, 107–113.

Smith, T. B., Clegg, S. M., Kimura, M., Ruegg, K. C., Mila, B. and Lovette, I. M. (2005) Molecular genetic approaches to linking breeding and overwintering areas in five neotropical migrant passerines. In: *Birds of Two Worlds: the ecology and evolution of migration* (eds. Greenberg, R. and Marra, P. P.), 222–234. The Johns Hopkins University Press, Baltimore, MD.

Snow, B. K. and Snow, D. W. (1972) Feeding niches of hummingbirds in a Trinidad valley. *Journal of Animal Ecology* **41**, 471-&.

Snow, D. W. (1981) Tropical frugivorous birds and their food plants – a world survey. *Biotropica* **13**, 1–14.

Snyder, N. and Snyder, H. (2000) *The California Condor: a saga of natural history and conservation*. Academic Press, San Diego, CA.

Soares-Filho, B. S., Nepstad, D. C., Curran, L. M., Cerqueira, G. C., Garcia, R. A., Ramos, C. A., Voll, E., McDonald, A., Lefebvre, P. and Schlesinger, P. (2006) Modelling conservation in the Amazon basin. *Nature* **440**, 520–523

Sodhi, N. S. (2010). Birds. In: *Encyclopedia of Invasive Introduced Species* (eds. Simberloff, D. and Rejmánek, M.), 70–74. University of California Press, Berkeley, CA.

Sodhi, N. S. and Brook, B. W. (2006) *Southeast Asian biodiversity in crisis*. Cambridge University Press, Cambridge, UK.

Sodhi, N. S. and Er, K. B. H. (2000) Conservation meets consumption. *Trends in Ecology and Evolution* **15**, 431.

Sodhi, N. S. and Sharp, I. (2006) *Winged Invaders: pest birds of the Asia Pacific*. SNP Reference, Singapore.

Sodhi, N. S. and Smith, K. G. (2007) Conservation of tropical birds: Mission possible? *Journal of Ornithology* **148**(Suppl 2), 305–309.

Sodhi, N. S., Didiuk, A. and Oliphant, L. W. (1990) Differences in bird abundance in relation to proximity of merlin nests. *Canadian Journal of Zoology* **68**, 852–854.

Sodhi, N. S., Briffet, C., Kong, L. and Yuen, B. (1999) Bird use of linear areas of a tropical city: implications for park connector design and management. *Landscape and Urban Planning* **45**, 123–130.

Sodhi, N. S., Peh, K. S. H., Lee, T. M., Turner, I. M., Tan, H. T. W., Prawiradilaga, D. M. and Darjono (2003) Artificial nest and seed predation experiments on tropical southeast Asian islands. *Biodiversity and Conservation* **12**, 2415–2433.

Sodhi, N. S., Koh, L. P., Brook, B. W. and Ng, P. K. L. (2004a) Southeast Asian biodiversity: an impending disaster. *Trends in Ecology and Evolution* **19**, 654–660.

Sodhi, N. S., Liow, L. H. and Bazzaz, F. A. (2004b) Avian extinctions from tropical and subtropical forests. *Annual Review of Ecology, Evolution and Systematics* **35**, 323–345.

Sodhi, N. S., Koh, L. P., Prawiradilaga, D. M., Tinulele, I., Putra, D. D. and Tan, T. H. T. (2005a) Land use and conservation value for forest birds in Central Sulawesi (Indonesia). *Biological Conservation* **122**, 547–558.

Sodhi, N. S., Lee, T. M., Koh, L. P. and Dunn, R. R. (2005b) A century of avifaunal turnover in a small tropical rainforest fragment. *Animal Conservation* **8**, 217–222.

Sodhi, N. S., Koh, L. P. and Brook, B. W. (2006a) Southeast Asian birds in peril. *Auk* **123**, 275–277.

Sodhi, N. S., Lee, T. M., Koh, L. P. and Prawiradilaga, D. M. (2006b) Long-term avifaunal impoverishment in an isolated tropical woodlot. *Conservation Biology* **20**, 772–779.

Sodhi, N. S., Brooks, T. M., Koh, L. P., Acciaioli, G., Erb, M., Tan, A. K. J., Curran, L. M., Brosius, P., Lee, T. M., Patlis, J. M., Gumal, M. and Lee, R. J. (2006c) Biodiversity and human livelihood crises in the Malay Archipelago. *Conservation Biology* **20**, 1811–1813.

Sodhi, N. S., Brook, B. W. and Bradshaw, C. J. A. (2007) *Tropical Conservation Biology*. Blackwell Publishing, Oxford, UK.

Sodhi, N. S., Posa, M. R. C., Lee, T. M. and Warkentin, I. G. (2008) Effects of disturbance or loss of tropical rainforest birds. *Auk* **125**, 511–519.

Sodhi, N. S., Brook, B. W. and Bradshaw, C. A. J. (2009a) Causes and consequences of species extinctions. In: *Princeton Guide to Ecology* (eds. Levin, S. A.), 514–520. Princeton University Press, Princeton, NJ.

Sodhi, N. S., Lee, T. M., Koh, L. P. and Brook, B. W. (2009b) A meta-analysis of the impact of anthropogenic forest disturbance on Southeast Asia's biotas. *Biotropica* **41**, 103–109.

Sodhi, N. S., Wilcove, D. S., Lee, T. M., Sekercioglu, C. H., Subaraj, R., Bernard, H., Yong, D. L., Lim, S. L. H., Prawiradilaga, D. M. and Brook, B. W. (2010) Deforestation and avian extinction on tropical landbridge islands. *Conservation Biology* **24**, 1290–1298.

Soh, M. C. K., Sodhi, N. S., Seoh, R. K. H. and Brook, B. W. (2002) Nest site selection of the house crow *Corvus splendens*, an urban invasive bird species in Singapore and implications for its management. *Landscape and Urban Planning* **59**, 217–226.

Soh, M. C. K., Sodhi, N. S. and Lim, S. L. H. (2006) High sensitivity of montane bird communities to habitat disturbance in Peninsular Malaysia. *Biological Conservation* **129**, 149–166.

Sol, D., Duncan, R. P., Blackburn, T. M., Cassey, P. and Lefebvre, L. (2005) Big brains, enhanced cognition, and response of birds to novel environments. *Proceedings of the National Academy of Sciences of the United States of America* **102**, 5460–5465.

Sol, D., Bacher, S., Reader, S. M. and Lefebvre, L. (2008) Brain size predicts the success of mammal species introduced into novel environments. *American Naturalist* **172**, s63–s71.

Soria-Auza, R. W., Kessler, M. B., K., Barajas-Barbosa, P. M., Lehnert, M., Herzog, S. K. and Bohner, J. (2010) Impact of the quality of climate models for modelling species occurrences in countries with poor climatic documentation: a case study from Bolivia. *Ecological Modelling* **221**, 1221–1229.

Sousa, W. P., Kennedy, P. G. and Mitchell, B. J. (2003) Propagule size and predispersal damage by insects affect establishment and early growth of mangrove seedlings. *Oecologia* **135**, 564–575.

Sparks, T. and Crick, H. (1999) Opinion – the times they are a-changing? *Bird Conservation International* **9**, 1–7.

Spielman, D., Brook, B. W. and Frankham, R. (2004) Most species are not driven to extinction before genetic factors impact them. *Proceedings of the National Academy of Sciences of the United States of America* **101**, 15261–15264.

Staicer, C. A. (1992) Social behavior of the Northern Parula, Cape May Warbler, and Prairie Warbler wintering in second-growth forest in southwestern Puerto Rico. In: *Ecology and Conservation of Neotropical Migrant Landbirds* (eds. Hagan, J. M. III. and Johnston, D. W.), 308–320. Smithsonian Institution Press, Washington, DC.

Stainforth, D. A., Aina, T., Christensen, C., Collins, M., Faull, N., Frame, D. J., Kettleborough, J. A., Knight, S., Martin, A., Murphy, J. M., Piani, C., Sexton, D., Smith, L. A., Spicer, R. A., Thorpe, A. J. and Allen, M. R. (2005) Uncertainty in predictions of the climate response to rising levels of greenhouse gases. *Nature* **433**, 403–406.

Stanley, M. C. and Lill, A. (2002) Importance of seed ingestion to an avian frugivore: an experimental approach to fruit choice based on seed load. *Auk* **119**, 175–184.

Stapp, P. and Polis, G. A. (2003) Marine resources subsidize insular rodent populations in the Gulf of California, Mexico. *Oecologia* **134**, 496–504.

Stapp, P., Polis, G. A. and Pinero, F. S. (1999) Stable isotopes reveal strong marine and El Niño effects on island food webs. *Nature* **401**, 467–469.

Stattersfield, A. J., Crosby, M. J., Long, A. J. and Wege, D. C. (1998) *Endemic Bird Areas of the World: priorities for biodiversity conservation*. BirdLife International, Cambridge, UK.

Steadman, D. W. (1995) Prehistoric extinctions of Pacific island birds – biodiversity meets zooarchaeology. *Science* **267**, 1123–1131.

Steadman, D. W. (1997a) The historic biogeography and community ecology of Polynesian pigeons and doves. *Journal of Biogeography* **24**, 737–753.

Steadman, D. W. (1997b) Extinction of Polynesian birds: reciprocal impacts of birds and people. In: *Historical Ecology in the Pacific Islands* (eds. Kirch, P. V. and Hunt, T. L.), 51–79. Yale University Press, New Haven, CT.

Steadman, D. W. (2005) The paleoecology and fossil history of migratory landbirds. In: *Birds of Two Worlds: the ecology and evolution of migration* (eds. Greenberg, R. and Marra, P. P.), 5–17. The Johns Hopkins University Press, Baltimore, MD.

Stevens, P. F. (1976) Altitudinal and geographical distributions of flower types in *Rhododendron* section *Vireya*, especially in Papuasian species, and their significance. *Botanical Journal of the Linnean Society* 72, 1–33.

Stevenson, T and Fanshawe, J. (2002) *Field Guide to the Birds of East Africa*. T & AD Poyser, London, UK.

Stiles, F. G. (1978) Ecological and evolutionary implications of bird pollination. *American Zoologist* 18, 715–727.

Stiles, F. G. (1981) Geographical aspects of bird-flower coevolution, with particular reference to Central America. *Annals of the Missouri Botanical Garden* 68, 323–351.

Stiles, F. G. (1985) On the role of birds in the dynamics of Neotropical forests. In: *Conservation of Tropical Forest Birds* (eds. Diamond, A. W. and Lovejoy, T. E.), 49–59. International Council for Bird Preservation, Cambridge, UK.

Stocker, G. C. and Irvine, A. K. (1983) Seed dispersal by cassowaries (*Casuarius casuarius*) in north Queensland (Australia) rainforests. *Biotropica* 15, 170–176.

Stork, N. E. (2010) Re-assessing current extinction rates. *Biodiversity and Conservation* 19, 357–371.

Stork, N. E., Coddington, J. A., Colwell, R. K., Chazdon, R. L., Dick, C. W., Peres, C. A., Sloan, S. and Willis, K. (2009) Vulnerability and resilience of tropical forest species to land-use change. *Conservation Biology* 23, 1438–1447.

Stotz, D. F., Fitzpatrick, J. W., Parker, T. A. III. and Moskovits, D. K. (1996) *Neotropical Birds: ecology and conservation*. University of Chicago Press, Chicao, Illinois, USA.

Stouffer, P. C. and Bierregaard, R. O. (1995) Use of Amazonian forest fragments by understory insectivorous birds. *Ecology* 76, 2429–2445.

Stouffer, P. C., Strong, C. and Naka, L. N. (2009) Twenty years of understory bird extinctions from Amazonian rain forest fragments: consistent trends and landscape-mediated dynamics. *Diversity and Distributions* 15, 88–97.

Stratford, J. A. and Robinson, W. D. (2005) Gulliver travels to the fragmented tropics: geographic variation in mechanisms of avian extinction. *Frontiers in Ecology and Environment* 3, 85–92.

Strode, P. K. (2003) Implications of climate change for North American wood warblers (Parulidae). *Global Change Biology* 9, 1137–1144.

Strong, A. M. and Sherry, T. W. (2001a) Habitat-specific effects of food abundance on the condition of ovenbirds wintering in Jamaica. *Journal of Animal Ecology* 69, 883–895.

Strong, A. M. and Sherry, T. W. (2001b) Body condition of Swainson's Warblers wintering in Jamaica and the conservation value of Caribbean dry forest. *Wilson Bulletin* 113, 410–418.

Struhsaker, T. T. and Leakey, M. (1990) Prey selectivity by Crowned Hawk-eagles on monkeys in the Kibale Forest, Uganda. *Behavioral Ecology and Sociobiology* 26, 435–443.

Studds, C. E. and Marra, P. P. (2001) Linking fluctuations in rainfall to nonbreeding season performance in a long-distance migratory songbird *Setophaga ruticilla*. *Climate Research* 35, 115–122.

Studds, C. E. and Marra, P. P. (2005) Nonbreeding habitat occupancy and population processes: an upgrade experiment with a migratory bird. *Ecology* 86, 2380–2385.

Studds, C. E. and Marra, P. P. (2007) Linking fluctuations in rainfall to nonbreeding season performance in a long-distance migratory bird, *Setophaga ruticilla*. *Climate Research* 35, 115–122.

Studds, C. E., Kyser, T. K. and Marra, P. P. (2008) Natal dispersal driven by environmental conditions interacting across the annual cycle of a migratory songbird. *Proceedings of the National Academy of Sciences of the United States of America* 105, 2929–2933.

Stutchbury, B. (2007) *Silence of the Songbirds*. Walker and Company, New York, NY.

Stutchbury, B. J. M., Tarof, S. A., Done, T., Gow, E., Kramer, P. M., Tautin, J., Fox, J. W. and Afanasyev, V. (2009) Tracking long-distance songbird migration by using geolocators. *Science* 323, 896.

Subramanya, S. and Radhamani, T. R. (1993) Pollination by birds and bats. *Current Science* 65, 201–209.

Sun, C., Ives, A. R., Kraeuter, H. J. and Moermond, T. C. (1997) Effectiveness of three turacos as seed dispersers in a tropical montane forest. *Oecologia* **112**, 94–103.

Swarup, D., Patra, R. C., Prakash, V., Cuthbert, R., Das, D., Avari, P., Pain, D. J., Green, R. E., Sharma, A. K., Saini, M. and Taggart, M. A. (2007) Safety of meloxicam to critically endangered *Gyps* vultures and other scavenging birds in India. *Animal Conservation* **10**, 192–198.

Tabarelli, M. and Peres, C. A. (2002) Abiotic and vertebrate seed dispersal in the Brazilian Atlantic forest: implications for forest regeneration. *Biological Conservation* **106**, 165–176.

Tabarelli, M., Lopes, A. V. and Peres, C. A. (2008) Edge-effects drive tropical forest fragments towards an early-successional system. *Biotropica* **40**, 657–661.

Takekawa, J. Y. and Garton, E. O. (1984) How much is an Evening Grosbeak worth? *Journal of Forestry* **82**, 426–428.

Takekawa, J. Y., Garton, E. O. and Langelier, L. A. (1982) Biological control of forest insect outbreaks – the use of avian predators. *Transactions of the North American Wildlife and Natural Resources Conference* **47**, 393–409.

Tanner, E. V. J., Kapos, V. and Healey, J. R. (1991) Hurricane effects on forest ecosystems in the Caribbean. *Biotropica* **23**, 513–521.

Teixeira, D. M. and Negret, A. (1984) The Dwarf Tinamou (*Taoniscus nanus*) of Central Brazil. *Auk* **101**, 188–189.

Tejeda-Cruz, C. and Sutherland, W. J. (2005) Cloud forest bird responses to unusually sever storm damage. *Biotropica* **37**, 88–95.

Temple, S. A. (1977) Plant-animal mutualism – coevolution with Dodo leads to near extinction of plant. *Science* **197**, 885–886.

Terborgh, J. and Janson, C. H. (1986) The socioecology of primate groups. *Annual Review of Ecology and Systematics* **17**, 111–135.

Terborgh, J., Robinson, S. K., Parker, T. A. I., Munn, C. A. and Pierpont, N. (1990) Structure and organization of an Amazonian forest bird community. *Ecological Monographs* **60**, 213–238.

Terborgh, J. W. (1980) The conservation status of Neotropical migrants: present and future. In: *Migrant Birds in the Neotropics: ecology, behavior, distribution, and conservation* (eds. Keast, A. and Morton, E. S.), 21–30. Smithsonian Institution Press, Washington, DC.

Terborgh, J. W. (1989) *Where have all the birds gone?*, Princeton University Press, Princeton, NJ.

Terborgh, J. and Weske, J. S. (1969) Colonization of secondary habitats by Peruvian birds. *Ecology* **50**, 765–782.

Terborgh, J. and Winter, B. (1980) Some causes of extinction. In: *Conservation Biology: an evolutionary-ecological perspective* (eds. Soulé, M. A. and Wicox, B. A.), 119–133. Sinauer, New York, NJ.

Terborgh, J., Lopez, L., Nunez, P., Rao, M., Shahabuddin, G., Orihuela, G., Riveros, M., Ascanio, R., Adler, G. H., Lambert, T. D. and Balbas, L. (2001) Ecological meltdown in predator-free forest fragments. *Science* **294**, 1923–1926.

Terborgh, J., Pitman, N., Silman, M., Schichter, H. and Nuñez, P. V. (2002) Maintenance of tree diversity in tropical forests. In: *Seed Dispersal and Frugivory: ecology, evolution and conservation* (eds. Levey, D. J., Silva, W. R. and Galetti, M.), 1–18. CABI Publishing, Wallingford, UK.

Tewes, M. E. (1984) Opportunistic feeding by white-tailed hawks at prescribed burns. *Wilson Bulletin* **96**, 135–136.

Thibault, J. C. and Meyer, J. Y. (2001) Contemporary extinctions and population declines of the monarchs (*Pomarea* spp.) in French Polynesia, South Pacific. *Oryx* **35**, 73–80.

Thibault, J. C. M., J. L., Penloup, A. and Meyer, J. Y. (2002) Understanding the decline and extinction of monarchs (Aves) in Polynesian Islands. *Biological Conservation* **108**, 161–174.

Thiollay, J.-M. (1992) Influence of selective logging on bird species-diversity in a Guianan rain forest. *Conservation Biology* **6**, 47–63.

Thiollay, J.-M. (1995) The role of traditional agroforests in the conservation of rain forest bird diversity in Sumatra. *Conservation Biology* **9**, 335–353.

Thiollay, J.-M. (1997a) Disturbance, selective logging and bird diversity: a Neotropical forest study. *Biodiversity and Conservation* **6**, 1155–1173.

Thiollay, J. M. (1997b) Distribution and abundance patterns of bird community and raptor populations in the Andaman archipelago. *Ecography* **20**, 67–82.

Thiollay, J.-M. (1999) Responses of an avian community to rain forest degradation. *Biodiversity and Conservation* **8**, 513–534.

Thiollay, J.-M. (2005) Effects of hunting on Guinean forest game birds. *Biodiversity and Conservation* **14**, 1121–1135.

Thiollay, J.-M. and Rondeau, G. (2004) West African vulture decline. *Vulture News* **51**, 13–33.

Thomas, D. W., Blondel, J., Perret, P., Lambrechts, L. M. and Spearman, J. R. (2001) Energetic and fitness costs of mismatching resource supply and demand in seasonally breeding birds. *Science* **291**, 2598–2600.

Thompson, J. N. (1996) Evolutionary ecology and the conservation of biodiversity. *Trends in Ecology and Evolution* **11**, 300–303.

Tieleman, B. I. (2009) High and low, fast or slow: the complementary contributions of altitude and latitude to understand life-history variation. *Journal of Animal Ecology* **78**, 293–295.

Tiffney, B. H. and Mazer, S. J. (1995) Angiosperm growth habit, dispersal and diversification reconsidered. *Evolutionary Ecology* **9**, 93–117.

Tilman, D., May, R. M., Lehman, C. L. and Nowak, M. A. (1994) Habitat destruction and the extinction debt. *Nature* **371**, 65–66.

Timmermann, A., Oberhuber, J., Bacher, A., Esch, M., Latif, M. and Roeckner, E. (1999) Increased El Niño frequency in a climate model forced by future greenhouse warming. *Nature* **398**, 694–697.

Toh, I., Gillespie, M. and Lamb, D. (1999) The role of isolated trees in facilitating tree seedling recruitment at a degraded sub-tropical rainforest site. *Restoration Ecology* **7**, 288–297.

Tompkins, D. M. (1999) Impact of nest-harvesting on the reproductive success of black-nest swiftlets *Aerodramus maximus*. *Wildlife Biology* **5**, 33–36.

Tonioli, M., Escarre, J., Lepart, J. and Speranza, M. (2001) Facilitation and competition affecting the regeneration of *Quercus pubescens* Willd. *Ecoscience* **8**, 381–391.

Torchin, M. E., Lafferty, K. D., Dobson, A. P., McKenzie, V. J. and Kuris, A. M. (2003) Introduced species and their missing parasites. *Nature* **421**, 628–630.

Townsend, J. M., Rimmer, C. C., Brocca, J., McFarland, K. P. and Townsend, A. K. (2009) Predation of a wintering migratory songbirds by introduced rats: can nocturnal roosting behavior serve as predator avoidance. *Condor* **111**, 565–569.

Trail, P. W. (2007) African hornbills: keystone species threatened by habitat loss, hunting and international trade. *Ostrich* **78**, 609–613.

Traill, L. W., Bradshaw, C. J. A., Field, H. E. and Brook, B. W. (2009) Climate change enhances the potential impact of infectious disease and harvest on tropical waterfowl. *Biotropica* **41**, 414–423.

Traill, L., Lim, M., Sodhi, N. S. and Bradshaw, C. J. A. (2010) Mechanisms driving change: altered ecosystem functions and species interactions from global warming. *Journal of Animal Ecology* **79**, 937–947.

Trainor, C. R. (2007) Changes in bird species composition on a remote and well-forested Wallacean Island, South-East Asia. *Biological Conservation* **140**, 373–385.

Traveset, A. (2002) Consequences of the disruption of plant-animal mutualisms for the distribution of plant species in the Balearic Islands. *Revista Chilena De Historia Natural* **75**, 117–125.

Traveset, A. and Verdú, M. (2002) A meta-analysis of the effect of gut treatment on seed germination. In: *Seed Dispersal and Frugivory: ecology, evolution, and conservation* (eds. Levey, D. J., Silva, W. R and M. Galetti, M.), 339–350. CABI Publishing, Wallingford, UK.

Traveset, A., Riera, N. and Mas, R. E. (2001) Passage through bird guts causes interspecific differences in seed germination characteristics. *Functional Ecology* **15**, 669–675.

Treca, B. and Tamba, S. (1997) Role of birds on the regeneration of the woody *Boscia senegalensis* (Pers.) Lam. in sahelian savanna in north Senegal. *Revue d Ecologie-La Terre Et La Vie* **52**, 259–260.

Tscharntke, T. (1992) Cascade effects among four trophic levels – bird predation on galls affects density-dependent parasitism. *Ecology* **73**, 1689–1698.

Tscharntke, T., Sekercioglu, C. H., Dietsch, T. V., Sodhi, N. S., Hoehn, P. and Tylianakis, J. M. (2008) Landscape constraints on functional diversity of birds and insects in tropical agroecosystems. *Ecology* **89**, 944–951.

Tucker, N. I. J. and Murphy, T. M. (1997) The effects of ecological rehabilitation on vegetation recruitment: some observations from the wet tropics of north Queensland. *Forest Ecology and Management* **99**, 133–152.

Turcq, B., Sifeddine, A., Martin, L., Absy, M. L., Soubies, F., Suguio, K. and Volkmer-Ribeiro, C. (1998) Amazonia rainforest fires: a lacustrine record of 7000 years. *Ambio* **27**, 139–142.

Turner, I. M. and Corlett, R. T. (1996) The conservation value of small, isolated fragments of lowland tropical rain forest. *Trends in Ecology and Evolution* **11**, 330–333.

Tye, H. (1992) Reversal of breeding-season by lowland birds at higher altitudes in western Cameroon. *Ibis* **134**, 154–163.

Ueta, M. (2001) Azure-winged Magpies avoid nest predation by breeding synchronously with Japanese Lesser Sparrowhawks. *Animal Behaviour* **61**, 1007–1012.

Uhl, C. and Buschbacher, R. (1985) A disturbing synergism between cattle ranch burning practices and selective tree harvesting in the Eastern Amazon. *Biotropica* **17**, 265–268.

Uhl, C. and Kauffman, J. B. (1990) Deforestation, fire susceptibility, and potential tree responses to fire in the Eastern Amazon. *Ecology* **71**, 437–449.

UNEP (United Nations Environmental Programme) (2002) *Global Environmental Outlook 3: past, present and future perspectives*. Earthscan, London, UK.

Valdivia-Hoeflich, T., Rivera, J. H. V. and Stoner, K. E. (2005) The Citreoline Trogon as an ecosystem engineer. *Biotropica* **37**, 465–467.

Valentine, L. E., Schwarzkopf, L., Johnson, C. N. and Grice, A. C. (2007) Burning season influences the response of bird assemblages to fire in tropical savannas. *Biological Conservation* **137**, 90–101.

Valli, E. and Summers, D. (1990) Nest gatherers of Tiger Cave. *National Geographic* **177**, 107–133.

Vall-llosera, M. and Sol, D. (2009) A global risk assessment for the success of bird introductions. *Journal of Applied Ecology* **46**, 787–795.

Van Bael, S. A. and Brawn, J. D. (2005) The direct and indirect effects of insectivory by birds in two contrasting neotropical forests. *Oecologia* **145**, 658–668.

Van Bael, S. A., Brawn, J. D. and Robinson, S. K. (2003) Birds defend trees from herbivores in a neotropical forest canopy. *Proceedings of the National Academy of Sciences of the United States of America* **100**, 8304–8307.

Van Bael, S. A., Aiello, A., Valderrama, A., Medianero, E., Samaniego, M. and Wright, S. J. (2004) General herbivore outbreak following an El Niño-related drought in a lowland Panamanian forest. *Journal of Tropical Ecology* **20**, 625–633.

Van Bael, S. A., Bichier, P., Ochoa, I. and Greenberg, R. (2007) Bird diversity on cacao farms and forest fragments in western Panama. *Biodiversity and Conservation* **16**, 2245–2256.

Van Bael, S. A., Philpott, S. M., Greenberg, R., Bichier, P., Barber, N. A., Mooney, K. A. and Gruner, D. S. (2008) Birds as predators in tropical agroforestry systems. *Ecology* **89**, 928–934.

Van Balen, B. (1999) Differential extinction patterns in Javan forest birds. *Tropical Resource Management Papers* **30**, 39–57.

Van Euler, F. (2001) Selective extinction and rapid loss of evolutionary history in bird fauna. *Proceedings of the Royal Society of London Series B* **268**, 127–130.

Van Houtan, K. S., Pimm, S. L., Halley, J. M., Bierregaard, R. O., Jr. and Lovejoy, T. E. (2007) Dispersal of Amazonian birds in continuous and fragmented forests. *Ecology Letters* **10**, 219–229.

Van Langevelde, F., Van de Vijver, C. A. D. M., Kumar, L., Van de Koppel, J., de Ridder, N., Van Andel, J., Skidmore, A.K., Hearne, J. W., Stroosnijder, L., Bond, W. J., Prins, H. H. T. and Rietkerk, M. (2003) Effects of fire and herbivory on the stability of Savanna ecosystems. *Ecology* **84**, 337–350.

Van Riper III, C., Van Riper, S. G., Goff, M. L. and Laird, M. (1986) The epizootiology and ecological significance of malaria in Hawaiian landbirds. *Ecological Monographs* **56**, 327–344.

Vander Wall, S. B. (1992) The role of animals in dispersing a wind-dispersed pine. *Ecology* **73**, 614–621.

Vera, C., Silvestri, G., Liebmann, B. and Gonzalez, P. (2006) Climate change scenarios for seasonal precipitation in South America from IPCC-AR4 models. *Geophysical Research Letters* **33**, L13707–13710.

Veran, S., Gimenez, O., Flint, E., Kendall, W. L., Doherty, P. F. and Lebreton, J. D. (2007) Quantifying the impact of longline fisheries on adult survival in the Black-footed Albatross. *Journal of Applied Ecology* **44**, 942–952.

Verdu, M. and Garcia-Fayos, P. (1996) Nucleation processes in a Mediterranean bird-dispersed plant. *Functional Ecology* **10**, 275–280.

Vidal, E., Medail, F., Tatoni, T., Vidal, P. and Roche, P. (1998) Functional analysis of the newly established plants induced by nesting gulls on Riou archipelago (Marseille, France). *Acta Oecologica* **19**, 241–250.

Vidal, E., Medail, F., Tatoni, T. and Bonnet, V. (2000) Seabirds drive plant species turnover on small Mediterranean islands at the expense of native taxa. *Oecologia* **122**, 427–434.

Vidal, E., Jouventin, P. and Frenot, Y. (2003) Contribution of alien and indigenous species to plant-community assemblages near penguin rookeries at Crozet archipelago. *Polar Biology* **26**, 432–437.

Viitala, J., Korpimaki, E., Palokangas, P. and Koivula, M. (1995) Attraction of kestrels to vole scent marks visible in ultraviolet light. *Nature* **373**, 425.

Visser, M. E. (2008) Keeping up with a warming world; assessing the rate of adaptation to climate change. *Proceedings of the Royal Society of London Series B* **275**, 649–659.

Voelker, G., Outlaw, R. K. and Bowie, R. C. K. (2010) Pliocene forest dynamics as a primary driver of African bird speciation. *Global Ecology and Biogeography* **19**, 111–121.

Walsberg, G. E. (1993) Thermal consequences of diurnal microhabitat selection in a small bird. *Ornis Scandinavica* **24**, 174–182.

Waltert, M., Mardiastuti, A. and Mühlenberg, M. (2004) Effects of land use on bird species richness in Sulawesi, Indonesia. *Conservation Biology* **18**, 1339–1346.

Wanger, T. C., Rauf, A. and Schwarze, S. (2010) Pesticides and tropical biodiversity. *Frontiers in Ecology and the Environment* **4**, 178–179.

Wanless, R. M., Angel, A., Cuthbert, R. J., Hilton, G. M. and Ryan, P. G. (2007) Can predation by invasive mice drive seabird extinction? *Biology Letters* **3**, 241–244.

Warburton, N. H. (1997) Structure and conservation of forest avifauna in isolated rainforest remnants in tropical Australia. In: *Tropical Forest Remnants: ecology, management, and conservation of fragmented communities* (eds. Laurance, W. F. and Bierregaard, R. O. Jr.), 190–206. Chicago University Press, Chicago, IL.

Warkentin, I. G. and Hernandez, D. (1996) Conservation implications of site fidelity: a case study involving nearctic-Neotropical migrant songbirds wintering in a Costa Rican mangrove swamp. *Biological Conservation* **77**, 143–150.

Warkentin, I. G., Greenberg, R. and Salgado-Ortiz, J. (1995) Songbird use of gallery woodlands in recently cleared and older settled landscapes of the Selva Lacandona, Chiapas, Mexico. *Conservation Biology* **9**, 1095–1106.

Waser, N. M., Chittka, L., Price, M. V., Williams, N. M. and Ollerton, J. (1996) Generalization in pollination systems and why it matters. *Ecology* **77**, 1043–1060.

Watanuki, Y., Ito, M., Deguchi, T. and Minobe, S. (2009) Climate-forced seasonal mismatch between the hatching of rhinoceros auklets and the availability of anchovy. *Marine Ecology-Progress Series* **393**, 259–271.

Watling, D. (1978) Observations on the naturalized distribution of the Red-vented Bulbul in the Pacific, with special reference to the Fiji islands. *Notornis* **25**, 109–117.

Watson, D. M. (2001) Mistletoe – a keystone resource in forests and woodlands worldwide. *Annual Review of Ecology and Systematics* **32**, 219–249.

Watson, D. M. (2002) Effects of mistletoe on diversity: a case-study from southern New South Wales. *Emu* **102**, 275–281.

Watson, J. E. M., Whittaker, R. J. and Dawson, T. P. (2004) Habitat structure and proximity to forest edge affect the abundance and distribution of forest-dependent birds in tropical coastal forests of southeastern Madagascar. *Biological Conservation* **120**, 311–327.

Waycott, M., Duarte, C. M., Carruthers, T. J. B., Orth, R. J., Dennison, W. C., Olyarnik, S., Calladine, A., Fourqurean, J. W., Heck, K. L., Hughes, A. R., Kendrick, G. A., Kenworthy, W. J., Short, F. T. and Williams, S. L. (2009) Accelerating loss of seagrasses across the globe threatens coastal ecosystems. *Proceedings of the National Academy of Sciences of the United States of America* **106**, 12377–12381.

Weatherhead, P. J., Tinker, S. and Greenwood, H. (1982) Indirect assessment of avian damage to agriculture. *Journal of Applied Ecology* **19**, 773–782.

Weathers, K. C., Cadenasso, M. L. and Pickett, S. T. A. (2001a) Forest edges as nutrient and pollutant concentrators: potential synergisms between fragmentation, forest canopies, and the atmosphere. *Conservation Biology* **15**, 1506–1514.

Weathers, W. W. (1997) Energetics and thermoregulation by small passerines of the humid, lowland tropics. *Auk* **114**, 341–353.

Weathers, W. W. and Greene, E. (1998) Thermoregulatory responses of bridled and juniper titmice to high temperature. *Condor* **100**, 365–372.

Weathers, W. W., Hodum, P. J. and Blakesley, J. A. (2001b) Thermal ecology and ecological energetics of California Spotted Owls. *Condor* **103**, 678–690.

Webb, C. O. and Peart, D. R. (2001) High seed dispersal rates in faunally intact tropical rain forest: theoretical and conservation implications. *Ecology Letters* **4**, 491–499.

Webber, B. L. and Woodrow, I. E. (2004) Cassowary frugivory, seed defleshing and fruit fly infestation influence the transition from seed to seedling in the rare Australian rainforest tree, *Ryparosa* sp nov 1 (Achariaceae). *Functional Plant Biology* **31**, 505–516.

Weber, W. J. (1979) *Health hazards from pigeons, starlings and English sparrows*. Thomson Publications, Fresno, CA.

Webster, M. S. and Marra, P. P. (2005) The importance of understanding migratory connectivity and seasonal interaction. In: *Birds of Two Worlds: the ecology and evolution of migration* (eds. Greenberg, R. and Marra, P. P.), 199–209. The Johns Hopkins University Press, Baltimore, MD.

Webster, M. S., Marra, P. P., Haig, S. M., Bensch, S. and Holmes, R. T. (2002) Links between worlds: unraveling migratory connectivity. *Trends in Ecology and Evolution* **17**, 76–83.

Weeks, P. (1999) Interactions between Red-billed Oxpeckers, *Buphagus erythrorhynchus*, and domestic cattle, *Bos taurus*, in Zimbabwe. *Animal Behaviour* **58**, 1253–1259.

Weeks, P. (2000) Red-billed Oxpeckers: vampires or tickbirds? *Behavioral Ecology* **11**, 154–160.

Wenny, D. G. (2000) Seed dispersal, seed predation, and seedling recruitment of a Neotropical montane tree. *Ecological Monographs* **70**, 331–351.

Wenny, D. G. (2001) Advantages of seed dispersal: a re-evaluation of directed dispersal. *Evolutionary Ecology Research* **3**, 51–74.

Wenny, D. G. and Levey, D. J. (1998) Directed seed dispersal by bellbirds in a tropical cloud forest. *Proceedings of the National Academy of Sciences of the United States of America* **95**, 6204–6207.

West, B. and Zhou, B.-X. (1989) Did chickens go north? New evidence for domestication. *World's Poultry Science Journal* **45**, 205–218.

Westcott, D. A. and Graham, D. L. (2000) Patterns of movement and seed dispersal of a tropical frugivore. *Oecologia* **122**, 249–257.

Wheelwright, N. T. (1985) Fruit size, gape width and the diets of fruit-eating birds. *Ecology* **66**, 808–818.

Wheelwright, N. T. and Orians, G. H. (1982) Seed dispersal by animals – contrasts with pollen dispersal, problems of terminology, and constraints on coevolution. *American Naturalist* **119**, 402–413.

Whitcomb, R. F., Robbins, C. S., Lynch, J. F., Whitcomb, B. L., Klimkiewicz, M. K. and Bystrak, D. (1981) Effects of forest fragmentation on avifauna of the eastern deciduous forest. In: *Forest Island Dynamics in Man-dominated Landscapes* (eds. Burgess, R. L. and Sharpe, D. M.), 125–205. Springer-Verlag, New York, NY.

Whittaker, R. J. and Turner, B. D. (1994) Dispersal, fruit utilization and seed predation of *Dysoxylum gaudichaudianum* in early successional rain-forest, Krakatau, Indonesia. *Journal of Tropical Ecology* **10**, 167–181.

Whitten, A. J., Mustafa, M. and Henderson, G. S. (1987) *The Ecology of Sulawesi*. Gadjah Mada University Press, Yogyakarta, Indonesia.

Wiemeyer, S. N., Lamont, T. G., Bunck, C. M., Sindelar, C. R., Gramlich, F. J., Fraser, J. D. and Byrd, M. A. (1984) Organochlorine pesticide, polychorobiphenyl, and mercury residues in bald eagle eggs – 1969–79 – and their relationships to shell thinning and reproduction. *Archives of Environmental Contamination and Toxicology* **13**, 529–549.

Wiemeyer, S. N., Bunck, C. M. and Krynitsky, A. J. (1988) Organochlorine pesticides, polychlorinated biphenyls, and mercury in Osprey eggs – 1970–79 – and their relationships to shell thinning and productivity. *Archives of Environmental Contamination and Toxicology* **17**, 767–787.

Wiens, J. A. (1973) Pattern and process in grassland bird communities. *Ecological Monographs* **43**, 237–270.

Wiersma, P., Munoz-Garcia, A., Walker, A. and Williams, J. B. (2007) Tropical birds have a slow pace of life. *Proceedings of the National Academy of Sciences of the United States of America* **104**, 9340–9345.

Wikelski, M., Hau, M. and Wingfield, J. C. (2000) Seasonality of reproduction in a neotropical rain forest bird. *Ecology* **81**, 2458–2472.

Wikelski, M., Hau, M., Robinson, W. D. and Wingfield, J. C. (2003a) Reproductive seasonality of seven neotropical passerine species. *Condor* **105**, 683–695.

Wikelski, M., Tarlow, E. M., Raim, A., Diehl, R. H., Larkin, R. P. and Visser, G. H. (2003b) Costs of migration in free-flying songbirds. *Nature* **423**, 704.

Wikelski, M., Foufopoulos, J., Vargas, H. and Snell, H. (2004) Galápagos birds and disease: invasive pathogens as threats for island species. *Ecology and Society* **9**, 5.

Wikelski, M., Kays, R. W., Kasdin, N. J., Thorup, K., Smith, J. A. and Swenson, G. W. J. (2007) Going wild: what a global small-animal tracking system could do for experimental biologists. *Journal of Experimental Biology* **210**, 181–186.

Wilcove, D. S. (1985) Nest predation in forest tracts and the decline of migratory songbirds. *Ecology* **66**, 1211–1214.

Wilcove, D. S. (2008) *No Way Home: the decline of the world's great animal migrations.* Island Press, Washington, DC.

Wilcove, D. S., McLellan, C. J. and Dobson, A. P. (1986) Habitat fragmentation in the temperate zone. In: *Conservation Biology, The Science of Scarcity and Diversity* (eds. Soulé, M. E.), 237–256. Sinauer Associates, Sunderland, MA.

Wiles, G. J., Bart, J., Beck, R. E. and Aguon, C. F. (2003) Impacts of the brown tree snake: patterns of decline and species persistence in Guam's avifauna. *Conservation Biology* **17**, 1350–1360.

Will, T. (1991) Birds of a severely hurricane-damaged Atlantic Coast rain-forest in Nicaragua. *Biotropica* **23**, 497–507.

Williams, N. (2010) Malaria climbs the mountain. *Current Biology* **20**, R37–R38.

Williams, S. E. and Middleton, J. (2008) Climatic seasonality, resource bottlenecks, and abundance of rainforest birds: implications for global climate change. *Diversity and Distributions* **14**, 69–77.

Williams, S. E., Williams, Y. M., van Der Wal, J., Isaac, J. L., Shoo, L. P. and Johnson, C. N. (2009) Ecological specialization and population sizes in a biodiversity hotspot: how rare species avoid extinction. *Proceedings of the National Academy of Sciences of the United States of America* **106**, 19737–19741.

Williamson, M. (1996) *Biological Invasions.* Chapman & Hall, London, UK.

Williamson, M. (2006) Explaining and predicting the success of invading species at different stages of invasion. *Biological Invasions* **8**, 1561–1568.

Williamson, M. and Brown, K. C. (1986) The analysis and modelling of British invasions. *Philosophical Transactions of the Royal Society London B* **314**, 505–522.

Willis, C. G., Ruhfel, B., Primack, R. B., Miller-Rushing, A. J. and Davis, C. C. (2008) Phylogenetic patterns of species loss in Thoreau's woods are driven by climate change. *Proceedings of the National Academy of Sciences of the United States of America* **105**, 17029–17033.

Willis, E. O. (1966) The role of migrant birds at swarms of army ants. *Living Bird* **5**, 187–231.

Willis, E. O. (1974) Populations and local extinctions of birds on Barro Colorado Island, Panama. *Ecological Monographs* **44**, 153–169.

Willis, E. O. and Oniki, Y. (1978) Birds and army ants. *Annual Review of Ecology and Systematics* **9**, 243–263.

Willis, S. G., Hole, D. G., Collingham, Y. C., Hilton, G., Rahbek, C. and Huntley, B. (2009) Assessing the impacts of future climate change on protected area netwrks: a method to simulate individual species' responses. *Environmental Management* **43**, 836–845.

Wilkinson, D. M. (1997) Plant colonization: are wind dispersed seeds really dispersed by birds at larger spatial and temporal scales? *Journal of Biogeography* **24**, 61–65.

Williams, R. J., Berlow, E. L., Dunne, J. A., Barabasi, A.-L. and Martinez, N. D. (2002) Two degrees of separation in complex food webs. *Proceedings of the National Academy of Sciences of the United States of America* **99**, 12913–12916.

Willson, M. F. and Crome, F. H. J. (1989) Patterns of seed rain at the edge of a tropical Queensland rain forest (Australia). *Journal of Tropical Ecology* **5**, 301–308.

Willson, M. F. and Traveset, A. (2000) The ecology of seed dispersal. In: *Seeds: the ecology of regeneration in plant communities* (ed. M. Fenner), 85–110. CAB International, New York NY.

Wilson, C. and Tisdell, C. (2005) Knowledge of birds and willingness to support their conservation: an Australian case study. *Bird Conservation International* **15**, 225–235.

Winker, K., Rappole, J. H. and Ramos, M. A. (1990) Population dynamics of the wood thrush in southern Veracruz, Mexico. *Condor* **92**, 444–460.

Winkler, H. and Leisler, B. (2005) To be a migrant. In: *Birds of Two Worlds: the ecology and evolution of migration* (eds. Greenberg, R. and Marra, P. P.), 79–86. The Johns Hopkins University Press, Baltimore, MD.

With, K. A. and King, A. W. (2001) Analysis of landscape sources and sinks: the effect of spatial pattern on avian demography. *Biological Conservation* **100**, 75–88.

Witmer, M. C. and Cheke, A. S. (1991) The Dodo and the Tambalacoque Tree – an obligate mutualism reconsidered. *Oikos* **61**, 133–137.

Wolfe, K. M., Mills, H. R., Garkaklis, M. J. and Bencini, R. (2004) Post-mating survival in a small marsupial is associated with nutrient inputs from seabirds. *Ecology* **85**, 1740–1746.

Woinarski, J. C. Z. and Catterall, C. P. (2004) Historical changes in the bird fauna at Coomooboolaroo, northeastern Australia, from the early years of pastoral settlement (1873) to 1999. *Biological Conservation* **116**, 379–401.

Woinarski, J. C. Z., Risler, J. and Kean, L. (2004) Response of vegetation and vertebrate fauna to 23 years of fire exclusion in a tropical Eucalyptus open forest, Northern Territory, Australia. *Austral Ecology* **29**, 156–176.

Wong, T. C. M., Sodhi, N. S. and Turner, I. M. (1998) Artificial nest and seed predation experiments in tropical lowland rainforest remnants of Singapore. *Biological Conservation* **85**, 97–104.

Woodbridge, B., Finley, K. K. and Seager, S. T. (1995) An investigation of the Swainson's Hawk in Argentina. *Journal of Raptor Research* **29**, 202–204.

Woodworth, B. L., Atkinson, C. T., LaPointe, D. A., Hart, P. J., Spiegel, C. S., Tweed, E. J., Henneman, C., LeBrun, J., Denette, T., DeMots, R., Kozar, K. L., Triglia, D., Lease, D., Gregor, A. Smith, T. and Duffy, D. (2005) Host population persistence in the face of introduced vector-borne diseases: Hawaii amakihi and avian malaria. *Proceedings of the National Academy of Sciences of the United States of America* **102**, 1531–1536.

Wootton, J. T. (1994) The nature and consequences of indirect effects in ecological communities. *Annual Review of Ecology and Systematics* **25**, 443–466.

Wootton, J. T. (1995) Effects of birds on sea urchins and algae: a lower-intertidal trophic cascade. *Ecoscience* **2**, 321–328.

World Health Organization (2010) Rabies. Fact Sheet No. 99. http://www.who.int/mediacentre/factsheets/fs099/en/

Wright, S. J. (2005) Tropical forests in a changing environment. *Trends in Ecology and Evolution* **20**, 553–560.

Wright, S. J., Stoner, K. E., Beckman, N., Corlett, R. T., Dirzo, R., Muller-Landau, H. C., Nuñez-Iturri, G., Peres, C. A. and Wang, B. C. (2007) The plight of large animals in tropical forests and the consequences for plant regeneration. *Biotropica* **39**, 289–291.

Wright, T. F., Toft, C. A., Enkerlin-Hoeflich, E., González-Elizondo, J., Albornoz, M., Rodríguez-Ferraro, A., Rojas-Suárez, F., Sanz, V., A., T., Beissinger, S. R., Berovides, V., Gálvez, X., Brice, A. T., Joyner, K., Eberhard, J., Gilardi, J., Koenig, S. E., Stoleson, S., Martuscelli, P., Meyers, J. M., Renton, K., Rodríguez, A. M., Sosa-Asanza, A. C., Vilella, F. J. and Wiley, J. W. (2001) Nest poaching in Neotropical parrots. *Conservation Biology* **15**, 710–720.

Wunderle, J. M. (1991) Responses of bird populations in a Puerto-Rican forest to Hurricane Hugo – the first 18 months. *Condor* **97**, 879–896.

Wunderle, J. M. (1997) The role of animal seed dispersal in accelerating native forest regeneration on degraded tropical lands. *Forest Ecology and Management* **99**, 223–235.

Wunderle, J. M. J. and Latta, S. C. (2000) Winter site fidelity of Nearctic migrants in shade coffee plantations of different sizes in the Dominican Republic. *Auk* **117**, 596–614.

Wunderle, J. M. J., Currie, D., Helmer, E. H., Ewert, D. N., White, J. D., Ruzycki, T. S., Parresol, B. and Kwit, S. (2010) The Kirtland's Warbler in anthropogenically disturbed early-successional habitats on Eleuthera, The Bahamas. *The Condor* **112**, 123–137.

Yamaura, Y., Amano, T., Koizumi, T., Mitsuda, Y., Taki, H. and Okabe, K. (2009) Does land-use change affect biodiversity dynamics at a macroecological scale? A case study of birds over the past 20 years in Japan. *Animal Conservation* **12**, 110–119.

Yap, C. A. M. and Sodhi, N. S. (2004) Southeast Asian invasive birds: ecology, impact and management. *Ornithological Science* **3**, 57–67.

Yap, C. A. M., Sodhi, N. S. and Brook, B. W. (2002) Roost characteristics of invasive mynas in Singapore. *Journal of Wildlife Management* **66**, 1118–1127.

Yap, C. A.-M., Sodhi, N. S. and Peh, K. S.-H. (2007) Phenology of tropical birds in Peninsular Malaysia effects of selective logging and food resources. *Auk* **124**, 945–961.

Yibarbuk, D., Whitehead, P. J., Russell-Smith, J., Jackson, D., Godjuwa, C., Fisher, A., Cooke, F., Choquenot, D. and Bowman, D. (2001) Fire ecology and Aboriginal land management in central Arnhem Land, northern Australia: a tradition of ecosystem management. *Journal of Biogeography* **28**, 325–343.

Yumoto, T. (2000) Bird-pollination of three *Durio* species (Bombacaceae) in a tropical rainforest in Sarawak, Malaysia. *American Journal of Botany* 87, 1181–1188.

Zhijun, W. and Young, S. S. (2003) Differences in birds diversity between two swidden agricultural sites in mountainous terrain, Xishuangbanna, Yunnan, China. *Biological Conservation* **110**, 231–24[illegible].

Zurita, G. A. and Bellocq, M. I. (2010) Spatial patterns of bird community similarity: bird responses to landscape composition and configuration in the Atlantic forest. *Landscape Ecology* **25**, 147–158.

Zydelis, R., Bellebaum, J., Osterblom, H., Vetemaa, M., Schirmeister, B., Stipniece, A., Dagys, M., van Eerden, M. and Garthe, S. (2009) Bycatch in gillnet fisheries – an overlooked threat to waterbird populations. *Biological Conservation* **142**, 1269–1281.

Index

Page numbers in *italics* refer to figures, those in **bold** refer to tables and boxes.

Conservation of Tropical Birds, First Edition. Navjot S. Sodhi, Çağan H. Şekercioğlu, Jos Barlow and Scott K. Robinson.

Published 2011 by Blackwell Publishing Ltd.